THE CHANGING NATURE OF THE MAINE WOODS

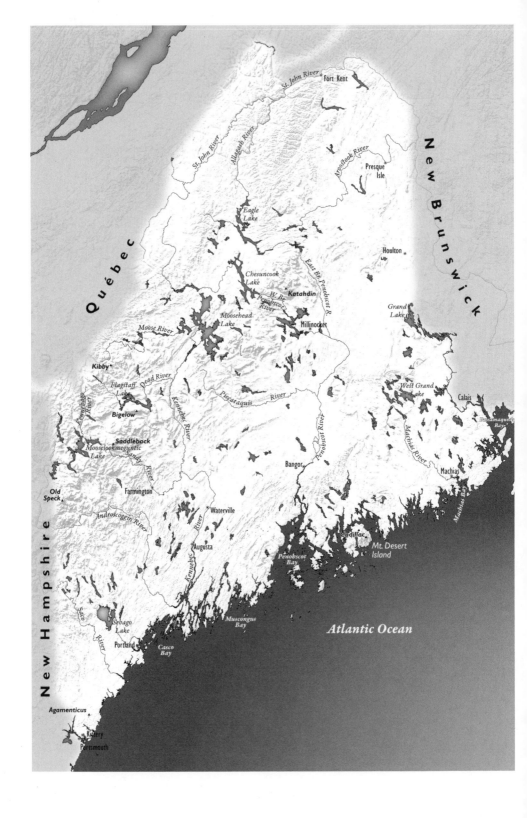

The
Changing
Nature
of the
Maine
Woods

Andrew M. Barton

with Alan S. White and Charles V. Cogbill

UNIVERSITY OF NEW HAMPSHIRE PRESS

DURHAM, NEW HAMPSHIRE

University of New Hampshire Press
An imprint of University Press of New England
www.upne.com

Manufactured in the United States of America
Designed by April Leidig
Typeset in Arno by Copperline Book Services, Inc.

University Press of New England is a member of the
Green Press Initiative. The paper used in this book meets
their minimum requirement for recycled paper.

For permission to reproduce any of the material in this book, contact:
Permissions
University Press of New England
One Court Street, Suite 250
Lebanon NH 03766
or visit www.upne.com

*This book was supported by a grant from
Furthermore: a program of the J. M. Kaplan Fund.*

Library of Congress Cataloging-in-Publication Data
Barton, Andrew M.
The changing nature of the Maine woods / Andrew M. Barton ;
with Alan S. White and Charles V. Cogbill. — 1st ed.
p. cm.
Includes bibliographical references and index.
ISBN 978-1-58465-832-0 (pbk. : alk. paper)
ISBN 978-1-61168-295-3 (ebook)
1. Forests and forestry — Maine — History. 2. Forest
ecology — Maine. I. White, Alan S. II. Cogbill, Charles V.
III. Title.
SD144.M2B37 2012
577.3 — dc23   2011053107

Chapter-opening and back cover art showing white pine branch
(*Pinus strobus*) courtesy of © iStockphoto.com/Linas Lebeliunas.

5 4 3 2 1

To Alan Barton,
who taught his son to love mountains and forests.

To Cindy White,
whose encouragement and support were so important
to ASW's involvement in the book.

To B.A. Cogbill,
who spent his youth in the now vanished forests of the
Mississippi River bottoms and inspired his son to
appreciate their relatives, the "puny" forests of the north.

Past where the last
gang of signs

comes out of the dark
to wave you back

and past telephone
wires lengthening

with the light of someone
beyond the next hill

just returning,
a slow single line

will take the eye
of your high beam. Around you

will be jewels
of the fox-watch.

Great trees will rise up
to see you passing by

all by yourself,
riding on light.

—Wesley McNair, Poet Laureate of Maine,
"Driving to Dark Country," *The New Criterion* 8 (1990): 43.
Courtesy of David R. Godine, Publisher

Contents

xi   Preface

1   One | The Maine Woods: Through Time and Across Space

     An Ancient Forest in Portland 1
     Moosehead Lake 5
     Maine Forests in Time and Space 6
     About This Book 7
     Appendix 1.1: Common and Scientific Names of Species Used in the Text 8

15   Two | From Rocks to Ice to Forests:
       How Did the Land and Forests of Maine Form
       and Change over Geological Time?

     The Bog in the Basin 15
     The Dynamic Earth 16
     Origin of the Bedrock of Maine 17
     Climate Change 20
     Deglaciation and the Emergence of Land in Maine 24
     How Glaciation Shaped the Maine Landscape 26
     Life in a New World: The Forest Reconstituted 29
     15,000 Years: Changing Environments, Changing Forests 32
     Postglacial Animals of Maine 40
     Lessons Learned 44

48   Three | The Presettlement Forest of Maine:
       What Were the Forests Like before European Settlement?

     Explorers: Observers of Maine's Presettlement Forest 53
     Land Surveys of Maine's Presettlement Forest 56
     Paleoecology: Remains from the Past 66
     The Remnant Old-Growth Landscape: "Still Stands the
       Forest Primeval" 68
     Remnant Old-Growth Stands: Museums of the Past 81
     Conclusion 84
     Appendix 3.1: Observations by Early Explorers of the Forests
       along the Maine Coast 90
     Appendix 3.2: Observations of Forests by Travelers and Surveyors
       in the Interior of Maine before Division into Towns 93
     Appendix 3.3: Equivalent Vernacular Names for Maine Trees in
       Old Records 99

100 Four | From European Settlement to Modern Times:
     How Did Human Culture Transform Maine Forests?

The Perham Farm, Sandy River Valley, Maine 100
Changes in the Land, Changes in the Forest 103
Before 1781: Colonial Maine and Its Forests 104
1781–1840: Settlers Come, White Pine Goes 107
1840–1880: Into the Interior—Forest Clearing, Tree Harvesting 112
1880–1920: The Remaking of the Maine Forests 117
1920–1970: Well-Worn Patterns 123
1970–2010: Conflict and Change 125
Maine Forests Today 131
Appendix 4.1: Chronology of Important Events in the History of the
     Maine Woods from European Settlement to Today 140

142 Five | The Length and Breadth and Height of Maine:
     How and Why Do Forests Vary across the State?

Six Maine Nature Snapshots 144
Making Sense of the Diversity of Maine Forests 155
Maine's Place in the Natural Geography of North America 160
Dividing Up the State: The Ecoregions of Maine 164
The Natural Communities of Maine 169
Why Is Maine Ecologically Diverse? 172
Putting Classification into Practice: Maine's Conservation Network 175
Appendix 5.1: Ecoregional Frameworks in Maine 180
Appendix 5.2: Seven Ecoregions of Maine 182
Appendix 5.3: The Natural Community System of Maine 184

189 Six | The Future of the Maine Woods:
     What Will Maine Forests Be Like in the Year 2100?

View from a Maine Woodlot 189
The Future: Changing Environments, Changing Forests 190
Changing Nutrient Cycles: Acid Rain and Nitrogen Saturation 191
Changing Enemies: Invasive Exotic Insects, Diseases, and Plants 197
Changing Disturbance: Land Use–Forestry and Housing
     Development 204
Changing Climate 211
Interacting Stresses and Forest Dieback 221
The Future of the Maine Woods: Forecasts and Uncertainty 230

235 Notes
289 Bibliography
331 Illustration Credits
339 Index

Preface

Maine is the most forested state in the country. Nearly half of its total land area falls within the boundaries of the Unorganized Territories, a region with no formal municipalities, covered in forest, broken mainly by logging roads. Forests are Maine's chief commodity, supporting a tourism economy and a receding but still substantial forest products industry. Maine is one of the few states where forest issues are front-page news. Forests and nature are at the very heart of the self-identity of Mainers.

We set out to synthesize what is known about these vast and important forests, which at least since Thoreau have been referred to in aggregate as the "Maine Woods." Forests are exceedingly complex, heterogeneous across the landscape, and changing over time. No book can hope to capture these attributes in any comprehensive way. And so it is with ours. We take on an ambitious ambit, examining Maine's forests over millennia and from border to border, but our view is necessarily circumscribed by time, book space, and our own biases as ecologists who pay attention mainly to trees in terrestrial habitats. When trying to understand forests, focusing on trees makes sense, for they provide the structure and resources upon which nearly all other organisms depend. Scientists also know a lot about trees, because they are large, abundant, important economically, and possess parts (pollen, wood) that can resist decay over long periods of time, leaving behind a trace of their distant past. Despite our bias, we embrace the equal standing of all organisms, and have included a wide range of nature's characters in addition to trees. This book is not aimed directly at the full array of current issues, some controversial, that confront the management and conservation of Maine's forests. Our highest hope, however, is that our work, in addition to educating and entertaining readers, will contribute to the scientific basis for debates and decisions about how to sustainably use and conserve the ecological diversity of the Maine Woods. We fully recognize that it is difficult to avoid writing about nature without implying certain perspectives and perhaps even management directions. We hope that the patterns and principles discussed in this book speak largely for themselves.

The book presents new scientific contributions, including an expanded presettlement land survey analysis of Maine, the collation of attributes for old-growth remnants, updated calculations of disturbance rates for presettlement old-growth, and a revised chronology of deglaciation — contributions that have not been fully presented elsewhere at the time of this book's completion. But the book is largely a synthesis of work that has already been published in journals, books, websites, and the "gray literature." Some of that research is our own, but the vast majority is the product of others. Our review of that literature has given us deep respect for the years of hard work and the successes of these many scientists, historians, land managers, and conservationists. We are honored to present their results here. We also greatly appreciate the generosity of many friends and colleagues who have critically evaluated the writing and ideas in many drafts of the book; their contribution has been immense, and some of the words and perspectives you will read are theirs. We have no doubt made mistakes, for which we take full responsibility and offer our sincerest apology.

We owe many deeply felt thanks. For permission to use unpublished material: Bill Livingston, Woody Thompson, Cees van Staal, Jeffrey Bain, and Molly O'Guinness Carlson. For help with the creation of illustrations: Woody Thompson, Dan Coker (and The Nature Conservancy in Maine), Sarah O'Blenes, Matt McCourt, and Community GIS (Ken Gross and Stephen Engle). For going out of their way to contribute their art or illustration: Catherine V. Schmitt, Gary Hoyle, Woody Thompson, Dana Moos, Cees van Staal, Robert Hatcher, Jr., Molly O'Guinness Carlson, The Nature Conservancy in Maine, Eric Zelz, the USDA Forest Service Northern Research Station (Anantha Prasad, L. R. Iverson, S. Matthews, M. Peters), Laura Conkey, David Cappaert, Eric White, Don Bassett, Trevor Persons, Maryellen Chiasson (for the late Bill Silliker), Guoping Tang and Brian Beckage, Parker Shuerman, and Dan Grenier, For assistance in finding information: University of Maine at Farmington Mantor Library, Peter Goodwin at the Sagadahoc History & Genealogy Room (Patten Free Library, Bath), the Maine State Archives, the Maine Historical Society, and the many, many registry of deed offices around the state visited by Charles V. Cogbill. For critical reading of parts of the book: Dean Bennett, Bill Roorbach, Ann Dieffenbacher-Krall, Andrea Nurse, Tom Charles, Bill Haslam, Doug Reusch, Julia Daly, David Struble, Ann Gibbs, Allison Kanoti, Herb Wilson, Bill Livingston, Janet McMahon, Dan Grenier, Nancy Sferra, Barbara Vickery, Andy Cutko, Woody Thompson, Richard Judd, Jeffrey Bain, and the 2011 UMF Forest Ecology and Conservation class. A very special thanks to those who read multiple chapters or the entire book, investing many of their precious hours: Cyndy Stancioff, Valerie Huebner,

Dave Gorchov, David Foster, and Mac Hunter. Our sincerest gratitude to those who encouraged the genesis of this book: Bill Roorbach; Dean Bennett; and Sarah Sloane; and the generous support of our funders, the Furthermore Foundation, a program of the J. M. Kaplan Fund; the University of Maine at Farmington; the Maine Agricultural and Forest Research Station; and the School of Forest Resources, University of Maine. Many thanks to our editor, Richard Pult, and the entire staff at University Press of New England. And, finally, infinite gratitude to our families for their support and forbearance. Yes, dear, the book is finished!

THE CHANGING NATURE OF THE MAINE WOODS

1 | The Maine Woods

Through Time and Across Space

From this elevation, just on the skirts of the clouds, we could overlook the
country, west and south, for a hundred miles. There it was, the State of Maine,
which we had seen on the map, but not much like that—immeasurable forest
for the sun to shine on, that eastern stuff we hear of in Massachusetts. No
clearing, no house. It did not look as if a solitary traveler had cut so much as
a walking-stick there.
—Henry David Thoreau, during his 1846 journey to Ktaadn (Katahdin)

An Ancient Forest in Portland

From the I-295 Fore River bridge into Portland, the new Mercy Hospital
stands out, east across the water, below Bramhall Hill and the Western
Promenade. In 2007, when the land was being prepared for construction,
excavators struck a layer of ancient tree trunks (figure 1.1). A team of scientists was quickly assembled. Led by Woody Thompson of the Maine Geological Survey, the crew uncovered not just trunks and branches, but also
remnants of other plants, insects, marine bivalves, and, quite remarkably,
green conifer needles still attached to the preserved twigs of some long-
dead tree. Reconstructing this prehistoric scene was not easy. It required
the work of geologists, paleoecologists, and a tree-ring expert, who assessed

Figure 1.1 White spruce tree trunk, 13,500 years old, excavated in 2007 near the new Mercy Hospital site in Portland, Maine. Note the bent trunk, thought to be the result of a landslide.

sediment layers, identified species, dated fossils, and unearthed clues about events leading to the deposition of this prehistoric forest into the sea.[1]

In their recent article, published in the journal *Quaternary Research*, the scientific team concluded that these plants and insects belonged to a white spruce–balsam poplar woodland that thrived some 13,500 years ago on the slope of Bramhall Hill. Fossil pollen in lake sediments from across Maine tell us that this woodland type was widespread in the state after the massive continental glacier melted away and retreated north into Canada. During that time, the position of the sea relative to Maine's coast was as much as 125 feet above its present level and consequently much farther inland. Bramhall Hill was an island and the Fore River part of the ocean proper. The contemporaneous deaths of the excavated trees, their broken trunks, and the mixed sediment layers suggest that the slope of Bramhall Hill gave way suddenly and the forest plummeted into the sea, mingling with clay and sea shells of the Presumpscot marine sediments. The remnants of the ancient forest and the ocean floor were preserved *in situ* in the sediment layers, now above sea level.[2]

Southern Maine was colonized by Paleoindians[3] not long after the Bramhall Hill spruce forest met its demise.[4] Millennia before their first contact

THE CHANGING NATURE OF THE MAINE WOODS

Figure 1.2 Map of Maine, highlighting Portland and Moosehead Lake.

with Europeans, they had established rich cultures across the entire state, influencing the ecosystems within which they lived in ways that have been largely obscured by time. The first successful European settlements in southern Maine (English) were established in the 1620s and 1630s.[5] Extensive harvesting of big white pines for ship masts and lumber followed soon thereafter, along the lower parts of rivers such as the Piscataqua, Saco, and Presumpscot. But settlement and clearing of the presettlement forest was slow and sporadic, held back by warfare with the Abenakis and the French, until the ultimate English victories in the late 1750s. When the trickle of settlement became a river in the early 1800s, the impact on the forests of southern Maine was profound. Land was rapidly cleared for farms and timber extracted for export, shipbuilding, local construction, and fuelwood. By the mid-1800s, about one-third of southern Maine was open farmland and the logging frontier had moved Down East along the coast and north into the interior along the Kennebec, Penobscot, and St. John rivers.

Figure 1.3 Pitch pine woodland in southern Maine, showing a stone wall, evidence of past agriculture. Location: The Nature Conservancy Basin Preserve in Phippsburg.

By the mid-nineteenth century, external conditions again brought changes to southern Maine. Developing cities in Massachusetts and elsewhere, the advent of railroads for transporting food, and better land in the Midwest lured people away from farms and rural towns throughout the state. The population of Gorham, for example, only ten miles inland from Portland, declined from 3,351 in 1870 to 2,541 in 1900.[6] Urban folks from the burgeoning cities along the Atlantic seaboard began to venture to the Maine Woods for vacations and respite from the cities. By the late nineteenth century, farms were being abandoned, resorts were springing up, and the forest began to recover — part of a widespread retransformation in the Northeast of farmland to forest, described as an "explosion of green" by Bill McKibben.[7]

Southern Maine today is the most populated and developed part of the state. In between the towns, however, is a rural forest, and also some of the rarest ecosystems and highest levels of biological diversity in Maine. Development pressure has led to the most recent land use change: the purchase of conservation lands and easements. Current projects for the Mount Agamenticus conservation region, for example, will soon protect more than

14,000 acres through the efforts of two towns, two local land trusts, a water district, the Maine Department of Inland Fisheries and Wildlife, and The Nature Conservancy.[8] Most of this land is long-abandoned farmland. Although vehicles and roads are never far away, the southern Maine landscape today has more forest than it has had since the early 1800s.

Moosehead Lake

Moosehead Lake is 130 miles north of Portland (figure 1.4). The region feeds the headwaters of three of Maine's major rivers, the Kennebec, Penobscot, and St. John. These 130 miles make a big difference in terms of ecology and history. The extra two degrees of northern latitude, the additional 1,000 feet of elevation, and the inland location of Moosehead give it a pronounced northern climate. In fact, whereas the southern Maine coastal area is well ensconced in plant hardiness zone 5b (minimum winter low of –10 to –15°F), the Moosehead region is 4a (–30°F).[9] This chilly climate supports a northern forest of spruces, fir, northern white cedar, birches, maples, and boreal pines, which contrasts strikingly with the temperate plant cover of southern Maine.

The history of the Moosehead region also differs from that of southern Maine, starting long before the arrival of European settlers. When the

Figure 1.4 Aerial photograph, showing Moosehead Lake and alternating and mixing northern hardwood and spruce-fir forests.

white spruce–balsam poplar woodland was thriving on the side of Bramhall Hill 13,500 years ago, Moosehead Lake was just being freed of ice. The presettlement forest of the Moosehead region remained intact until much later than in southern Maine. European settlers did not move into the area until the early 1800s, and then in very small numbers, two centuries after the Euro-American colonization of southern Maine.[10] Even by 1857, during his second visit to the lake, Henry David Thoreau described the area as "boundless forest undulating away from its shores on every side, as densely packed as a rye-field, and enveloping nameless mountains in succession."[11] In fact, logging had already begun in the region, and soon the forests supported massive timber harvests, making the city of Bangor and the state of Maine world leaders in the timber industry. Unlike southern Maine, however, very little of the land was ever cleared for agriculture and other development. Repeated harvesting has altered the Moosehead forests over the past 150 years, from a little-disturbed forest dominated by large, shade-tolerant tree species (e.g., spruce, beech, maple) to stands with smaller trees and more fast-growing, light-demanding species (e.g., aspens, birches). Even today, however, the region continues to be overwhelmingly forested, and timber harvesting remains a major economic engine. As in southern Maine, nature tourism, which began not long after the establishment of the first towns, has also increasingly become an important source of livelihood in the region. This has led to new development pressures and an increasing emphasis on conservation lands. A highly controversial proposal for real estate and resort complexes on Plum Creek Timber Company land around Moosehead Lake, approved by the Maine Land Use Regulation Commission in 2009 but still in the courts, promises to magnify these trends.

Maine Forests in Time and Space

These two areas — southern Maine and the Moosehead Lake region — provide a microcosm of Maine's ecological diversity. Differences in climate, differences in topography, differences in history — from all of these have sprung divergent forested landscapes. Pull the lens back and the picture becomes an even richer mosaic. Maine is 320 miles from north to south, 190 miles east to west, and 5,267 feet from sea level to the top of Mt. Katahdin.[12] These 19.8 million acres (33,204 square miles) of land contain 17.7 million acres of forests — the highest percentage of any state in the United States.[13] The land supports coniferous forests resembling those stretching up to northern Canada but also temperate oak-hickory woods like those in the Carolinas; large, wet peatlands but also dry pine barrens; treeless grass-lands but also treeless alpine tundra; and northern hardwood forest but also southern softwood swamps. Canada lynx and jack pine, boreal icons, live here, but so do southerners such as the flowering dogwood and the black racer snake. Maine bounds seven plant hardiness zones (3a to 6a), so even backyard gardens express remarkable differences in growing potential.[14]

These differences in forest structure and species have developed over 15,000 years, since the land emerged from under the massive continental glacier that stretched from the Arctic to as far south as Ohio and Long Island. Maine's diversity of natural communities is the product of two phenomena. First, the broad range of climates and landforms encompassed by Maine's borders naturally gives rise to an accordingly diverse range of forest types. Second, each part of Maine has experienced a distinct history that has shaped the forests in a unique way. Some areas were cleared during European colonization and remain so; others were cleared and now have reverted to forest; and some — such as much of the vast Unorganized Territories — have been continuously covered by forest, despite multiple rounds of harvesting since European settlement. Even the timing of the retreat of the glaciers differed across the state, with land in the southwest emerging 14,000 to 15,000 years ago and some far northern areas not for another two millennia.[15] These two processes then — differences in the climate-topography template and differences in land-use history — have given rise to and maintain tremendous forest heterogeneity across the state.

About This Book

The goal of this book is to explore these forest patterns across space and through time. We approach this by addressing five fundamental ecological questions about Maine forests, each allocated to a separate chapter. How was the land formed over geological time, and, after the most recent glacial retreat, how did the forests recover and then change over the millennia (chapter 2)? What were Maine forests like before the profound influence exerted by Euro-Americans (chapter 3)? In what ways were the forests then transformed by four centuries of settlement and forest exploitation (chapter 4)? How and why do Maine forests vary so much from one place to another across the length and breadth and elevation of the state (chapter 5)? Finally, what will the Maine Woods be like in the future (chapter 6)? In other words, have we reached a point of stability, or are processes at work now that will continue to alter forests? The short answer to this last question is that the forces that shape forests — climate, insects and diseases, nonnative organisms, land use — are shifting in ways that are likely to considerably transform the Maine Woods. These changes will, in turn, profoundly influence the people who live in, visit, and cherish the natural environment of the state. A comprehensive analysis of these alterations will require an integration of answers developed in chapters 2 through 5. Answering the questions posed here is challenging. It requires multiple perspectives, knowledge developed in many overlapping disciplines, and scientific debate. Fortunately, the last two decades, even the last few years, have seen remarkable advances in our understanding of the relationships between forests and the environment. By applying these modern ecological principles and historical knowledge, we hope to illuminate the diversity and changing nature of the Maine Woods.

APPENDIX 1.1
Common and Scientific Names of Species Used in the Text

Names from Arthur Haines and Thomas F. Vining, *Flora of Maine* (Bar Harbor, ME: V.F. Thomas Co., 1998); Maine Forest Service, *Forest Trees of Maine*, Centennial Edition (Augusta: Maine Forest Service, 2008); Malcolm Hunter Jr., Aram J.K. Calhoun, and Mark McCollough, eds., *Maine Amphibians and Reptiles* (Orono: University of Maine Press, 1999); Susan Gawler and Andrew Cutko, *Natural Landscapes of Maine: A Guide to Natural Communities and Ecosystems* (Augusta: Maine Natural Areas Program, 2010); The American Ornithologist's Union, Checklist of North American Birds (retrieved on August 10, 2011, at www.aou.org/checklist/north); Don E. Wilson and DeeAnn M. Reeder, eds., *Mammal Species of the World: A Taxonomic and Geographic Reference*, 3rd edition (Baltimore, MD: Johns Hopkins University Press, 2005) (taxonomy retrieved on August 10, 2011, at www.bucknell.edu/msw3).

Alder (*Alnus*)
Alder, Speckled (*Alnus incana*)
Alewife (*Alosa pseudoharengus*)
Alpine Bearberry (*Arctostaphylos alpina*)
Apatosaurus or Brontosaurus
Apple (*Malus domestica*)
Armillaria Root Rot or Honey Fungus (*Armillaria* species)
Ash (*Fraxinus*)
Ash, Black (*Fraxinus nigra*)
Ash, Green (*Fraxinus pennsylvanica*)
Ash, White (*Fraxinus americana*)

Asian Longhorned Beetle (*Anoplophora glabripennis*)
Aspen (*Populus*)
Aspen, Big-Tooth or Popple (*Populus grandidentata*)
Aspen, Quaking or Popple (*Populus tremuloides*)
Auk, Great (*Pinguinus impennis*)
Auricled Twayblade (*Listera auriculata*)
Balm-of-Gilead (see Poplar, Balsam)
Balsam Woolly Adelgid (*Adelges piceae*)
Banded Bog Skimmer Dragonfly (*Williamsonia linteri*)
Barberry, Japanese (*Berberis thunbergii*)
Basswood, American or Linden (*Tilia americana*)
Bat, Eastern Pipistrelle (*Perimyotis subflavus*)
Bear, American Black (*Ursus americanus*)
Bear, Brown (*Ursus arctos*)
Bear, Grizzly (*Ursus arctos horribilis*)
Bear, Polar (*Ursus maritimus*)
Bear, Short-Faced (*Arctodus simus*)
Beaver, Canadian (*Castor canadensis*)
Beaver, Giant (*Castoroides ohioensis*)
Beech Bark–*Neonectria* disease (see Beech scale insect and *Neonectria*)
Beech Scale Insect (*Cryptococcus fagisuga*)
Beech, American (*Fagus grandifolia*)
Beetle, Japanese (*Popillia japonica*)
Birch (*Betula*)
Birch, Gray (*Betula populifolia*)
Birch, Heart-Leaved (*Betula papyrifera* var. *cordifolia*)
Birch, Paper or White (*Betula papyrifera*)
Birch, Sweet or Black (*Betula lenta*)
Birch, Yellow (*Betula alleghaniensis*)
Bittersweet, Asian (*Celastrus orbiculatus*)
Black Racer (*Coluber constrictor*)
Blanding's Turtle (*Emydoidea blandingii*)
Blueberry (*Vaccinium*)
Bobcat (*Lynx rufus*)
Bobolink (*Dolichonyx oryzivorus*)
Bog Fritillary (*Boloria eunomia*)
Bog Lemming, Northern (*Synaptomes borealis*)
Brown Spruce Longhorned Beetle (*Tetropium fuscum*)
Browntail Moth (*Euproctis chrysorrhoea*)
Buckthorn, Common (*Rhamnus cathartica*)
Butternut or White Walnut (*Juglans cinerea*)
Buttonwood (see Sycamore, Eastern)

Camel, Extinct (*Camelops* species)
Cardinal, Northern (*Cardinalis cardinalis*)
Caribou (*Rangifer tarandus*)
Cattail, Common (*Typha latifolia*)
Cedar, Atlantic White (*Chamaecyparis thyoides*)
Cedar, Northern White (*Thuja occidentalis*)
Cheetah (*Acinonyx jubatus*)
Cherry (*Prunus*)
Cherry, Black (*Prunus serotina*)
Cherry, Pin or Fire Cherry (*Prunus pensylvanica*)
Chestnut Blight (*Cryphonectria parasitica*)
Chestnut, American (*Castanea dentata*)
Chickadee, Black-Capped (*Poecile atricapillus*)
Cinnamon Fern (*Osmunda cinnamomea*)
Cobweb Skipper Butterfly (*Hesperia metea*)
Coltsfoot (*Tussilago farfara*)
Common Hairgrass (*Deschampsia flexuosa*)
Tachinid Fly (*Compsilura concinnata*)
Cottontail, New England (*Sylvilagus floridanus*)
Crabapple (*Malus*)
Crowberry Blue Butterfly (*Lycaeides idas empetri*)
Cut-Leaved Anemone (*Anemone multifida*)
Deer-Hair Sedge (*Trichophorum cespitosum*)
Deer, White-Tailed (*Odocoileus virginianus*)
Dicranum Moss (*Dicranum* spp.)
Dogwood, Flowering (*Cornus florida*)
Duck, Labrador (*Camptorhynchus labradorius*)
Dutch Elm Disease (*Ophiostoma ulmi*)
Dwarf Rattlesnake Root (*Prenanthes nana*)
Eagle, Bald (*Haliaeetus leucocephalus*)
Eastern Tiger Swallowtail (*Papilio glaucus*)
Elephant, African (*Loxodonta africana*)
Elm (*Ulmus*)
Elm, American (*Ulmus americana*)
Elongate Hemlock Scale (*Fiorinia externa*)
Emerald Ash Borer (*Agrilus planipennis*)
English Plantain (*Plantago lanceolata*)
Fir, Balsam (*Abies balsamea*)
Fisher (*Martes pennanti*)
Flycatcher, Alder (*Empidonax alnorum*)
Fox (Red Fox, *Vulpes vulpes,* and Gray Fox, *Urocyon cinereoargenteus*)
Furbish's Lousewort (*Pedicularis furbishiae*)

Grasses (family Poaceae)
Grouse, Ruffed (*Bonasa umbellus*)
Grouse, Spruce (*Falcipennis canadensis*)
Gypsy Moth (*Lymantria dispar*)
Hare, Snowshoe (*Lepus americanus*)
Hawthorn (*Crataegus*)
Hazelnut (*Corylus*)
Hemlock Woolly Adelgid (*Adelges tsugae*)
Hemlock, Eastern (*Tsuga canadensis*)
Heron, Great Blue (*Ardea herodias*)
Heron, Little Blue (*Egretta caerulea*)
Hickory (*Carya*)
Hickory, Pignut (*Carya glabra*)
Hickory, Shagbark (*Carya ovata*)
Holly (*Ilex*)
Honeysuckle (*Lonicera*)
Honeysuckle, Morrow (*Lonicera morrowii*)
Honeysuckle, Tatarian (*Lonicera tatarica*)
Hophornbeam, Eastern, Ironwood, or Hornbeam (*Ostrya virginiana*)
Hornbeam, American, Musclewood, Blue-Beech, or Ironwood
    (*Carpinus caroliniana*)
Horse, North American (*Equus ferus caballus*)
Huckleberry, Black (*Gaylusaccia baccata*)
Ips Bark Beetle (*Ips* species)
Ironwood (see Hornbeam or Hophornbeam)
Jelly Lichen (*Collema*)
Juniper, Common (*Juniperus communis*)
Katahdin Arctic Butterfly (*Oeneis polixenes katahdin*)
Knotweed, Japanese (*Fallopia japonica*)
Lapland Rosebay (*Rhododendron lapponicum*)
Lark, Horned (*Pooecetes gramineus*)
Leafy Lungwort (*Lobaria pulmonaria*)
Lion, African (*Panthera leo*)
Little Bluestem (*Schizachyrium scoparium*)
Loosestrife, Purple (*Lythrum salicaria*)
Lynx, Canada (*Lynx canadensis*)
Mallard (*Anas platyrhynchos*)
Mammoth, Woolly (*Mammuthus primigenius*)
Maple (*Acer*)
Maple, Norway (*Acer platanoides*)
Maple, Red (*Acer rubrum*)
Maple, Silver (*Acer saccharinum*)

Maple, Striped or Moosewood (*Acer pensylvanicum*)
Maple, Sugar (*Acer saccharum*)
Marginal Woodfern (*Dryopteris marginalis*)
Marten, American or Pine (*Martes martes*)
Mastodon (*Mammut americanum*)
Merlin (*Falco columbarius*)
Mile-a-Minute Vine (*Persicaria perfoliata*)
Mink Frog (*Rana septentrionalis*)
Mink, Sea (*Neovison macrodon*)
Mistassini Primrose (*Primula mistassinica*)
Moose, Eastern (*Alces alces*)
Moosewood (see Maple, Striped)
Mountain Ash (*Sorbus americana*)
Mountain Ash, Showy or Northern Mountain Ash (*Sorbus decora*)
Mountain Avens (*Dryas drummondii*)
Mountain Avens (*Dryas integrifolia*)
Mountain Laurel (*Kalmia latifolia*)
Mountain Lion (*Puma concolor*)
Mountain Holly (*Nemopanthus mucronata*)
Musk ox (*Ovibos moschatus*)
Muskrat (*Ondatra zibethicus*)
Naked Bladderwort (*Utricularia cornuta*)
Neonectria fungi (*Neonectria faginata* and *N. ditissima*)
Northern Blazing Star (*Liatris scariosa var. novae-angliae*)
Northern Comandra (*Geocaulon lividum*)
Northern Painted Cup (*Castilleja septentrionalis*)
Northern Woodsia (*Woodsia alpina*)
Oak (*Quercus*)
Oak, Chestnut (*Quercus montana*)
Oak, Northern Red (*Quercus rubra*)
Oak, Scrub or Bear Oak (*Quercus ilicifolia*)
Oak, White (*Quercus alba*)
Opossum, North American or Possum (*Didelphis virginiana*)
Orchid, White-Fringed (*Platanthera blephariglottis*)
Osprey (*Pandion haliaetus*)
Otter, North American River (*Lontra canadensis*)
Ovenbird (*Seiurus aurocapilla*)
Passenger Pigeon (*Ectopistes migratorius*)
Pine, Eastern White (*Pinus strobus*)
Pine, Jack (*Pinus banksiana*)
Pine, Pitch (*Pinus rigida*)
Pine, Red (*Pinus resinosa*)

Pine, Rocky Mountain Pinyon (*Pinus edulis*)
Pipit, American (*Anthus rubecens*)
Pitcher Plant (*Sarracenia purpurea*)
Plum, Canada (*Prunus nigra*)
Poplar (*Populus*)
Poplar, Balsam or Balm-of-Gilead (*Populus balsamifera*)
Popple (see Big-Tooth or Quaking Aspen)
Porcupine, North America (*Erethizon dorsatum*)
Puffin (*Fratercula*)
Pygmy Snaketail Dragonfly (*Ophiogomphus howei*)
Raccoon (*Procyon lotor*)
Ragweed (*Ambrosia artemsiifolia*)
Rand's Goldenrod (*Solidago simplex* var. *randii*)
Raspberry, Red (*Rubus idaeus*)
Raven, Common (*Corvus corax*)
Red-Winged Sallow Moth (*Xystopeplus rufago*)
Reindeer Lichen (*Cladina* species)
Rocktripe Lichen (*Umbilicaria* species)
Rose Family (Rosaceae)
Rose Pogonia Orchid (*Pogonia ophioglossoides*)
Round-Leaved Sundew (*Drosera rotundifolia*)
Salmon (*Salmo salar*)
Sandpiper, Upland (*Bartramia longicauda*)
Sapsucker, Yellow-Bellied (*Sphyrapicus varius*)
Sarsaparilla (*Aralia nudicaulis*)
Sassafrass (*Sassafrass albidum*)
Seal, Bearded (*Erignathus barbatus*)
Seal, Harp (*Pagophilus groenlandicus*)
Seal, Ringed (*Pusa hispida*)
Sensitive Fern (*Onoclea sensibilis*)
Shadbush or Serviceberry (*Amelanchier* species)
Sheep (*Ovis aries*)
Sheep Laurel (*Kalmia angustifolia*)
Sheep Sorrel (*Rumex acetosella*)
Skunk, Striped (*Mephitis mephitis*)
Sloth, Giant Ground (*Megatherium americanum*)
Sparrow, Grasshopper (*Ammodramus savannarum*)
Sparrow, Vesper (*Pooecetes gramineus*)
Spruce (*Picea*)
Spruce Budworm (*Choristoneura fumiferana*)
Spruce, Black (*Picea mariana*)
Spruce, Red (*Picea rubens*)

Spruce, White (*Picea glauca*)
Sudden Oak Death or ramorum blight (*Phytophthora ramorum*)
Sundew, Round-Leaved (*Drosera rotundifolia*)
Sycamore, Eastern (*Platanus occidentalis*)
Tall Cotton Grass (*Eriophorum angustifolium*)
Tamarack, Larch, or Hackmatack (*Larix laricina*)
Three-Toothed Cinquefoil (*Sibbaldiopsis tridentata*)
Thrush, Bicknell's (*Catharus bicknelli*)
Titmouse, Tufted (*Baeolophus bicolor*)
Tomato (*Solanum lycopersicum*)
Tulip Tree or Yellow Poplar (*Liriodendron tulipifera*)
Tupelo, Pepperidge, or Black Gum (*Nyssa sylvatica*)
Turkey, Wild (*Meleagris gallopavo*)
Turtle, Blanding's (*Emydoidea blandingii*)
Tussock Cotton Grass (*Eriophorum vaginatum* var. *spissum*)
Walrus (*Odobenus rosmarus*)
Warbler, Magnolia (*Dendroica magnolia*)
Warbler, Prairie (*Dendroica discolor*)
Warbler, Wilson's (*Wilsonia pusilla*)
Western Pine Beetle (*Dendroctonus brevicomis*)
Whale, Beluga (*Delphinapterus leucas*)
Whip-Poor-Will (*Caprimulgus vociferous*)
White Pine Blister Rust (*Cronartium ribicola*)
Willow (*Salix*)
Wolf, Dire (*Canis dirus*)
Wolf, Eastern Gray (*Canis lupus lycaon*)
Wood Sorrel, European (*Oxalis acetosela*)
Wood Turtle (*Glyptemys insculpta*)
Woodpecker, Pileated (*Dryocopus pileatus*)
Woodpecker, Red-Bellied (*Melanerpes carolinus*)
Wren, Carolina (*Thryothorus ludovicianus*)
Zigzag Darner Dragonfly (*Aeshna sitchensis*)

2 | From Rocks to Ice to Forests

How Did the Land and Forests of Maine Form
and Change over Geological Time?

Are these the end times? Yes. And they have been this way since the beginning.
Welcome to planet Earth, a wonderful but not entirely stable place to live.
— Craig Childs

The Bog in the Basin

We're standing in the middle of a wet black spruce bog in The Nature Conservancy's Basin Preserve on Phippsburg peninsula. I am here with Andrea
Nurse, from the University of Maine's Climate Change Institute, and our
three research assistants, Tommy Hannington, Bobby Harrington, and
Laura Lalemand. Our mission is to push a long metal cylinder through the
muck to the hard bottom, extracting a vertical section of the bog. Thousands of years ago, this place was simply a water-filled depression surrounded by sand and silt and stones discarded by the melting ice sheet. But
then plants invaded the land, and the basin began to fill with organic matter. Pollen, leaves, and seeds from the surrounding vegetation floated onto
the surface and sank to the bottom, depositing layer upon layer, year after
year — creating what is now a multimillennial time capsule of the changing
forests of the Basin Preserve.

Andrea and Tommy stand the Russian peat corer up on its pointy end, attach the T-bar handle, and begin pushing it into the sediments. The metal slashes quickly through the top few inches corresponding to recent decades of secondary forest, then through a couple centuries of farmland now long abandoned, back to the beginning of European settlement in the seventeenth century, into the remains of the mixed spruce and oak forest before European colonization, when Indians lived off the woods and nearby ocean. The corer's three-foot chamber is full, so we give it the heave-ho and out it comes with a great sucking sound. We carefully empty the contents — a four-inch-diameter tube of organic matter — onto a semi-circular piece of rigid plastic, just the right size to accommodate the core. Andrea and Tommy add an extension to the borer and drive it back into the bog. Four, five, six feet into the past, the forest constantly changing: hemlock and spruce, then hemlock is gone, replaced by oaks, beech, and birches. Another full chamber; we repeat the process. Hemlock again and then it disappears for good, replaced this time by white pine and oaks, dominating the core for many inches and centuries. Now it's just spruce and some tundra plants; we're almost back to glacial times. The sediments suddenly resist, and Andrea, breathless, says, "We've hit the bottom." We extract the borer and lay out this bottom core. It looks different than previous ones — more solid, little identifiable organic matter. At the top is a dark smooth substance: plant matter that has been transformed by compaction and downward oozing of humic acid. Below that a grayish clay. And finally at the bottom: seashells. As much as we expected that — Phippsburg was under the ocean just after the ice melted — it's still wondrous to see. On this day, scientific discovery is not a dry enterprise undertaken by the dispassionate, but an endeavor, full of surprise and inspiration, pursued by a wet, motley crew.

The Dynamic Earth

Our nearly twelve-foot core is but a small piece — less than 15,000 years — of the long timeline of the land we call Maine. Before that, the ground was under ice for nearly 100,000 years. Before that, there was warmth and forests and life. And before that, ice again. Over millions of years, the very fabric of the Earth's surface has dramatically transformed, as continental plates were newly created and eventually descended back into the hot interior of the planet. In this chapter, we will begin to describe how the environment of Maine came to be what it is today. In so doing, we will discuss some of the fundamentals of ecology and earth science, including the newest ideas about the cycles that govern change over time. Two main types of geological forces have created and modified the Maine landscape. First, the raw material of the land was formed over hundreds of millions of years by creation, movement, alteration, and destruction of the Earth's crust — *tectonics*. Second, this raw material was reshaped by periodic *glaciation*, a product of *climate cycles* that drive the Earth from greenhouse warmth to icehouse cold. We will examine each of these sets of processes in turn. Then, we will focus on the last 15,000 years, the time during which Maine recovered from the most recent cataclysm: the massive continental glacier that covered much of northern North America. We will look in detail at how the forest colonized the newly liberated land and how it changed in response to continuing climate fluctuation and other ecological factors. This exploration will take us to the beginning of the period — about 400 years ago — when Euro-American settlers began to remake the forests of Maine.

Origin of the Bedrock of Maine

"Utter, damned rot," said the president of the American Philosophical Society, one of the most eminent scholarly institutions of its day.[1] He was speaking about Alfred Wegener's new theory of *continental drift*, which proposed that landmasses moved around the Earth, constantly reshuffling its surface. Wegener explicated the theory in his 1922 book *The Origin of Continents and Oceans*, but it wasn't until the 1960s that evidence supporting this remarkable theory convinced the majority of geologists.[2] Part of their reluctance was the lack of a clear, known mechanism that could propel the continents across the face of the Earth. In the 1960s, Harry Hess hypothesized that an elevated seafloor ridge in the middle of the Atlantic Ocean was created by upwelling hot rock from the mantle, leaking through the crust to the surface. He argued that this new material spreads out in each direction, east and west, creating new sea floor, and, as a consequence, a widening basin. He suggested further that crustal and upper-mantle material was eventually recycled in places where one moving continental plate converged with another. At these so-called *subduction* zones, one plate would dive under the other, rejoining the mantle. Here was a powerful mechanism that could explain how oceans widen and contract, continents combine and separate, and landmasses move around the lithosphere, the rigid outer portion of the Earth.

Unlike the common usage of the word to mean "speculation," *theory* is the pinnacle, the Holy Grail of science. The term is attached only to those major ideas that have a solid logical cause-and-effect basis, explain a large number of observed events and phenomena in nature, and are supported by repeated tests and multiple lines of evidence. Examples include the theory of gravity, biological cell theory, the theory of evolution, and the atomic theory of matter. By the 1970s, *plate tectonics theory* — as it came to be known — fulfilled these criteria and became the new framework for the earth sciences. The establishment of theories is rarely as cut-and-dried as this sounds, however, usually requiring years of struggle, debate, and controversy before full development and acceptance, as in the case of plate tectonics.

The land we call Maine was born in the proto-Atlantic zone of plate tectonics between what is now North America and Europe (figure 2.1). Approximately 500 million years ago, the Iapetus[3] Ocean separated ancestral North America, called Laurentia, from the huge landmass of Gondwana, centered over the South Pole. Two microcontinents, Ganderia and Avalonia, peeled off of this southern land mass and became source material for a future New England.[4] Over the next 100 million years or so, the two large continents first spread farther apart, enlarging the Iapetus Ocean, and then reversed direction, eventually colliding, erasing the ocean and creating the giant supercontinent Pangaea (meaning "all land") by around 400 million years ago. This collision rammed Ganderia and Avalonia into the developing North American continent, pushing up the Appalachian Mountain chain. Another consequence of this collision was the intrusion of large quantities of magma that crystallized into granite, eventually forming features such as Mount Katahdin. A new spreading rift formed around 150 million years ago, creating the incipient Atlantic Ocean, with Maine left on the North American side. Since then, the Atlantic Ocean has gradually widened as the seafloor spreads from the mid-oceanic ridge. Meanwhile, Maine has moved latitudinally from an equatorial location of around five degrees north to its present location at about 45 degrees north. In places such as Phillips and Hurricane Mt. (near Moosehead Lake), geological sutures, with contrasting rocks on opposite sides, provide modern-day traces of the ancient closure of the Iapetus Ocean.[5]

Weathering of the surface over these vast amounts of time played just as important a role in shaping the land as movements and upheavals of plates. The rocks of today's Maine were at one time buried miles under the ground, and have been exposed by millions of years of gradual, persistent erosion. The state's bumpy topography of hills and valleys is the product

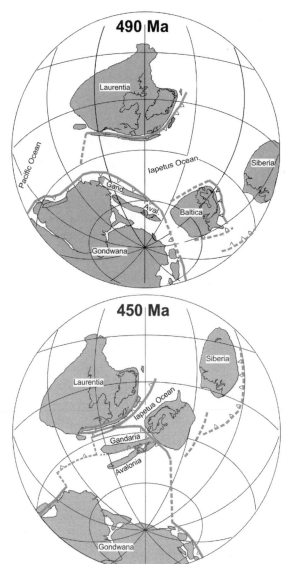

Figure 2.1 Location of continents and microcontinents 490 and 450 million years ago. The land that today makes up the state of Maine had its origins in two microcontinents, Ganderia ("Gand") and Avalonia ("Aval"). Solid lines represent established subduction zones; dashed lines represent possible additional subduction zones.

of differential erosion, which occurs because some rocks, like the granite of Mt. Katahdin, are harder and more resistant to this process, but also because of spatial variability in the action of surface erosion.

The complex bedrock geology of Maine we observe today, then, is the product of hundreds of millions of years of plate construction and destruction, significant latitudinal shifts, tilting and folding, and weathering and erosion. This deep geological history has created tremendous heterogeneity in landforms and contributed to the diversity of soils across the state, which in turn has promoted a corresponding diversity of forest communities, a subject we will take up in chapter 5.

Climate Change

In a recent article, James Zachos and colleagues wrote, "It has become increasingly apparent that during much of the last sixty-five million years and beyond, Earth's climate system has experienced continuous change, drifting from extremes of expansive warmth with ice-free poles, to extremes of cold with massive continental ice-sheets and polar ice caps"[6] (figure 2.2). Fluctuations in climate over the vast time spans of the geological record might seem unspectacular. But recent research has revealed that climate change characterizes not just periods of millions and thousands of years, but also centuries and even decades — a scale perceptible within a human lifespan.[7]

The profound climate swings of the past 20,000 years have the most bearing on the structure and composition of Maine forests today. An understanding of the development of those forests is enriched, however, by an examination first of the general processes that underlie climate change. This is one of the most active areas of scientific research today, in no small part because of its application to current and future human-caused climate change. Because of the enormous complexity of climate, the vast time scales involved, and the profound relevance to the fate of the planet, climate science is rife with new discoveries and debate.[8]

Weather refers to conditions (temperature, humidity, precipitation, wind, etc.) at a particular time and place, whereas climate is the long-term average of weather for a given area.[9] People expect weather to vary, but why does climate fluctuate so dramatically over time? Although we don't have a complete answer, scientists have identified three main "knobs" that *directly* control the Earth's climate system: insolation, albedo, and greenhouse gases (see figure 2.3).[10] *Insolation* is the amount of energy reaching the Earth from the Sun. Orbital cycles, explained below, influence climate by altering the amount of insolation striking the Earth (usually measured as watts per m^2 of land surface) and its latitudinal and seasonal distribution. *Albedo*, which is the reflectivity of the Earth's surface, determines how much of the Sun's insolation is actually absorbed by the planet and how much gets reflected back to space. For example, more ice means higher albedo and lower absorption of insolation. *Greenhouse gases* in the atmosphere

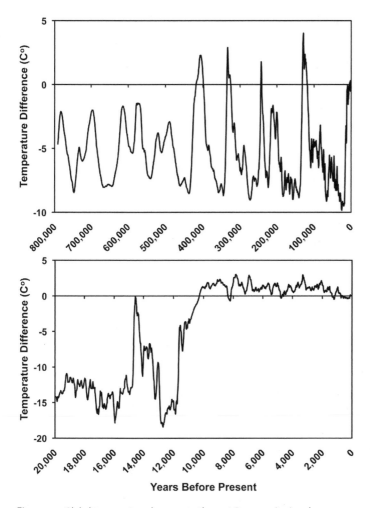

Figure 2.2 Global temperature change over the past 800,000 (*top*) and 20,000 years (*below*). Data, expressed in comparison to average temperature over the past 1,000 years, are from hydrogen isotope ratios of water from Antarctica (EPICA Dome C) and Greenland (GISP2), respectively.

allow the Sun's energy to pass through largely unobstructed on its way to the Earth's surface, but then trap a portion of the heat energy that gets re-radiated back toward space. It's not surprising, then, that conditions were relatively warm during periods in the geological record when atmospheric greenhouse gas concentrations were elevated. Tectonic processes also strongly alter climate over long periods of time, acting indirectly through greenhouse gases and albedo — by governing the amount of rock weathering that removes carbon dioxide from the air and deposits it on the ocean

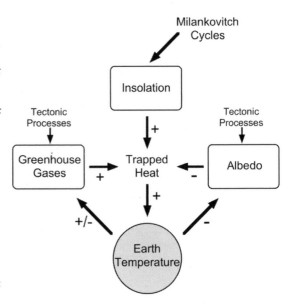

Figure 2.3 Three main physical drivers of global temperature over long periods of time. *Insolation:* radiative energy received from the Sun. *Albedo:* degree to which Earth's surface reflects energy back toward space. *Greenhouse Gases:* atmospheric concentration of heat-trapping gases. *Heat Trapped:* amount of heat retained by Earth's atmosphere. *Milankovitch cycles:* variations in the Earth's orbital cycles. *Tectonic processes:* volcanoes, continental plate movement, and plate construction and destruction. Arrows signify that one factor influences the magnitude of another, with "+" indicating a positive effect and "−" a negative effect.

floor, by determining the abundance of carbon dioxide–producing volcanoes, and by rearranging the continents, which influences ocean and air currents, the growth of ice sheets, and the abundance of carbon dioxide–absorbing plants (figure 2.3). In fact, tectonic processes seem to drive the major planetary shifts from ice ages to greenhouse conditions lasting tens of millions of years.[11]

The climate has been especially volatile over the past 2.6 million years, a geological time span known as the Quaternary period, consisting of the Pleistocene epoch and the Holocene epoch, which began only 12,000 years ago.[12] The Quaternary has been subject to repeating glaciations, approximately 100,000 years apart, evidence for which comes mainly from cores extracted from glaciers and ocean sediments. Each cycle is characterized by roughly 90,000 years of glacial conditions and only 10,000 years of interglacial warmth, such as the current time, during which human civilization has developed. These regular climate fluctuations appear to be driven in part by three orbital cycles (the *Milankovitch cycles*) that govern the amount and seasonal distribution of energy received by the Earth (figure 2.3). The Earth spins around a tilted axis each day and orbits around the Sun each year. The Milankovitch cycles occur because aspects of these parameters — orbital shape, tilt of the Earth, and trueness of the spin — are not stable over time.

First, the Earth's orbit describes not a circle but an ellipse. The degree of departure from circularity is known as *eccentricity*. Over about a 100,000-year period, the eccentricity rises, reaches a peak, and then declines back to its original state. Increased orbital eccentricity increases seasonal variation in the amount of energy the Earth receives.[13] A second cycle involves changes in *obliquity*, the tilt of the Earth on its axis. Over a cycle of approximately 41,000 years, the Earth's tilt varies by about 2.4 degrees. Given that the very existence of seasons north and south of the equator depends on the axial tilt, it should not be surprising that more tilt leads to warmer summers and cooler winters (i.e., enhanced seasonality). Finally, the Earth does not spin true on its axis but instead wobbles as it turns, like a top. Over a cycle of about 20,000 years, this wobble fluctuates from smaller to larger and then back. The degree of wobble — called *precession* — also affects the difference in seasonality between the northern and southern hemispheres. At the current point in the precession cycle, seasonality in the Southern Hemisphere tends to be more extreme than for the Northern Hemisphere. Here's the punch line: these three cycles combine about every 100,000 years to produce a nudge in insolation sufficient to induce a dramatic change in climate.[14]

When Milutin Milankovitch proposed this idea in 1941,[15] many scientists objected because the changes in energy input to the Earth seemed too small to lead to the drastic alterations apparent in the climate record. Current thinking actually agrees with this notion by viewing the Milankovitch cycles as merely a trigger that sets in motion a series of *positive feedbacks* (where one effect magnifies another) that together lead to climate change. The most important triggering effect of the cycles is often not the change in total energy input, but instead the alteration of the degree of seasonality and the onset of seasons. For example, even modestly warmer summers can reduce ice cover, lowering the planet's albedo, leading to more absorption of the Sun's energy, resulting in further rise of global temperature, melting even more ice, further increasing temperature — in other words, cascading positive feedbacks and global warming. Over longer periods of time, natural rises in greenhouse gases produced by changing tectonic processes can play a similar role, warming the atmosphere, triggering a set of other alterations that leads to drastic climate change. A full understanding of the causes of climate change requires accounting for both the trigger and the entire cascade of positive feedbacks, a subject to which we will return in chapter 6. Even such accounting has not fully convinced all climate scientists about a primary role for Milankovitch cycles in glacial cycles, and many questions remain in this rapidly evolving area of climate science.[16]

Deglaciation and the Emergence of Land in Maine

Twenty-five thousand years ago, Maine was covered by glacial ice, part of the massive Laurentide Ice Sheet that stretched from the Arctic to as far south as present-day New York City. Triggered by the Milankovitch cycles, the Earth then began to warm. By about 12,000 years ago, Maine had been liberated of ice, ending the Pleistocene epoch and ushering in the current Holocene epoch. The process of deglaciation was complex, involving rapid temperature rise as well as abrupt temporary declines, and also changes in both global sea level and the elevation of the landmass that was to become Maine. Our summary below (and see figure 2.4) is from the most detailed chronology to date, based on the work of Harold Borns (University of Maine), Woodrow Thompson (Maine Geological Survey), and their many colleagues.[17]

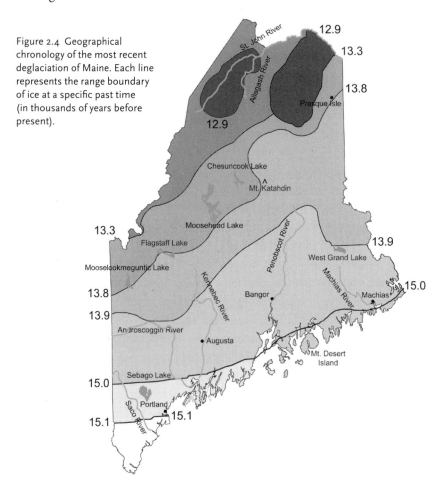

Figure 2.4 Geographical chronology of the most recent deglaciation of Maine. Each line represents the range boundary of ice at a specific past time (in thousands of years before present).

During maximum glaciation, about 25,000 years ago, ice covered not only the land of present-day Maine but also extended as far as 180 miles eastward to the present location of the Georges Bank in the Atlantic Ocean. The ice sheet began melting back about 21,000 years ago, and, by 15,000 years ago, had retreated to near the present coast of Maine. Thereafter, the ice withdrew steadily inland, but the appearance of coastal areas was delayed by a period of marine submergence. The weight of the ice sheet residing in northern New England was so great that it actually depressed the continental crust hundreds of feet below today's level. A little math can be very convincing on this issue. Ice weighs about 62.4 pounds per cubic foot.[18] The land area of Maine is about 35,000 square miles or 975,000,000,000 square feet. One and a half miles of ice over this area produces a volume of 7,727,000,000,000,000 cubic feet. That much ice would weigh 482,000,000,000,000,000 pounds, or a bit more than 240 quadrillion tons! The Earth's crust remained depressed for a time as the ice retreated, enabling the sea to flood onto land that today is coastal and central Maine. Dating of marine fossils extracted from deep lake sediments suggests that about 14,000 years ago, the sea reached as far inland as Livermore Falls, Farmington, Bingham, and Medway, nearly 100 miles from the present coast. (Ocean-front property in Jay, Maine!) As the ice sheet melted, its mass declined, and the crust began to rebound, elevating and draining the land. Eventually, rebound progressed very rapidly — at least in the context of geological phenomena — causing sea level to drop at a rate as much as 100 feet per thousand years.

Between 14 and 15,000 years ago, the first dry land was exposed. By 14,000 years ago, glaciers had retreated northward nearly halfway up the state. By 13,300 years ago, only far northern Maine remained covered in ice and the retreating sea had liberated most of southern Maine. By about 13,000 years ago, the ice was apparently reduced to a remnant ice cap near the Quebec border. Between 12 and 13,000 years ago, with continued crustal uplift, the coastline actually retreated seaward of where it is today, but then advanced slowly back toward its present location, as global sea level rose in response to continued release of water from glaciers.

In the movie *The Day After Tomorrow*, a globally warmed northern hemisphere is suddenly thrust into an ice age, with temperatures dropping to −150°F in a matter of hours (which does not discourage the hero scientist from hiking from Philadelphia to New York!). Sound unlikely? This scenario is actually very loosely based on a real climate event. Just when ice had nearly released its hold on Maine, about 12,900 years ago, the climate did a dramatic about-face, plunging back into near-glacial conditions.[19] This period, called the *Younger Dryas* (after *Dryas drummondii*, a tundra wild-flower), lasted about 1,300 years and occurred across much of the world. Remarkably, the cooling took hold over only one or two decades — not quite on par with *The Day after Tomorrow* scenario, but nonetheless very rapid. One hypothesis for this event lays blame on a catastrophic breach of an ice dam impounding Lake Agassiz in Canada. The sudden influx of fresh water into the North Atlantic could have disrupted the North Atlantic current moving warm equatorial water to northern latitudes. The Younger Dryas period led to readvances of the ice sheet in parts of Europe and the Canadian Maritimes. For Maine, evidence suggests a resurgence of ice in the Aroostook River valley. The Younger Dryas ended about 11,600 years ago even more abruptly than it began, with temperatures warming by 5 to 6.5°F in Maine and as much as 15°F in Greenland in a decade or less, after the apparent reestablishment of the North Atlantic current. Soon after, Maine was ice-free.[20]

How Glaciation Shaped the Maine Landscape

If the tectonic processes of continental construction and destruction can be thought of as a sculptor's rough-cut of a block of rock, then the action of glaciers is analogous to the fine sculpting that finishes the object. It might be a stretch to imagine a chunk of ice, one and a half miles thick, quadrillions of tons in weight, as a chisel. In fact, as it flowed, the ice sheet and its embedded rocks gouged, plucked,[21] excavated, flattened, smoothed, graded, scratched, and shaped the existing raw material of land created over millions of years. This metaphor takes us only so far, however, for the "waste products" of the sculpting — the enormous load of debris eventually released with melting — were then dumped directly on top of the land, obscuring many of the contours.

The effects of glaciers, then, can be divided into bedrock shaping and debris deposition. As ice sheets moved south,[22] they smoothed the northern sides of exposed mountains and plucked the bedrock on the south face, creating the impressive cliffs of Mt. Kineo (figure 2.5) and other peaks. This action also produced the landforms that now provide diverse habitats for plants and animals thriving in open, rocky environments, in places like the coastal Camden Hills, western Maine's Tumbledown Mountain, and Mt. Katahdin. Elsewhere, glaciers scoured out broad U-shaped valleys, for in-

Figure 2.5 Evidence that glaciers shaped the Maine landscape (*clockwise from top left*): plucked cliff on Mt. Kineo on Moosehead Lake; road running along five-mile-long Chesterville Esker with ponds on each side; till showing a heterogeneous mix of different-size rocks and rock particles; and U-shaped valley in Grafton Notch, scooped out by glacial action.

stance, in Grafton Notch in western Maine (figure 2.5) and Somes Sound on Mount Desert Island.[23]

The debris left over from glacial action is known as *drift*, the bits picked up or scraped off as the ice advanced and retreated, much of it frozen in the ice itself, and then dumped as the ice melted. Drift comes in two forms. *Stratified drift* has been sorted by water and wind into homogeneous layers of a single particle size, for example, sand or gravel. Sorting occurs as heavier particle sizes (e.g., rocks) fall out of the water or wind, while lighter ones (e.g., sand) get transported farther before deposition. Sorted layers formed when melting glaciers dumped enormous loads of accumulated sand, gravel, and rock at their retreating margins or in meltwater tunnels within the glacier, creating a wide variety of landforms. Moraines, which are ridges or mounds of debris dropped at a glacier's edge, are especially common near the coast. Eskers (figure 2.5), long, narrow, steep ridges formed as stream deposits in ice tunnels, occur across most of the state, for instance, the Whalesback in Hancock County. Drumlins, which are common in southwestern Maine, are smoothed elevated hummocks smeared across the landscape under actively flowing glaciers and elongated in the direction of glacial movement. Where meltwater moved off the ice sheet onto exposed land in southern and coastal Maine, the sorting action of meltwater produced sand deltas, which gave rise to the sand plains common in those areas, such as Kennebunk Plains, described in chapter 5 ("Six Maine Nature Snapshots"; figure 5.2).

Unsorted drift is called *till*: loose, heterogeneous mixtures of mud, sand, gravel, cobbles, and boulders (figure 2.5). Despite the dramatic imagery of glacial-shaped bedrock and steep-sided landforms of sorted drift, till has had the most ubiquitous impact on the postglacial Maine landscape. This is quickly borne out by the map of the state's geological surface, which shows eskers, exposed bedrock, stream and wetland deposits, glacial lake sediments, and glacio-marine silt and clay where coastal and central areas were inundated. Till, however, is by far the dominant map feature, covering most of the state, profoundly influencing soil conditions and drainage.[24]

Like all deposits in Maine, till has had relatively little time to develop into soil since deglaciation, and much of the source parent rock was intrinsically acidic. As a result, most Maine soils exhibit only modest fertility. On the other hand, in some places, highly compacted *lodgment till*, created by compression of till that has melted out of basal ice, "was a gift to the pioneers.... It provided a physical barrier that blocked ... seepage into the earth. This kept the soil moist, gave rise to perennial springs, and trapped the water in small ponds needed to water livestock."[25] Many of the valleys of Maine were also plugged by this glacial debris, retarding water drainage, and creating today's vast systems of lakes, ponds, meandering water courses, swamps, marshes, fens, and bogs.

Here was the land of Maine, some 12,000 to 15,000 years ago: exposed bedrock, sculpted mountains, stratified drift organized into topographically coherent shapes, marine sands and clays near the coast, myriad raw lakes and ponds and streams, and a layer of loose glacial till covering just about everything else. Millions of years of tectonic processes and the advance and retreat of the ice sheet had created a complex landscape that would guide the reestablishment of forests and govern the diversity of ecosystems in Maine.

Life in a New World: The Forest Reconstituted

What did newly deglaciated Maine look like? Raw soil. No green life. Mud, silt, sand, gravel, stones, boulders — in some places sorted by size, elsewhere thoroughly mixed. Imagine the floor of a gravel pit. These postglacial environments pose serious challenges to plants: undeveloped soil, scarce nitrogen vital to plant growth, little anchorage for roots, and stark exposure to the vagaries of weather. Even for plants adapted to such stresses, there's the matter of actually getting their seeds or spores from established plants to these newly available sites. As a result, establishment of vegetation on such land occurs slowly, sometimes requiring hundreds of years.

By analyzing preserved plant pollen and macrofossils (seeds, needles, etc.) in sediments (sidebar 2.1), paleoecologists have discovered that most of the land exposed in Maine by retreating glaciers between 12,000 and 15,000 years ago[26] at first supported a low tundra vegetation (figures 2.7, 2.8) consisting of grasses, sedges, mosses, herbs, and woody shrubs such as dwarf alders, willows, and birches (scientific names are given in appendix 1.1). Not long after, trees such as spruce, poplars (aspens), and white birch appeared.[27] It should come as no surprise that these *pioneer species* are well adapted to the rigors of a postglacial environment. The earliest species tend to grow slowly, have modest nutrient requirements, and maintain a prostrate growth form, taking advantage of the warmth of the boundary layer near the ground. The wood tissue of postglacial pioneers, including the trees, is composed of small tubes, which reduce the incidence of bubble formation during thawing, preventing embolisms that can block water flow up the trunk. Most pioneer species have small, wind-dispersed seeds, or spores in the case of mosses and ferns, capable of traveling to distant, recently exposed land. According to pollen chronologies, these early species were soon joined by alder and pines (jack and red[28]), with later additions of balsam fir, larch, ash, and elm. The assemblages first formed open woodlands, which later coalesced into spruce-dominated, closed canopy forests, indicated by declining pollen abundance from herbs, shrubs, and tree species such as poplars, which perform poorly in shaded conditions (figures 2.7, 2.9).

This development of forest from lifeless land was driven in part by the passage of time and the presence of pioneer plants, which slowly built up the soil through weathering, trapping of wayward particles, and the trans-

Past Worlds in a Grain of Pollen

Wouldn't it be great if ecologists could go back in time to sample past environments? Well, they can—sort of. After deglaciation, Maine was dotted with lakes and ponds, onto which all manner of windblown flotsam blew and sank to the bottom: detritus such as pollen, seeds, cones, and leaves of plants growing in the area, and even bits of ash and charcoal from fires. These aquatic habitats filled slowly with this organic matter; some remained as lakes, while others developed into bogs or fens.[1] Low oxygen levels kept decomposition rates minimal in these sediments, and the embedded plant parts were preserved.

Because each year's sediment forms a layer on top of past years', the entire vertical column represents a chronological time capsule of vegetation and environment. Moreover, the date of deposition of each slice of the column can be determined using radiocarbon dating. Carbon comes in two natural forms: a stable form (^{12}c) and a radioactive form (^{14}c), which naturally decays into the stable form over time at a known rate. Thus, by measuring the ratio of ^{12}c to ^{14}c in preserved organic matter, scientists can estimate the time since deposition (i.e., when that matter was part of a living organism).[2]

The concept of peat core analysis is simple but the work isn't. A vertical core must be collected, by forcefully inserting and then extracting a hollow metal cylinder into the lake bottom or peat bog (see figure 2.6). Very small subsamples are sectioned from multiple levels (i.e., depths) along the core. Some material is sent off to specialized labs for radiocarbon dating; other samples are chemically prepared for examination under a microscope. Researchers spend many hours identifying and quantifying charcoal fragments, macrofossils (seeds, cones, needles, etc.), and pollen in the samples from each slice in time. Finally, the numbers are crunched to develop a graphical view of the changing relative abundance of different plant species and environmental conditions over time, called a *pollen chronosequence* or *profile* (figure 2.7).

An ecologist's fantasy fulfilled? In some ways yes. This technique allows paleoecologists to reconstruct past environments and vegetation with sufficient resolution to determine the relative abundance of individual tree species in an area at some fairly specific time in the distant past. By using modern information about the moisture requirements of plant species or certain fossilized microorganisms, these methods can even allow estimates of past regional lake and precipitation levels.[3] There are important biases and limitations, however: resolution is only about fifty years at best (mainly due to the costs of sampling more finely), species that produce large quantities of pollen or plant parts can be overrepresented in samples, species without wind-dispersed pollen are hardly represented at all, and species within the same genus can be difficult to distinguish. Despite these drawbacks, this method is an extraordinarily powerful tool for deciphering the past.[4]

Figure 2.6 Peat core, about 4" in diameter, extracted from a bog in The Nature Conservancy's Basin Preserve, Phippsburg, Maine, as described at beginning of the chapter.

NOTES

1. Wetlands have been classified as follows, from Ronald B. Davis and Dennis S. Anderson, "Classification and distribution of freshwater peatlands in Maine," *Northeastern Naturalist* 8 (2001): 1–50. Bogs are characterized by a dense layer of peat, low nutrients derived only from the atmosphere, highly acidic conditions, low productivity, and low decomposition rates. Fens are covered in peat and exhibit a higher level of nutrients derived both from the atmosphere and ground water and higher rates of productivity and decomposition compared to bogs.

2. For details on these methods, see Owen K. Davis, "Palynology: an important tool for discovering historic ecosystems," in Dave Egan and Evelyn A. Howell, eds., *The Historical Ecology Handbook* (Washington, DC: Island Press, 2001), 229–255.

3. Two examples from Maine: Heather Almquist et al., "The Holocene record of lake levels of Mansell Pond, central Maine, USA," *The Holocene* 11 (2001): 189–201; Ann M. Dieffenbacher-Krall and Andrea M. Nurse, "Late-glacial and Holocene record of lake levels of Mathews Pond and Whitehead Lake, northern Maine, USA," *Journal of Paleolimnology* 34 (2005): 283–310.

4. This sidebar is but a sample of the methods of paleoecology. For example, in the Southwest, where natural bodies of water are uncommon, packrats collect plant parts (and other treasures) and cache them in a pile, which can be sorted into layers and dated, providing a chronology of past vegetation and environment.

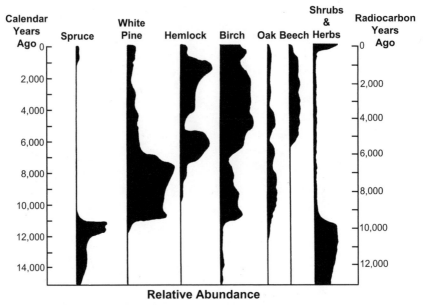

Figure 2.7 Generalized graph for central Maine of changes over time in relative abundance of tree species, indicated by fossil pollen. Bar thickness represents abundance of pollen type in lake sediments. Both calendar age (left side) and uncalibrated radiocarbon dates are provided.

formation of living matter into soil organic matter generation after generation. Enhanced nitrogen availability, water-holding capacity, and anchorage would then have promoted the success of a broader range of species, less able to tolerate the rigorous early conditions. This process, called *primary succession*, would have eventually led to a complex layer of vegetation and an organic soil into which plants were rooted. As important as these processes might have been in the early revegetation of the land, the transformation from open tundra to closed spruce forest was probably driven largely by the warming climate, which eventually favored boreal over subarctic vegetation.[29]

15,000 Years: Changing Environments, Changing Forests

By at least 13,500 years ago in southern and 12,000 years ago in northern Maine, boreal woodlands and forests covered the land. This was only the beginning of the story of late Pleistocene and Holocene vegetation change in the state. Driven by continuing climate fluctuations, as well as further migrations of trees from the south, the species composition and physical structure of Maine forests went through several major transformations

Figure 2.8 Tundra with sparse white spruce woodland, from Denali National Park, Alaska.

Figure 2.9 Spruce-fir forest in western Maine.

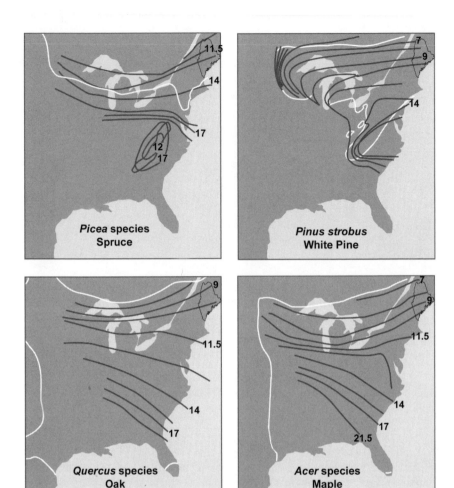

Figure 2.10 Postglacial migration of spruces, maples, oaks, and white pine into Maine. White lines show current ranges in eastern North America. Dark lines represent northern range boundaries at specific times in the past, and each graph tracks migration in response to warming. Dates are thousands of calendar years before present. Outline of Maine is shown.

during the following millennia, right up to the time of contact between Europeans and Indians (figure 2.7).

As described previously, the rise in temperature leading to deglaciation was abruptly interrupted about 12,900 years ago, ushering in the 1,300-year Younger Dryas period. This cooling temporarily retarded the progression toward more temperate forests, favoring tundra in northern Maine and the Maritimes and open spruce woodlands in Vermont, New Hampshire, and central and southern Maine. In southern New England, boreal spruce forests returned and temperate oak forests declined.[30]

Figure 2.11 Red oak–white pine forest, common in southern Maine. This photo shows oaks in the foreground without their leaves.

Sudden warming at the end of the Younger Dryas completely reshaped the vegetation of Maine.[31] Some of the most familiar tree species, such as oaks, maples, and white pine, made their first postglacial appearance at that time (figure 2.10). White pine (sidebar 2.2) and oaks were so successful in their new setting that they and other temperate species largely replaced the spruce forests across the state and were, until about 8,000 years ago, the most abundant tree species (figures 2.10, 2.11). Compared to the rapid climate and forest changes associated with the Younger Dryas, this was a long, relatively stable period of nearly 4,000 years.

Several lines of evidence suggest that this transformation to pine and oak was driven by higher temperature, drier conditions, and a resulting increase in lightning-caused wildfire. Specifically, studies have shown for New England that summers warmed, lake levels dropped, temperate tree species advanced up the White Mountains,[32] and the abundance of charcoal in corresponding layers of lake sediments increased.[33] Compared to spruce and fir, oaks and white pine are temperate species with low tolerance of extreme cold but high tolerance of the drier conditions that ensued during

Figure 2.12 Eastern white pine (*Pinus strobus*), showing its tendency to tower above other species—so-called *supercanopy* or *emergent* trees.

this period. Furthermore, most of the species replacing spruce and fir are well adapted to a regime of frequent fire through their ability to establish from seed after fire (white pine, birches, aspens) or resprout from roots or stumps (aspen, oaks). White pine has not since been as abundant in Maine as it was during that period between about 12,000 and 8,000 years ago, and spruces did not again become widespread until about 1,000 years ago.

By around 8,000 years ago, eastern hemlock, American beech, and yellow birch had all finally migrated northward into Maine during a time in which the presence of charcoal in sediments declined, suggesting a reduction in fire frequency, apparently as a result of cooler, moister conditions (figure 2.7).[34] These three species are less tolerant of fire and drought than are white pine and oaks. They slowly replaced those drier, relatively open woods with what was probably a moist, dense, heavily shaded forest. These mixed temperate forests persisted until about 1,500 years ago, with one striking exception. About 5,400 years ago, across eastern North America, the abundance of hemlock pollen dropped abruptly, so much, so fast that some paleoecologists have hypothesized that this species was struck by a virulent pathogen or a massive insect outbreak that killed most individuals (figure 2.7). In sediments contemporaneous with the decline, in fact, Najat Bhiry and Louise Filion of the Université Laval found chewed hemlock needles and abundant fossils of the eastern hemlock looper, a major herbivore of hemlock. A contemporary analog to such a remarkable biotic event is the rapid decimation of American chestnut populations throughout eastern North America in the twentieth century as a result of the accidental release and spread of the nonnative chestnut blight fungus. Hemlock decline caused by a looper outbreak is a compelling story, but recent evidence suggests the possibility that plain vanilla climate change — decreased precipitation, to which hemlock is vulnerable — may have caused or at least contributed to the decline. A reasonable possibility is that climate change could have induced or exacerbated a pest or pathogen attack, leading to the documented rapid decline of hemlock. In chapter 6, we discuss the possible role of such climate-pest interactions in current and future ecological changes in Maine. Whatever the cause of its crash, hemlock recovered within 2,000 years, and hemlock-beech-birch forests again covered most of Maine.[35]

About 1,000–1,500 years ago, the balance of northern versus southern species shifted again. After persisting in low numbers for about 10,000 years, spruce suddenly increased dramatically (figure 2.7). This resurgence was apparently in response to very modest declines in temperature, perhaps less than 1°F, as well as apparent increases in moisture availability, indicated by lake level rise.[36] Recent studies have added two compelling details to this story. First, Molly Schauffler and George Jacobson, of University of Maine's Climate Change Institute, demonstrated that, despite low numbers across most of Maine before 1,500 years ago, spruce was actually quite abundant right along the cool, moist coast — *refugia* that probably provided the source for the eventual upsurge in the interior.[37] Second, new statistical tools developed by Matts Lindbladh, a Swedish ecologist visiting the Maine institute, revealed shifts in the balance of the three common eastern spruce species: white and black spruce dominated in the initial postglacial forests, but were later supplanted by red spruce, the dominant species today, which was biding its time in those coastal refugia or off the present coast on the exposed continental shelf.[38] By a thousand years ago, Maine supported its present-day unique combination of tree species, a mixing of southern and northern elements, part of the *Acadian Forest* of northern New England and southeastern Canada, sandwiched between two massive global biomes, the temperate deciduous forest to the south and the boreal forest to the north (see chapter 5).

Where in North America did the species that eventually colonized

Eastern White Pine (*Pinus strobus*)

The importance of the eastern white pine[1] to Maine is obvious: it's the state tree and its pollen tassel and cone together constitute the state "flower" (see Figure 2.12).[2] One could say that the tree is almost legendary, a Paul Bunyan of the plant kingdom. With a lifespan of up to 450 years, it is one of the largest tree species in eastern North America. The national champion, boasting a diameter just over 6 feet and a height of 132 feet, resides in Morrill, Maine.[3] White pine put the territory on the map during colonial times: it was so valuable for ship masts that trees greater than 24 inches in diameter and within three miles of water were blazed and reserved for the British Royal Navy. Less appreciated is that white pine during those times was not particularly abundant, especially compared to many other major tree species in the state (see chapter 3; table 3.1). Based on natural abundance, in fact, the "spruce tree state" would be a more fitting moniker for Maine than the "pine tree state" (see table 4.3).

This wasn't always true, according to the University of Maine Climate Change Institute's George Jacobson and Ann Dieffenbacher-Krall, who analyzed over 150 lake sediment pollen chronosequences from the public North American Pollen Database.[4] The earliest white pine fossil pollen dates from nearly 16,000 years ago in lake sediments in the Shenandoah Valley. It reached northern New England by about 12,000 years ago, the Great Lakes region by about 10,000 years ago, and western Ontario by around 8,000 years ago. White pine is sensitive to cold and wet conditions, mineral soil promotes its germination, and open conditions greatly accelerate its juvenile growth. Accordingly, the species reached the pinnacle of its abundance across its entire range during a period of relatively dry, warm, fire-prone conditions between 12,000 and 8,000 years ago. It even occurred at an elevation about a thousand feet above its current limit in the White Mountains during that time.[5] Since then, its overall abundance has declined and shifted to the Great Lakes region.

Although few New Englanders are aware of this distant history, there is general knowledge that white pine was very common in the twentieth century. How does this square with the modest abundance of this species in presettlement times? Nearly all of the huge white pines in the state were harvested by the mid-nineteenth century (see chapter 4: "1840–1880"). This was coincidentally the beginning of widespread abandonment of farmland in Maine, which led to a resurgence of white pine—so-called pasture pines—which thrived in these open environments. In more natural areas, such as Maine's Unorganized Territories, which have remained forested since long before colonial times, white pine is sufficiently uncommon that special care is taken to insure its regeneration and growth.

NOTES

1. Silvics information from G.W. Wendel and H. Clay Smith, "White Pine (*Pinus strobus* L.)," in Russell M. Burns and Barbara H. Honkala, tech. cords., *Silvics of Forest Trees of the United States*, Volume 1, Agriculture Handbook 654 (Washington, DC: U.S. Department of Agriculture, Forest Service, 1990); Maine Forest Service, *Forest Trees of Maine, Centennial Edition* (Augusta: Maine Forest Service, Department of Conservation, 2008). Paleoecological information from Jacobson and Dieffenbacher-Krall, "White pine and climate change."

2. Technically, the pine tassel, which bears the tree's pollen, is not a flower at all since pines are gymnosperms, not angiosperms (flowering plants), but the authors of this book embrace the inclusiveness of the state flower designation.

3. National Big Tree Register (retrieved on November 17, 2009, at www.americanforests.org/resources/bigtrees/index.php).

4. North American Pollen Database (retrieved on August 10, 2011, at www.ncdc.noaa.gov/paleo/napd.html).

5. Margaret B. Davis, Ray W. Spear, and Linda C.K. Shane, "Holocene climate of New England," *Quaternary Research* 14 (1980): 240–250.

Maine originate? In other words, where were they when Maine and much of the north were covered in ice? This is an issue with a long history of debate in ecology. E. Lucy Braun, one of the most prominent forest biogeographers of the twentieth century, hypothesized in her 1950 book, *Deciduous Forests of Eastern North America*, that most species resided in small refugia, with stable, ameliorated climate conditions, in the southern mountains.[39] A compelling idea, but one that turned out to be at least partially wrong. Later paleoecological studies of pollen have revealed that many of the prominent tree species of early Holocene Maine actually came from several different parts of the southern United States. Today, these studies, genetic research, and investigations of the evolutionary history of groups of plants suggest an even broader pedigree for the Maine flora, with species coming from the south, the eastern coastal plain, the west, and even the high Arctic.[40]

How fast did species migrate from their ice age ranges to Maine? Estimates vary from six to sixty miles per century.[41] There's been much discussion about how species might achieve such formidable rates of migration, including an emphasis on rare long-distance seed dispersal events via wind and animals.[42] A recent study of beech and red maple challenges these ideas. Molecular data revealing the history of geographic movements for these species point to considerably slower rates from small, isolated populations surviving closer to the glacial front than previously supposed, suggesting that the maps in figure 2.10 might be somewhat misleading.[43] The resolution of this debate has implications for the impact of current and future anthro-

Are Communities "Superorganisms"?

Does a community of plants, animals, fungi, and other creatures resemble an organism—with mutually dependent internal parts and natural homeostasis? Forcefully promulgated by Frederic Clements, one of the early titans of ecology, this *superorganism* idea reigned supreme in the discipline for nearly half a century. Not until the 1950s, did R.H. Whittaker working in the Great Smoky Mountains and John Curtis in Wisconsin convince other ecologists to view communities differently.[1] They presented dramatic evidence that species behaved independently and that communities were sets of interacting organisms occurring in the same geographic location, sharing similar environmental requirements, rather than as inseparable species produced from eons of mutual evolution. Daniel Simberloff has argued that this marked a philosophical abandonment in the field of ecology of a view of nature as unchanging, balanced, and deterministic ("essentialism").[2]

Margaret Bryan Davis, now professor emerita at the University of Minnesota, played a crucial role in this shift. She helped popularize the use of fossil pollen to reconstruct past migrations of tree species after deglaciation (described in sidebar 2.1), and Maine, with its northern location, was a key geographical linchpin in this work. Her 1976 paper, "The Pleistocene Biogeography of Temperate Deciduous Forests," was a decisive blow to the superorganism theory of natural communities.[3] By showing that the major tree species of eastern North America arrived at their eventual locations at very different times, that those species were often not found together over thousands of years, and that forest types common today did not exist over much of the past 20,000 years, she forcefully demonstrated that forest communities were not composed of tightly co-evolved, inextricably linked plant species.[4] This was to be a major influence on the way scientists (including the authors of this book) viewed nature.[5]

pogenic climate change on forests in Maine and beyond. Namely, do tree species have the capacity to keep pace as their climate zones move north (Northern Hemisphere) and south (Southern Hemisphere)? Perhaps not if their movements are restricted to the rates suggested by this new study.

Postglacial Animals of Maine

How about animals? We know much less about animal than plant species inhabiting Maine during early postglacial times, at least in part because

NOTES

1. Robert P. McIntosh, *The Background of Ecology, Concept and Theory* (New York: Cambridge University Press, 1985); John T. Curtis, *The Vegetation of Wisconsin* (Madison: University of Wisconsin Press, 1959); Robert H. Whittaker, "A consideration of climax theory, the climax as a population and pattern," Ecological Monographs 23 (1953): 41–78. Decades earlier, in the 1920s, Henry Gleason, of the New York Botanical Garden, published a strong rebuttal of the superorganism theory, which helped inspire the work of Curtis, Whittaker, and others.

2. Daniel Simberloff, "A succession of paradigms in ecology: essentialism to materialism and probabalism," *Synthese* 43 (1980): 3–39.

3. Davis, "Pleistocene biogeography of temperate deciduous forests," *Geoscience and Man* 13 (1976): 13–26.

4. This is not to say that species in communities don't interact intimately, which they do, sometimes competitively but also often in ways that benefit one or both species, an expanding research focus in ecology. Very close linkages have been found, for example, between plants and mycorrhizal fungi. See Rob W. Brooker and Ragan M. Callaway, "Facilitation in the conceptual melting pot," *Journal of Ecology* 97 (2009): 1117–1120.

5. Davis was also a pioneer in the emergence of women in ecology. Women made major contributions to the discipline in the early 1900s, but their participation (in all sciences) waned in the 1930s–1950s. Davis persisted, receiving her PhD from Harvard in 1957, becoming a leader in the field, including president of the Ecological Society of America. She helped mentor a new wave of women ecologists, many of whom have also taken up leadership roles in the field. Her thoughts on her success reveal much about the challenges of gender equality in science. "Now in my sixties, I find I have achieved the goal I was striving for all my life. I am a professor in a department with a strong graduate program. I have a group of excellent students and my research combines ecology and paleoecology. In this benign environment I spend relatively little time on women's issues, but a decade or so ago I added up all the time I had spent . . . maneuvering for laboratory space and a faculty position, fighting for equal wages . . . and it comes to an appalling 25% of my total investment in science. My experience isn't unusual, either." From Jean. H. Langenheim, "Early history and progress of women ecologists: emphasis upon research contributions," *Annual Review of Ecology and Systematics* 27 (1996): 1–53, herself a pioneer woman ecologist.

there is no equivalent zoological counterpart to the resistant tiny plant parts that accumulate in aquatic sediments.[44] Maine's most prominent late Pleistocene fossil is undoubtedly the woolly mammoth excavated in 1959 in Scarborough in southern Maine and reexamined recently (figure 2.13). The authors of a new report[45] reconstructed a gruesome scenario of a female mammoth dying in the winter and then floating down the proto-Saco River, possibly subject to disarticulation as a result of being scavenged during its transport, until it reached its resting place, which is now about 125 feet above sea level. Fossil bones and teeth of musk ox, bearded seal, and a

Figure 2.13 A depiction of woolly mammoths in early postglacial Maine.

now-extinct North American horse have also been found in Maine, dating from immediate postglacial times. Fossils from the wider region suggest that the Maine fauna of this time likely also included grizzly bear, polar bear, caribou, walrus, beluga whales, and ringed and harp seals.[46]

During the late Pleistocene, climate and plant communities were undergoing dramatic transformations across North America, as described previously. Not to be outdone, animal communities — specifically an assemblage of huge mammals, sometimes called *megafauna* — suffered a continent-wide mass extinction that continues to fascinate researchers. A giant ground sloth (*Megalonxy*), the dire wolf (larger than any wolves today), the short-faced bear (larger than the modern brown bear), a giant beaver (300 lbs!), huge camels, and the mastodon — these were just a few of the dozens of large mammal species abundant in the late Pleistocene fossil record that suddenly disappeared.[47] It is likely that some of these, in addition to the Scarborough mammoth, occupied the highly productive tundra and woodlands of Maine in postglacial times.

What led to this remarkable extinction event? Scientists initially focused on what seemed an obvious link: the demise of large mammals was contemporaneous with a rapidly changing climate at the end of the Pleistocene, which likely resulted in disruptions of habitat and food supply.[48] Others, most prominently Paul Martin, Professor of Geosciences at the University of Arizona, have argued vigorously for an anthropogenic cause — the *overkill hypothesis* — citing the coincidence of increasing human populations, the lack of extinctions in other groups of organisms, physical evidence of large mammal kills, and probable human-caused extinctions elsewhere.[49] A recent comprehensive review of this long-standing debate takes a middle ground, citing a key role for humans but a contributing effect of climate change.[50] On the other hand, Richard Firestone[51] of Lawrence Berkeley Laboratory and colleagues have developed a completely new hypothesis: that an extraterrestrial impact, possibly from a comet, led to the abrupt and dramatic cooling of the Younger Dryas, which then led to the mass mammalian megafaunal extinction. The most recent work, by Jacquelyn Gill and colleagues, casts doubt on aspects of Firestone's argument: fossilized indicators of climate, vegetation, and animals in lake sediments (including those of a fungus specializing in mammal dung!) suggest that the megafaunal collapse occurred before both the Younger Dryas and the hypothesized extraterrestrial impact.[52] Adding a new wrinkle to the already gripping argument about this unusual period of paleohistory, the authors argue that the loss of megafauna released hardwood tree species from herbivory, increased fuel for wildfires, and led to the development of novel plant communities that haven't occurred since that time.

The outcome of this debate is more than academic. Some advocates of the overkill hypothesis have argued that, if megafaunal extinctions were the result of human actions, then we have the responsibility to restore some measure of the faunal diversity that would otherwise exist today — by introducing to the United States species similar to those lost, such as the African lion, the African elephant, and the cheetah.[53] According to William Stolzenburg in *Conservation* magazine, response to this proposal has spanned the gamut: "some of the comments were congratulatory, a good many of them were disparaging, a handful of them were downright hateful."[54]

We know much more about the animals of Maine starting from the time of documented arrival of Paleoindians in the early Holocene, 12–13,000 years ago.[55] Like for Maine's flora, the warming at the end of the Younger Dryas period, about 11,600 years ago, triggered a dramatic transformation of the fauna, including the appearance of most of the familiar modern animal species: white-tailed deer, moose, caribou, raccoon, bobcat, fox, beaver, muskrat, skunk, and porcupine. Two species — the sea mink[56] and the great auk (a large flightless bird) — were also common along the coast during this time, but have since become extinct, from hunting and trapping pressure by Euro-Americans.

Like plant species, populations of animals continued to exhibit sensitivity to changes in climate and corresponding habitat transformation during the Holocene. Arthur Spiess, an archeologist at the Maine Historic Preservation Commission, illustrates this well with the abundance of moose, an indicator of northern coniferous forests, versus deer killed at the Turner Farm site over the past 4,000 years. As spruce-fir forest increased at the expense of hardwoods, the ratio of moose to deer increased correspondingly.[57] It is reasonable to expect that ancient people in Maine would also have been strongly affected by these multiple correlated changes in their environment. Heather Almquist-Jacobson and David Sanger, for example, have postulated that Late Archaic increases in human populations in Maine were stimulated in part by the development of a beech-birch forest, resulting from the decline of hemlock 5,400 years ago. They argue that increased hardwood forage led to a rise in beaver and muskrat populations, which provided native peoples with a highly productive resource.[58] Sidebar 2.4 describes a similarly compelling hypothesis about the connections between climate change, vegetation, and human populations in immediate postglacial times.

Lessons Learned

According to Pulitzer Prize–winning historian David McCullough, "History is who we are and why we are the way we are." Similarly, postglacial history illuminates the nature of Maine today and how it came to be that way. Moreover, like all good history, paleoecology has advanced our understanding well beyond a simple chronology of the comings and goings of climate events and species on the ecological stage. Embedded in these stories are general principles about the intrinsic workings of nature.

The first and perhaps most important lesson: nature changes, sometimes a lot, sometimes very fast. In a period of only 8,000 years, ice 8,000 feet deep melted off and exposed the bare landscape of Maine. Near the end of deglaciation, average temperatures abruptly dropped by nearly 10°F within a decade or two and then, 1,300 years later, rose at a similar rate. Just prior to that period, ecological communities were rocked by the rapid loss of many large mammal species. Over less than a century in the mid-Holocene, hemlock largely disappeared from the forests of the Northeast. Although from the perspective of humans, climate and forests were relatively stable during

A Paleoindian Response to Younger Dryas Climate Change

Paige Newby (Brown University), Jonathan Lothrop (New York State Museum), and their colleagues recently proposed an ambitious hypothesis connecting contemporaneous changes in climate, vegetation, large mammals, and Paleo-indians during the times surrounding the Younger Dryas.[1] As discussed previously, open spruce woodland and tundra spread across northern New England and the Maritimes in response to Younger Dryas cooling. Several lines of evidence suggest that these expanded subarctic habitats supported large herds of caribou, much like those found today under similar ecological conditions in far northern North America. Fluted projectile points, used in hunting large mammals, are found throughout New England and the Maritimes from before and during the Younger Dryas and provide the earliest evidence of Paleoindian presence in the region. Newby and Lothrop surmise that Paleoindians used these weapons to great advantage in hunting the large caribou herds and other large mammals adapted to the open habitats of the Younger Dryas. A key question in the archeology of New England and the Maritimes is why these fluted projectile points disappeared after the Younger Dryas. Abrupt warming at this time promoted the spread of closed canopy forests and a shift to more temperate tree species. Newby and Lothrop conclude that these habitat changes led to a major faunal shift from herd mammals (caribou) to more solitary ones, such as moose and deer, which in turn caused Paleoindians to abandon their fluted projectile points in favor of ones better adapted to the new milieu.

NOTES

1. P.E. Newby, J.B. Bradley, A. Spiess, B. Shuman, and P. Leduc, "A Paleoindian response to Younger Dryas climate change," *Quaternary Science Reviews* 24 (2005): 141–154; Jonathan C. Lothrop, Paige E. Newby, Arthur E. Spiess, and James W. Bradley, "Paleoindians and the Younger Dryas in the New England-Maritimes Region," *Quaternary International* 242 (2011): 546–569.

some periods (e.g., the pine-oak phase), these were inevitably perturbed by renewed pulses of physical and biotic change. In his book, the *Two-Mile Time Machine*, well-known climate scientist Richard Alley concluded that, "The Greenland ice cores and other records show that climate changes large enough and rapid enough to scare civilized peoples have occurred repeatedly in the past, and that our civilization has risen during an anomalously stable time."[59]

Second, climate change dramatically alters the abundance and distribution of species across the landscape. Fossil pollen chronologies from throughout eastern North America reveal consistently strong responses of tree species, even continental-scale migration, to changing temperature and moisture.[60] At the same time, animal communities substantially reconfigured in concert with changes in climate and vegetation. Even small climate alterations appear capable of producing large effects. We saw, for example, that spruce forests were uncommon in inland Maine until modest cooling 1,000–1,500 years ago. In their fine-scale analysis of pollen cores, Emily Russell and her colleagues have shown that even as European explorers were arriving in New England, spruce and fir were still increasing at the expense of beech and hemlock.[61] These transformations should raise concerns about the impacts of anthropogenic climate warming, which we will address in chapter 6.

Third, despite the ubiquitous power of climate in shaping forest communities, other environmental factors also play important ecological roles. Lightning-caused fire, associated with warm, dry conditions, appeared to favor some temperate species over northern tree species in New England during parts of the Holocene. The collapse of hemlock populations indicates the potential of pathogens and insect outbreaks to alter forests. These paleoecological revelations are corroborated by the ubiquitous roles of natural disturbances in modern ecosystems around the world.[62]

Finally, paleoecology profoundly questions assumptions, deeply embedded in our culture, about the degree to which nature is governed by timeless laws that maintain long-term stability in natural communities — a "balance of nature." This idea is untenable in the face of evidence of independent migrations of different tree species and the co-occurrence today of many tree species that did not overlap geographically for much of the Holocene (sidebar 2.3). In an article in the journal *Ecology*,[63] Jack Williams and his coauthors take this argument a step further: not only did plant communities exist in the past that have no analog today (so called *no-analog communities*), but these assemblages were apparently themselves the result of *no-analog climates* — unique combinations of temperature, precipitation, and seasonality with no modern counterpart. As climate changes, nature apparently does not simply rotate through identical environments and communities over time in clockwork fashion. This conclusion is perhaps obvious for longer periods of time, for the very pool of species in a region and on the Earth is constantly changing as new species are produced and others go extinct. But even over shorter periods, stability of communities is, at best, ephemeral, for individual species are constantly responding to natural disturbances, pest and pathogen outbreaks, shifting climate, and other phenomena, often independent of the response of other species.

3 | The Presettlement Forest of Maine

What Were the Forests Like before European Settlement?

The hemlock–white pine–northern hardwoods region . . . is characterized by the pronounced alternation of deciduous, coniferous, and mixed forest communities. Sugar maple, beech, basswood, yellow birch, hemlock, and white pine are climax dominants of the region as a whole . . . [and in the east] red spruce [is a] characteristic species. . . . The boundaries of the region are ill-defined, for this is a great tension zone between encroaching more southern species and retreating more northern species. — E. Lucy Braun

On August 18, 1607, captains George Popham and Raleigh Gilbert landed their ships, *Gift of God* and *Mary and John*, at Sabino Head at the mouth of the Kennebec River (figure 3.1). Two years earlier, George Waymouth had conducted the first detailed English exploration of what is now mid-coast Maine. On this second expedition, however, Ferdinando Gorges and other English investors sent Popham and Gilbert with a higher purpose. Its contingent of 120 colonists aimed to set up the first permanent English settlement in America north of Virginia (sidebar 3.1). The chronicler for Popham and Gilbert described the destination as "gallant Illands full of heigh and mighty trees of sundry sorts." This forested stretch of coastline must have seemed remote and alien to the Englishmen, for they had departed at the end of May from Plymouth, England, a busy port city on the Devon coast. The location did, however, satisfy their search for a land rich in timber and

Figure 3.1 Map of sites discussed in the text for Popham Colony, early explorers, pollen cores, and old-growth sites.

a supply of ship-building stores. Earlier that year, a sister colony was established at Jamestown in Virginia. The Pilgrim's Plymouth Colony was still thirteen years in the future.[1] Unlike these two early settlements, the Popham Colony survived for hardly more than a year, undone by "the winter . . . soe extreame and frosty,"[2] and the deaths of major expedition backers as well as the colony's leader, George Popham.[3] The survivors returned to England in the fall of 1608, some in the *Virginia*, the first English ship built

The First Forests Cut Down by European Settlers in Maine

On August 14, 1607, the chronicler for Popham and Gilbert described the coastal forest at Small Point, as "full of pyne trees and ocke and abundance of whorts [heaths] of (l)ower Sorts of them." Four days later and six miles to the east, they landed near what is now Popham Beach State Park in Phippsburg and immediately began to build their settlement. Surrounded by what was undoubtedly the same pine-oak forest, the colonists started by cutting down trees to erect a fort containing some dozen structures. During a 2005 archaeological excavation of Fort St. George, the name given to the settlement by the colonists, Jeffrey Brain, of the Peabody Essex Museum, found the remains of several posts from the fort's 1607 storehouse, including an amazingly preserved twenty-five-inch-long fragment of a pitch pine post, cut off on the butt end, squared on three sides, and rotten at the top (see figure 3.2). With bark still attached to one side, the tree contained ninety-three growth rings. The thirteen-inch squared dimensions indicate that the tree's original diameter must have been about eighteen inches. This artifact confirms the occurrence of relatively large pitch pine trees growing on Sabino Point in 1607. Furthermore, the colony's "smithy" excavated in 2010 contained charcoal derived from oak, with a trace of birch wood. Significantly, today there is a pitch pine dune forest in Popham Beach State Park (as at Small Point) with trees up to seventy-six years old, albeit less than ten inches in diameter. There is also a young red spruce and mixed hardwood forest, containing small white pine

in America. They also took back "many kinds of [trade] furrs" and a "good store of sarsaparilla" obtained in "New England." It would be two more decades before the English began successfully settling the Maine coast.

What was the nature of the presettlement Maine Woods seen by Popham and Gilbert and later European settlers? Was it all Josselyn's (1675) "daunting terrible . . . cloathed with infinite thick woods." Was Maine covered with Longfellow's "forest primeval. The murmuring pines and the hemlocks/ Bearded with moss, and in garments green, indistinct in the twilight"? Was it all a predictable hemlock–white pine–northern hardwoods in accordance with E. Lucy Braun's classification? Were the forests of the "Pine Tree State" dominated by huge white pines deserving the "Broad Arrow" mark (⬆),

Figure 3.2 Preserved fragment of storehouse post from 1607 Popham Colony, excavated at Fort St. George, Phippsburg peninsula. Note the axe marks of the Popham carpenters.

and red oak. Apparently this settlement was in the midst of a forest, from which they cut local timber to construct buildings and use for fuel.[1] At Popham they also constructed the ship *Virginia*, undoubtedly using native oak, pine, and spruce.

NOTES

1. Details from Popham, personal communication from Jeffrey Brain, e-mail on August 16, 2011. Post artifact details, personal communication from Molly O'Guinness Carlson.

which reserved them for the Royal Navy? Was the "North Maine Woods" an undisturbed spruce-fir forest waiting for the saw of the pulp and paper economy? The nature of the presettlement forest is shrouded by myths, anecdotal narratives, and monolithic generalities. In this chapter, we will tap a variety of historic and modern resources to go beyond the speculation and attempt to uncover the nature of the presettlement Maine forest.[4]

What do we mean by *presettlement forest*? The term *presettlement* refers to the time just before the wave of settlement in North America by people of European descent. Undoubtedly, Native Americans influenced the forests of Maine for millennia prior to contact with Europeans. Compared to other areas in North America, however, the number of Indians in Maine was relatively small, they pursued a predominantly low-impact hunter-gatherer lifestyle, and their ecological effects on forests were relatively less intense and widespread.[5] The level of impact of human culture on the Maine Woods changed dramatically with Euro-American settlement (chapter 4).[6] Exactly when this period occurred varied with location, generally being earliest along Maine's coast and latest far inland. For our purposes, the presettlement forest is what existed just prior to the period of 1600 to 1860 (depending on location), which brought rapid and dramatic change to Maine's forests.

Why devote a chapter to a long-forgotten landscape? The presettlement forest provides a unique benchmark for how forests function without substantial human influence — a baseline that can inform strategies for using, managing, and preserving forests. *Silvicultural* treatments in forests, for example, sometimes use these "original" conditions as a way to evaluate what works and doesn't work.[7] In some cases, the very goal of management is to restore or maintain forests with composition and structure similar to that of presettlement forests in order to promote the full range of ecological diversity.[8] Presettlement forests are also an important starting point for evaluating how Euro-American land use has affected the Maine Woods and how they might change in the future. Finally, an objective description of historical forests can enhance our understanding, enjoyment, connectedness, and reverence for Maine's forest heritage.

Reconstructing the nature of presettlement forests is not a straightforward process. It is difficult enough to describe modern forests, which are exceedingly complex and varied.[9] Here we are talking about ecosystems that existed 200 to 600 years ago, differed from one place to another across the state, and were themselves undergoing changes. No single approach or type of data will suffice; we must instead rely on multiple lines of evidence that operate at different geographic scales and times. This approach combines the disciplines of history, biogeography, archaeology, surveying, natural history, and ecology, and comprises its own discipline called *historical ecology*.[10]

We employ four methodological strategies. First, early explorers wrote descriptions of what they saw upon arriving in what would eventually become Maine. The earliest explorers wrote about coastal forests, mainly in the 1600s, whereas inland forests were described later, in some cases not until the 1800s. Second, after exploration, Maine lands were surveyed to establish the boundaries of towns, as well as lots within some towns, to subdivide private ownerships, etc. These surveys varied in techniques and information recorded, but provide broad coverage of the state and include a wealth of information about presettlement forests, including identification of *witness trees*. Third, we draw upon *paleoecological* research, which, as described in sidebar 2.1, uses preserved pollen and macrofossils to describe species composition and charcoal, insect remains, and other evidence to describe disturbances and environments in the past. Finally, modern old-growth forests, broadly defined as forests that have not been significantly modified by humans, are presumed to resemble presettlement forests and contain clues about forest composition, structure, and the roles of various disturbances in shaping their attributes. All these methodologies have their strengths and weaknesses, but in the aggregate yield a picture of presettlement forests that can help us understand the forests of today.

Explorers: Observers of Maine's Presettlement Forest

The earliest written evidence about the nature of Maine's forests comes from the first European visitors (see appendix 3.1 for details). Their anecdotal and sketchy narratives are from the sea voyages in the age of discovery and exploration of the New World. Few substantive observations of the coast survive, however, until 100 years after the first "discovery" of Maine by John Cabot in 1497. The most informative early explorers' comments on the vegetation include Giovanni da Verrazano's (1524) "nothing extraordinary except vast forests" and John Walker's (1580) observations from a hill near Camden that "(t)he country was most excellent both for the soyle, diuersity [diversity] of sweete woode and other trees" (figure 3.1).[11]

Even when identified, the common names given to species were ambiguous, such as the "firre" and "cypress" noted by Verrazano near Cape Cod, reflecting attempts by Europeans to use familiar names for foreign trees. Neither native fir (*Abies*), spruce (*Picea*), nor cedar grew in Britain, the home of many of the early explorers of the Maine coast. Furthermore, in the sixteenth century, European tree names were rudimentary and flexible. The common Scots pine was more often called "Riga fir" or even "Baltic spruce." "Cypress" was especially ambiguous, as it was definitely not an actual cypress; rather, it could have meant "cedar," which itself is a name applied to several genera in both Europe and North America. Thus, early usage of "fir," "spruce," or "cypress" could have indicated virtually any conifer.[12] George Waymouth's 1605 return voyage is thought to have first introduced white pine logs to England (figure 3.3), but the narration of the trip by Rosier never mentions "pine" directly, rather using "firre" or simply "high timber,

CAPT. WEYMOUTH SAILING UP THE PENOBSCOT.

Figure 3.3 Engraving of the voyage of Captain George Waymouth up the Penobscot Bay in 1605.

masts for ships." Thus the first certain identification of coastal vegetation was by Martin Pring in 1603, when he named four separate conifer species (cedar, spruce, pine, and firre) all together on the islands of Penobscot Bay.[13] Despite the challenges of species identity, the extensive descriptions of Pring (1603), Champlain (1604–5), Rosier (1605), and Popham and Gilbert (1607) provide a fairly clear description of the coastal presettlement forests of Maine: predominantly conifers east of and pine and mixed hardwoods (oaks, beech, and maple) southwest of Penobscot Bay.[14] Many of these references to pine on the southwest coast are probably for pitch rather than white pine, given descriptions of thick heath understory (typical of pitch pine forests), the identification of the preserved pitch pine at Popham, and the modern prevalence of coastal pitch pine.

On June 12, 1605, Waymouth anchored the *Archangell* on the shore of Penobscot Bay (figure 3.3) and "marched up into the country towards the mountains," the Camden Hills (in current day Knox County).[15] Here James Rosier, chronicler for the voyage, reports that they

passed over very good ground, pleasant and fertile, fit for pasture, for the space of some three miles having but little wood, that Oke like stands left in our pastures in England, good and great, fit timber for any use. Some small birch, Hazle and Brake . . . in many places are lowe thicks like our copoisses of small young wood. And surely it did all resemble a

THE CHANGING NATURE OF THE MAINE WOODS

stately parke, wherein appeare some old trees with high withered tops, and other flourishing with living greene boughs. Upon the hills grow notable high timber trees, masts for ships of 400 tun.

This description is fundamentally different from that of the forests of thick conifers on the nearby islands and the surrounding coast and leaves little doubt that the Camden Hills supported an open oak forest with some areas of younger shrubby thickets. In addition, there were tall pines higher on the hills, perhaps the source of the first white pine logs taken to England by Waymouth's expedition. The same impression of a thin, open oak forest is given by both Champlain near the Saco River and Popham and Gilbert near Cape Elizabeth for the southern coast (appendix 3.1).

We also get a glimpse of inland forests as the sea explorations sailed up the major rivers along the coast. Champlain (1604) describes the Penobscot River vegetation at Bangor as a few spruce with some pines on one side and all oaks on the other side of the river.[16] Rosier (1605) adds that the east side of the river is "very thin with oaks and small young birches." These observations and the memories of early Bangor residents indicate an oak-dominated forest 400 years ago.[17] Apparently the Brewer side of the river supported a younger mixed forest of oaks, pine, birch, and spruce. In 1607 Gilbert described the forests along the Kennebec River near Augusta as spruce trees good for masts with "oak, walnut [butternut], and pine-aple [pine]." Presently this location shows some historical continuity, with a prominent butternut forest on the Kennebec floodplain in Chelsea's "Butternut Park" and the forests across the river in Hallowell's Vaughan Woods containing white pine, beech, white birch, and red oak, albeit all young. These initial inland observations, then, suggest that pines and oaks occurred not just along the southern coast, but also inland, at least along the major rivers and in some of the coastal hills.[18]

Extensive inland exploration of Maine was delayed for more than a century because of periodic warfare among the English, French, and Indians (chapter 4: "Before 1781..."). In the late 1700s, taking up where the previous frontier had stalled, the interior of Maine was divided into individual ownerships and the exploration of the backcountry brought new perspectives of the forests. At least sixteen separate species (or genera) of prominent trees, such as hemlock, poplar, basswood, elm, and ash, are mentioned in narratives by surveyors, developers, town committees, and boundary commissions (appendix 3.2). Naming of trees became more consistent as settlers and surveyors developed their own vernacular terminology (appendix 3.3).

The forests in the interior were decidedly mixed in composition according to these early accounts. Many areas were dominated by northern hardwoods. For example, a description of Towns No. 5 and No. 6 (now Buckfield, Sumner, and Hartford in southeastern Oxford Co.), surveyed in 1786, indicates "growth in both Towns are . . . Birch, Beach, Mapple, &c the mountains & rivers and swamps are Hamlock Sprus with scattering pines & Ceder with Some ash and Red Oak."[19] To the north and in the mountains, the vegetation became more coniferous, with spruce, fir, and cedar dominating. In 1806, land surveyors Solomon Adams and Lemuel Perham (see chapter 4: "The Perham Farm . . .") describe Mt. Abraham as "primarily evergreen top mostly mosse and rock."[20] But even in the far north, forests also apparently supported hardwoods, revealed by surveys executed in response to conflicting claims about the location of the border between Maine and British Canada.[21] The land in New Brunswick just north of the St. John River at Matawaska (Aroostook County), for example, was described in 1818 by British surveyor, W.F. Odell, as "a mixture of hardwood, fir, and spruce, with some pine, the mountain ash is abundant and there are a very few wild cherry trees . . . the softwood grows mostly in the valleys, and the hardwoods on the tops of the hills."[22] Significantly, there was almost as much variety within a town as across the entire state. For example, a survey for a new road in 1794, now Route 3 in Belmont/ Morrill in Waldo County, passed through seven separate forest communities, including oak; "beach" (beech), birch, and rock (sugar) maple; and "black wood" (northern spruce). These forests contained ten dominant species mentioned in fourteen alternating patches over five miles.[23] Across almost every mile of survey line in the state there was a series of forest types and forest ages, demonstrating a fine-scale pattern to the presettlement forest.

Land Surveys of Maine's Presettlement Forest

On October 23, 1789, Ephraim Ballard arrived at a marked beech tree at the northwest corner of the Kennebec Purchase, now located on the New Vineyard–Industry town line in Franklin County. Over the next four days Ballard surveyed the current eastern line of Farmington and recorded in his field notes:

Run due S good land all the way Down hill 1st mile a Beech tree very good land 159 rod to a pond which we call Clear water pond. I made offsets & run around the pond on the W Side & found 3 mile to be out 17 rod before the S Course came out of the pond & still surveyed by the eye of the pond to where a SE course came out 59 rod & marked a stone in the Bank of the pond thus [Ballard's personal mark] & a Cedar tree EB there SE middling good land Beech Birch Cedar & hemlock 1st mile hemlock tree pretty good 226 rod a [brook] & good meadow mostly on the left hand very good land 2nd mile a hemlock tree all good Basswood maple & Birch 3d mile a yellow Birch tree all pretty good land 186 rod a brook 4th mile small beech tree middling good 155 rod a [brook] good swailly land 5th mile a small Rock maple & 2 hemlocks good swaly land 6th mile a great hemlock tree good land 292 rod to Sandy River an oak tree marked the same mark.

This narrative is typical of the numerous land surveys executed in anticipation of settlement as the owners (proprietors) began to develop their lands now contained within Maine.

Starting in the seventeenth century, the Province of Maine of the Massachusetts Bay Colony or, after 1783, the Commonwealth of Massachusetts' District of Maine, granted, chartered, or sold land to private "corporate" groups. The territory was divided into ownership of large tracts granted to individual "developers" ("great proprietors") or to small towns, ideally six miles square, owned in common by "town proprietors" or charitable institutions. Typically, the proprietors located the grant, surveyed its boundaries, mapped the interior of the town, subdivided the parcels, and sold or leased those lots to individual settlers (figure 3.5). For example, after a century of intermittent use, the original royal grants (1629–1652) to the Kennebec River region were revived by The Kennebeck Proprietors. In 1753, this "great proprietor" began selling parcels in the 1.5 million acres of accessible land 15 miles on either side of the Kennebec River.[24] In September 1789, the company finally got around to locating and marking the extent of their claim and hired Ephraim Ballard to survey the outlines of the grant, as described previously.[25]

These surveys contain remarkable data, for exact locations, on rating of the land for settlement, topography, water drainage, forests, and specific trees. Individual trees marked by surveyors to locate boundaries — at actual corners, near corners marked by stones or stakes, or at points along survey lines — are called *witness trees* (sidebar 3.2). If we assume that the witness trees were chosen in a consistent way, such as the nearest tree above

A Witness Tree in the Courthouse

In December 2010, we walked into the Penobscot County Registry of Deeds in Bangor, intending to peruse some surveyors' maps for new citations to witness trees. As is customary, we asked the Registrar of Deeds, Susan Bulay in this case, if she had any records that might contain "witness tree" references. Expecting the usual quizzical response, "What is a witness tree?" we were astonished as she replied, "Do you want to see an *actual* witness tree in the land records room?" Asserting that it was the only witness tree "filed" in any deed office in the country, she led us to a glass case beside shelves filled with 200 years of original deeds. Amazingly, the case contained several blocks of wood with large resin-filled scars deeply buried among the annual rings. These were the remains of an axe cut (or blaze) that sliced the bark and a strip of wood off the tree. It turns out that in 1878 there was a dispute over the ownership of recently logged land on the north line of Sebois Plantation (T3 R8 NWP), and a surveyor, Noah Barker, testified that the true line had been marked by Samuel Weston in 1794. He substantiated his declaration by cutting out three "chips" containing the blazes covered with overgrowth ("82–'3 gains growth over the old spot"). These wood blocks were entered as "evidence" in the case (figure 3.4). Putting the legalities aside, the "chips" from approximately two-foot-diameter trees vividly show the methodology surveyors used to mark witness trees on town lines when setting out towns in northern Maine.

Barker's deposition states that, searching for the old line, he located remains of blazes on several trees in small patches of "old growth" surrounded by second growth. Completing the archival trail, the original survey notes in the Maine State Archives[1] indicate that in 1820 Alexander Greenwood, surveying the same lines, blazed eight poplar (aspen) mile trees and found the region was "all hurricane and burnt, now young hardwood"—probably from the 1795 fire and 1815 hurricane. In 1831, Rufus Gilman described the same lines as second growth with some "40 to 50 acres of hardwood standing." Thus the blazes we observed in the Registry of

a certain size to the survey point, the frequency of these trees provides an estimate of the actual species composition of the forest at that time. Thus, the tally of trees marked during Ballard's survey gives a quantitative, and presumably unbiased, sample of the species growing along the eastern line of what is now Farmington in 1789: 45 percent hemlock, 18 percent beech, with some cedar, birch, maple, and oak. Moreover, based on the survey's actual description of the central segment, the presettlement condition of this

Figure 3.4 Witness tree blaze placed by Samuel Weston in 1794. Block cut out of approximately two-foot-diameter red spruce tree by Noah Barker in 1878 on T3 R8 NWP. Blaze and block now in the Penobscot County Registry of Deeds in Bangor.

Deeds had originally been placed on trees in 1794, and those trees had survived a severe fire and hurricane in undisturbed patches, then lived for 85 years, growing 3¼ inches of wood over the wounds.[2] These trees were large enough to be used as witness trees in 1794, had chunks removed in 1878, and have been permanently archived in Bangor ever since.

NOTES

1. Maine State Archives, Land Office Records, Field Notes, Gilman 1831, 13:16; Greenwood 1820, 15:20 ff.

2. Note the 1794 blaze was apparently placed after the 1794 growth year. The chip was removed in May before the 1878 growth ring, thus 83 rings over 85 elapsed years.

specific line was presumably "good Basswood, maple, birch," and hemlock. We will return to this location, settled two years earlier by Silas Perham in chapter 4 ("The Perham Farm . . .").

Such early land division surveys contain widespread eyewitness accounts of the forests and trees growing in the state's presettlement forest. Compared with written observations of early explorers, who described forests often with vested interests in pleasing their backers or attracting settlers, land surveyors generally had every incentive and a legal duty to describe trees and forests as objectively as possible. As the land was settled further, more detailed surveys expanded the spatial coverage and resolution of the estimates of forest composition, compared to the fragmentary extent of descriptions by early explorers and developers.

Across Maine three sources of early surveys incorporating trees growing in the original forest have been preserved in town records. Towns settled in the seventeenth and eighteenth centuries along the immediate coast and inland in southern Maine contain descriptions of parcels, often using specific corner trees, in the deeds for the earliest property transfers. Towns in the central interior settled before the mid-1800s were generally divided by surveys into grids of parcels with individual trees cited as the corners of the lots (figure 3.5). Towns in the western mountains and northern interior were seldom fully subdivided, but the outlines of these towns were surveyed and trees cited at the town corners and every half or full mile along the exterior lines of the town. These later northern surveys also included extensive narrative descriptions of the land and forest along the boundaries, for example, "burned," "blown down," "beaver meadows." Whatever the source, the cited trees constitute a quantitative sample at predetermined locations distributed around the towns and recorded in archival documents.[26] Al-

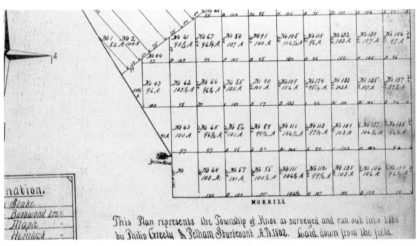

Figure 3.5 Detail of a lotting map for Town of Knox, part of the Waldo Patent (Waldo County). Note drawing of a yellow birch witness tree and letter codes to other witness tree or stakes (/) written at many of the lot corners. Compilation from the map indicates a composition of 76 witness trees in 11 species: 29 percent beech, 28 percent hemlock, 12 percent maple, 9 percent spruce, 8 percent birch, 7 percent pine. For Waldo Patent, see Taylor, *Liberty Men and Great Proprietors*.

though the dates of these surveys vary from 1662 on the southern coast to the 1850s in the northern interior, they all represent the prevailing presettlement forest, before it was altered by Euro-American land use.

Charles V. Cogbill, one of the authors of this book, has collated these presettlement land survey records from a wide variety of repositories, yielding citations of 23,490 witness trees in what is now Maine. Some thirty-six species or closely related groups of species were easily identified from among the English colloquial names used by surveyors.[27] We have assumed that a sample of more than 50 witness trees across a roughly town-sized area will yield an accurate estimate of the local presettlement abundance of common species (but not those of rare ones). Some 180 "towns" in Maine have adequate samples to determine the relative frequency or tree composition of that area. For example, the town of New Pennycook (now Rumford, Oxford Co.) has 888 witness trees from surveys in 1779–80 that indicate a beech dominated (28 percent) forest with about 15 percent each of maple, birch, and hemlock, less than 10 percent spruce, and less than 5 percent pine and oak. Sample towns are spread across the state and represent 58 percent of the land area of Maine (table 3.1).

Here's what the witness tree and associated evidence tell us for the state as a whole. The presettlement forest of Maine had a modest number of tree species, on average 14.6 species of witness trees per town, and numerical dominance by only 7–10 common species. Among all towns, spruces, beech, fir, yellow birch, hemlock, cedar, sugar maple, white birch, and white pine, in that order, were the most abundant trees in Maine's presettlement forest (table 3.1).[28] These prominent species were found in more than half of the towns and accounted for 88 percent of all trees. Spruces, fir, cedar, and yellow birch were dominant in the northern part of the state, while beech, hemlock, sugar maple, white pine, and red maple typified the southern regions. Minor species such as aspen, white ash, black ash, white birch, red maple, and tamarack were moderately widespread, but had low overall frequency. More restricted species such as red and white oaks and pitch pine were common only in the extreme south. Within each region of the state, there was a mosaic of several "types" of forest, as the early surveys all mention the constant alternation of hardwood, softwood, mixed wood, and swamp forests.

Lumping towns into groups of similar composition (technically a cluster analysis) shows a dramatic distribution of seven major forest "clusters" in the state (figure 3.6). These clusters can be roughly separated by their composition into two northern conifer types dominated by spruce, three

Table 3.1 Percentage abundance of tree species in the witness tree data set for four regional forest types and for all combined

	Southern Oak-Pine	Northern Hardwood	Coastal Conifer	Northern Conifer	Total
Spruces	5.3	**13.5**	**27.0**	**24.1**	20.2
Beech	**11.1**	**21.2**	8.1	9.6	12.2
Balsam fir	1.5	1.9	8.3	**17.6**	10.5
Yellow birch	3.5	8.4	6.2	**12.2**	9.4
Hemlock	**10.8**	**18.9**	**10.4**	2.9	8.9
Northern white cedar	0.6	3.3	3.1	**13.9**	8.0
Sugar maple	2.2	8.4	1.0	6.7	5.6
White birch	3.8	2.9	7.8	4.9	4.9
White pine	**11.1**	4.6	8.6	1.4	4.5
Red maple	8.8	2.4	7.1	1.7	3.6
Red oak	**12.3**	3.5	5.1	0.0	3.0
White oak	**11.2**	1.0	1.1	0.0	1.6
Aspens	2.3	1.8	2.0	0.7	1.4
Pitch pine	8.0	1.7	0.7	0.0	1.3
Black ash	2.0	0.6	0.6	1.3	1.1
White ash	1.2	1.9	1.2	0.3	0.9
Tamarack	0.2	0.4	0.7	1.1	0.8
Other species[a]	4.1	3.2	0.9	1.6	2.1
Number of Towns	18	43	33	86	180
Number of Trees	3,486	6,968	5,133	7,903	23,490
Dates	1662–1811	1687–1830	1662–1825	1792–1858	1662–1858

[a]Other species: elms, basswood, striped maple, ironwoods, red pine, mountain ash, cherries, butternut, chestnut, other oaks, poplar, hickories, shadbush, plum, crab-apple, hawthorn, tupelo, buttonwood, and Atlantic white cedar, each 0.5 percent or less and in descending order of abundance.

Source: Personal collection of Charles Cogbill.

Note: Bold denotes the most common species (at least 10 percent). Note that "Northern Hardwoods" occur in the central part of the state.

northern hardwood types in the central part of the state dominated by beech, and two southern types dominated by oak or pine. There is a gradual transition in composition between adjacent clusters, except for the division between the southern and central groups, which is remarkably abrupt. This boundary marks the approximate eastern North American limit of domi-nance of several species of both southern (white oak,[29] tupelo, hickory, pitch pine) and northern (northern white cedar, red spruce, white birch)

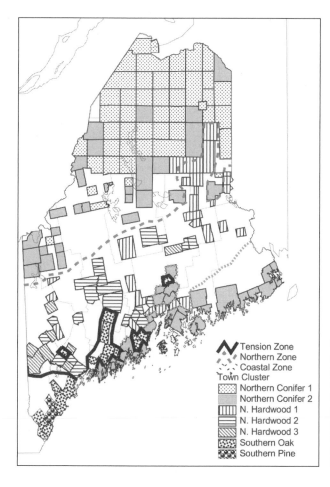

Figure 3.6 Regional clusters of similar tree species composition for the witness tree data set. White areas are for towns for which data are missing. Lines mark separation of the towns into four zones.

Legend:
- Tension Zone
- Northern Zone
- Coastal Zone
- Town Cluster
  - Northern Conifer 1
  - Northern Conifer 2
  - N. Hardwood 1
  - N. Hardwood 2
  - N. Hardwood 3
  - Southern Oak
  - Southern Pine

distribution. It also marks the rapid transition of oak-dominated to beech-dominated forests along the southern edge of Cumberland County, with an extension in the Kennebec River valley, outliers in the Camden Hills in Knox County, and two individual isolated towns (Bangor and Oxford). This dramatic change in vegetation composition (or *ecotone*) is a clear condensation of the forest geographer E. Lucy Braun's "great tension zone" (see quotation at beginning of this chapter). This "New England tension zone," however, should not be viewed as a titanic clash between homogeneous temperate versus northern forests, "locked" into regional climates, but instead as a reflection of changing proportions of southern and northern elements within towns (that is, more boreal moving north) due to multiple local independent factors such as climate, substrate, topography, disturbance, and postglacial history.

There is another latitudinal transition across the state as the northern hardwoods blend into the northern conifer forest. This is not as abrupt or dramatic as the "tension zone," but covers a broad swath over the northern half of the state. On the coast, the ecotone reverses, as a tongue of conifer along the Gulf of Maine results in a transition from hardwoods to conifers going from north (inland) to south (coastal). Thus, overall the presettlement Maine forests can be divided into four relatively recognizable zones: southern, central, northern, and coastal. Each of these zones had a distinct composition: the southern zone dominated by oaks, white pine, hemlock, and beech; the central zone by beech, hemlock, and spruces; the northern zone by spruces, fir, cedar, and yellow birch; and the coastal zone by spruces and hemlock (table 3.1).[30]

According to the witness tree data set, each tree species in the presettlement forest of Maine had a unique distribution across the state, independent from others. Smoothing the species abundances from the 180 towns displays several patterns from four selected species (figure 3.7). Beech was the most abundant species across a diagonal band through the central interior of the state. Oak was abundant only in the southern reaches of the state and confirms Aaron Young's 1848 observation that "[York County] is literally a land of oaks."[31] Pines were most abundant (28 percent of all trees) in southern Maine as well, but also occurred commonly along rivers near the coast. Nearly half (43 percent) of the pine south of Penobscot Bay was pitch pine, however. Significantly, white pine seldom exceeded 3 percent of the trees in towns in the northern interior — despite the oft-cited view of the presettlement abundance of the Maine state tree. Spruce was generally found in two bands: a broad northern swath, where it accounted for more than 25 percent of the trees, and a narrow coastal band with a maximum of 50 percent in Down East Washington County. Interestingly, spruce was also found in moderate abundance (12 percent) throughout most of the central diagonal band of northern hardwoods. These species distributions clearly display the New England boundaries between the spruce and beech to the north and pine and oak to the south of a line from southern Cumberland County to the mouth of Penobscot Bay (although right along the coast, spruce was common down to Casco Bay).

The detailed town line descriptions in northern Maine also provide insight into presettlement forest structure and disturbance. In the early 1800s, surveyors noted many young or "second growth" areas. For example, consider the following description from 1850 of T6 R19 WELS (Somerset County): "[T]he old growth is mostly killed by fire or wind." Compilation

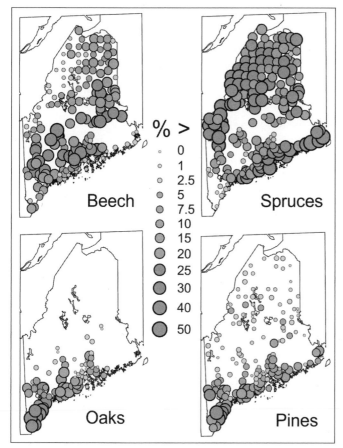

Figure 3.7 Percentage abundance in townships of beech, spruces, oaks, and pines in the presettlement forests of Maine.

of Maine presettlement surveys by Craig Lorimer indicates that 9.3 percent of the town lines of northeastern Maine had been recently burned, mostly by huge fires in 1795 and 1825.[32] Lori Mitchener, a graduate student at the University of Maine, extended the disturbance compilation using over 4,000 miles of presettlement boundary surveys, finding that 3.4 percent of northern Maine was noted as burned or supporting post-fire vegetation in original surveys.[33] These line descriptions also mention "windthrows," "fallen timber," or "hurricane land" over 0.76 percent of the lines. Assuming that such burned land remained recognizable for 50 years and windthrow for 25 years, these figures imply that the typical site in presettlement northern Maine burned on average every 1,461 years and was blown down by a major windstorm, such as the hurricane of 1815, on average every

3,289 years. The presettlement surveys indicate a stand-ending large disturbance with a joint expectation of every 1,012 years in the northern Maine Woods — in other words, infrequently. Softwoods burned more often at particular times and places, but this was balanced by the hardwoods burning very infrequently. Although surveyors noted fires and blowdowns as the most common disturbances, insect damage was also mentioned. Probably the earliest written evidence of a spruce budworm outbreak is given in the 1815 survey of Plymouth (Penobscot Co.) by Charles Hayden, who saw lots 3 and 4 "formerly black [conifer forest] but considerable killed, suposed by worms and fallen down."[34]

The most northwest corner of Maine had a very different composition and dynamics from other areas of the state. Presettlement surveys of the northern five towns' lines in The Nature Conservancy's St. John River Forest Preserve (Aroostook County) were described as "burnt," "young growth," "sapling pine," "white birch," "poplar," or "cherry." These areas covered roughly half of the land in 1844–1850, implying a return time of fire of 94 years[35] (sidebar 3.4). No areas were described as blown down or subject to insect ("budworm") damage. The heavy coniferous composition, including abundant black (rather than red) spruce, and fire disturbance regime make this area similar to the true boreal forest north of the St. Lawrence River in Canada, and unlike the rest of Maine (see chapter 5). Even in this area of spruce-fir dominance, however, there were towns that were decidedly mixed with scattered hardwood ridges. For example, the town at the northern tip of Maine (Big Twenty Twp., Aroostook County) was described in 1845 this way: "[G]rowth is principally fir, spruce, birch (both white and yellow marked), pine, maple, cedar, beech, juniper, and balm of Gilead and on two hills we saw some red oak."

Paleoecology: Remains from the Past

While land surveyors were recording witness trees, witness trees (and their many neighbors) were busy shedding pollen, leaves, cones, etc., some of which became preserved in the sediments of lakes, estuaries, and wetlands. Cores extracted from these sites (see chapter 2) provide an independent test of the occurrence of the four presettlement forest regions indicated by land surveys: southern oak-pine, centrally located northern hardwoods, northern spruce-fir, and coastal spruce-hemlock. We would expect the species composition of these samples to reflect the presettlement forest of the region from which they were taken.[36]

At the beginning of chapter 2, we described coring a black spruce bog in The Nature Conservancy's Basin Preserve on Phippsburg peninsula (Sagadahoc County), in the southern oak-pine region (figure 3.1). This pollen core documents a sudden surge in the abundance of ragweed around 1800, a marker for extensive land clearing and Euro-American settlement. Layers directly beneath that marker, corresponding to the presettlement period, document a forest consisting mainly of pines (pitch, red, and white), birches, hemlock, oak, beech, and spruce, with smaller amounts of poplars, maples, basswood, ironwood, and cedar — a diverse collection indeed, but very much consistent with the witness tree data from the region. The current forest at the Basin is also similar in composition. Charcoal, an indicator of fire, was common in the sediments from presettlement times, not surprising given the abundance of fire-adapted pines, oaks, and birches.[37]

The Gould Pond core in Dexter (Penobscot County), collected by Scott Anderson and colleagues, is from the northern hardwoods zone in the central part of the state (figure 3.1). The presettlement layer from just before 1780 indicates mainly hemlock and birch, with moderate amounts of beech, and small amounts of maple, white pine, fir, cedar, ironwood and oak. This same layer has macrofossils of hemlock, with small amounts of yellow birch and white pine. This mix of species is consistent with the diverse northern hardwoods forests described by land surveys. Charcoal is barely present throughout the core, with small peaks at 10,000 years ago and after settlement (since 1780), when plants associated with land clearing appear in abundance.[38]

The South Branch Pond core (T5 R9 WELS, Piscataquis Co.) is from the northern conifer zone (figure 3.1), and its pollen and macrofossils reflect that presettlement forest, documented by witness trees. The layer corresponding to about 1800, just before settlement, contains pollen of birch, spruce, fir, pine, and beech, with lesser amounts of hemlock, maple, and oak. Macrofossils include hemlock, cedar, spruce, fir, and white and yellow birch. This region was settled later and not as intensively as the other sites, which is reflected in a weak increase in herbs and decline in pine, hemlock, and beech pollen at the time of settlement. There has been an irregular input of charcoal with peaks at 9,000; 3,800; and 1,500 years ago and at and following settlement.[39]

Molly Schauffler and George Jacobson, of the University of Maine Climate Change Institute, collected five cores from small wetland hollows from a long stretch of the mid- and Down East coast (figure 3.1). The presettlement layer shows dominance by spruce, birches, and alder, with smaller amounts of pine, fir, maple, oak, beech, and hemlock. This closely resembles both the witness tree data and current vegetation composition. With a few exceptions, there were only minor amounts of charcoal.[40]

Sediments collect pollen from both the local and regional vegetation, and the larger the collecting surface of the lake or bog, the higher the proportion of regional pollen. Thus, although biased by surrounding trees, the four studies above provide a sample of the forests in their respective regions — the coastal small hollows because of the multiple cores scattered along the coast and the other three because of their larger collecting basins. The close similarity of the pollen and witness tree records corroborate the regional forest zonation revealed by land surveys: more oaks and pines to the south, mixed hardwoods with some conifers in central areas, and higher numbers of spruce in the north and on the coast. Macrofossils, which largely represent the local vegetation, show a similar pattern: hemlock and yellow birch occur at Gould and South Branch, but mix with white pine at the central Gould site and northern conifers and paper birch at the more northerly South Branch location.

The Remnant Old-Growth Landscape: "Still Stands the Forest Primeval"[41]

In 1989, two friends and colleagues (Mac Hunter and Bob Seymour) and I (ASW) from the University of Maine took a trip to the Big Reed Forest Reserve, a 5,000-acre old-growth forest in northern Maine (figure 3.1). This property had recently been purchased by The Nature Conservancy from the Pingree heirs, whose family had owned the property since 1844. This tract had escaped the axe because water transport, the cheapest and most common way to get wood out of the forest at the time, was not feasible.[42] Except for this difference in accessibility, the forest was typical of much of northern Maine. Its variability in topography and soils supported a variety of forest types and tree species, while its size allowed disturbances to occur naturally and to include multiple stands that represented the variation within forest types. These characteristics are very important when using old-growth forests to shed light on the composition, structure, and disturbance dynamics of presettlement forests. Unfortunately, most old-growth forests in Maine today exist as small, isolated stands and do not have the size or contain the diversity to give that kind of information. Reaching the Reserve is possible today via the network of private gravel roads that crisscross the northern Maine Woods. However, it is a long round trip from

Orono, our home base. Planning to spend at least one whole day exploring the Reserve, we stayed the previous night at an isolated camp, where we were treated to a spectacular display of northern lights and shooting stars, a fitting prelude to the upcoming day.

The next morning we got an early start, anticipating an hour's drive on logging roads. We were fortunate that Mac had been there before, as there were no signs directing us to the Reserve. In fact, access points may vary from year to year, depending on road conditions and logging operations outside the Reserve. Because of a lack of formal trails, getting around within the Reserve requires bushwhacking across rugged terrain. Over the course of the day, we hiked from the eastern boundary of the Reserve downhill to Big Reed Pond (figure 3.8), one of just a handful of lakes in Maine supporting the rare arctic charr (blueback trout), circled the pond, and then returned to our vehicle. This route did not take us into the western half of the Reserve, but by traversing some of the slopes surrounding the pond, we passed through several examples of the major forest types found in the Reserve. During the long ride back to Orono, we discussed what a gem the forest was and how much we could learn from it. Indeed, that trip marked the beginning of twenty-plus years of research in the Reserve with our graduate students (figures 3.8, 3.9).

Figure 3.8 Aerial view of Big Reed Forest Reserve, showing the intermixing of conifer and northern hardwoods stands.

Figure 3.9 Large sugar maple in mixed woods of the Big Reed Forest Reserve.

What have we learned about this forest? And what does it tell us about the presettlement forest?[43] Those familiar with Maine forests would find trees in the Reserve large—some as much as three feet in diameter at breast height (dbh[44])—but perhaps not as large as they might expect (table 3.3). Trees are certainly old for Maine, with a substantial number that established before 1800 (figure 3.10), but not ancient compared to old-growth areas elsewhere. Because trees are left to die in the Reserve, dead wood, including boles and snags larger than typically found in managed forests, is common in the Big Reed woods, adding structural complexity.

Table 3.2 lists the tree species found in sample plots in the Reserve and their relative abundance. This diverse array of species is typical of the *Acadian Forest* (Table 3.2), which lies between the boreal forest to the north and the temperate forest to the south (see chapter 5: "Maine's Place . . ."). Most of the species are common throughout Maine and will be familiar to readers. However, not all trees at Big Reed are readily identified because they sometimes look different due to their old age or large size. The typical shiny yellow, somewhat stringy bark of yellow birch trees, for example, is replaced by large, grayish, platy blocks on large trunks of this species. Such

THE CHANGING NATURE OF THE MAINE WOODS

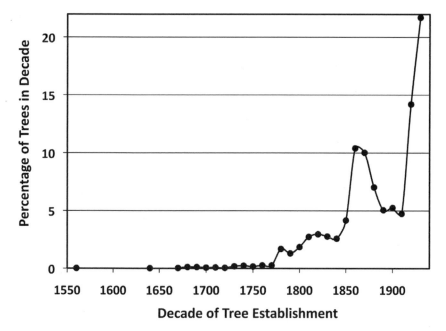

Figure 3.10 Age structure of trees establishing in Big Reed Forest Reserve from 1560 to 1950. Data are from increment cores of 6,909 trees ≥ 10 cm (about 4") diameter at breast height (dbh) taken from 1991 to 2001. A total of 63 trees date from at least before 1750. Many of the oldest cores give minimum ages due to trees with rotten centers, incomplete cores, and coring at breast height (rather than at ground level).

different appearances can make a forest of familiar composition appear somewhat unfamiliar at first glance. The only species on the list that is perhaps somewhat unusual is heartleaf birch (mountain paper birch), a close relative of paper birch and, in fact, easily mistaken for it, given its white, peeling bark. Some readers may recognize this as a taller, more robust form of the short, gnarly high-elevation birch. Five species were not found in our sample plots (table 3.2), but are present in the Reserve: tamarack is found in the wetlands, and red oak, paper birch, American elm, and hawthorn occur very rarely under special conditions, such as hill tops and seepage hollows. Combining these five species and the list from table 3.2 gives a total of twenty-six tree species in the Big Reed Forest Reserve.[45]

The richness of tree species in the Big Reed Forest Reserve is expressed in a variety of ways. First, the diversity sometimes manifests itself as many species occurring within a single stand; indeed, most classification schemes in this region use the term "mixed" in the names of some of their forest types, such as *mixed hardwoods* or *mixed softwood-hardwood*. Second, within any single forest type, species composition can differ from one stand to

Table 3.2 Relative importance (% basal area[a]) of tree species in four stand types and all stands combined at Big Reed Forest Reserve

Tree species	Hardwoods	Mixed Woods	Softwoods	Cedar Swamp	Total
Red spruce	7.8	26.2	70.4	14.4	40.1
Sugar maple	56.0	24.1	2.8	0.0	17.9
Northern white cedar	0.5	10.3	5.4	73.2	11.2
Beech	24.1	10.1	3.5	0.0	8.4
Balsam fir	3.5	11.1	5.5	2.9	7.8
Yellow birch	3.8	10.3	3.0	4.0	6.6
White spruce	0.0	2.4	1.4	2.7	1.8
Red maple	1.3	1.5	2.7	0.0	1.8
White pine	0.0	0.0	3.9	0.7	1.5
Hop-hornbeam	2.1	0.9	0.1	0.0	0.7
Other species[b]	1.0	3.0	1.1	2.1	2.0
Percent of land	10	48	37	6	100
Number of plots sampled	13	56	28	9	106
Number of tree species	12	16	15	10	16
Total basal area (ft²/acre)[a]	125	123	131	175	129

[a]Basal area: the total cross sectional area of trees at 4.5' height.
[b]Other species: hemlock, black ash, striped maple, heartleaf birch, white ash, and black spruce, each 0.5 percent or less and in descending order of abundance. Also present in plots but <4" dbh: shadbush, mountain maple, pin cherry, American mountain ash, showy mountain ash.

Source: Shawn Fraver, Unna Chokkalingam, Erika Rowland, Laura Conkey, Lissa Widoff, Erika Latty, John Hagen, Allison Dibble, Charles Cogbill, and Morton Mossewilde; compiled by Charles V. Cogbill.

another. The mixed hardwood type, for example, could include stands dominated by beech and red maple but also stands dominated by sugar maple and yellow birch. Third, this Acadian Forest is characterized by overlap in species among different forest types. For example, table 3.2 shows that the ten most common species, together comprising over 95 percent of the trees, all occur in three or all of the four forest types. In fact, rather than each forest type having a different list of species, it's the relative abundances of species that differ most among types. Finally, these major forest types also exhibit a fine-scale of spatial diversity, as they are distributed in alternating fashion across the Reserve, as shown in figure 3.11.

Despite their diversity, the species in Big Reed Forest Reserve share cer-

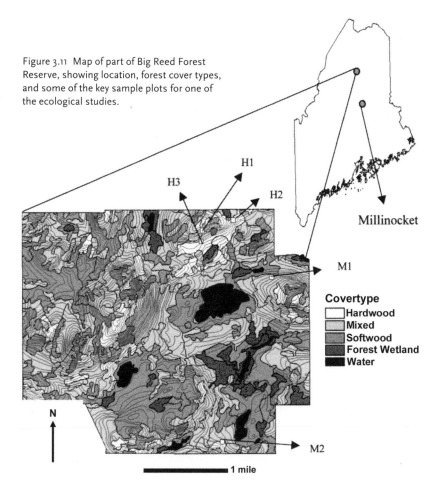

Figure 3.11 Map of part of Big Reed Forest Reserve, showing location, forest cover types, and some of the key sample plots for one of the ecological studies.

Covertype
Hardwood
Mixed
Softwood
Forest Wetland
Water

tain *life history* characteristics. For example, most are *tolerant* of shade; that is, they can regenerate under a forest canopy by utilizing the relatively small amounts of light and growing space associated with small gaps in the forest that are formed by the death of one or more trees. *Intolerant* species, in contrast, cannot regenerate under a forest canopy or in small gaps, relying instead on large disturbances to remove most of the forest canopy. It is worthy of note that there are virtually no intolerant tree species (such as aspen) in the Reserve, whereas they are common in the frequently harvested surrounding landscape. This suggests that the disturbances in the Reserve are relatively small, creating gaps that are sufficient to allow tolerant species to grow and make it into the canopy, but not big enough for intolerant species to survive. If there have been large disturbances in the Reserve, it has been a long time since the last one.

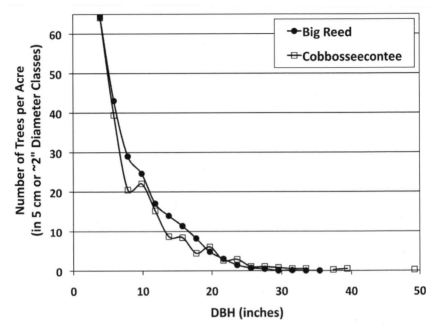

Figure 3.12 Size structure of trees of all species in sample plots covering 22.7 acres in Big Reed Forest Reserve and a complete survey of 3.7 acres in Cobbosseecontee old growth stand (see next section). Data are from trees > 10 cm (about 4") dbh, in 5 cm (about 2") increments.

What is the *size structure* of these forests? Like composition, structure is quite variable (figure 3.12). The distribution of tree diameters covers a wide range, and there are typically many more small-diameter trees than large ones. Tree heights are also diverse, due to differences in species height growth rates and to the gaps that are scattered throughout the forest. Being formed over time, the gaps give rise to a continual influx of new stems that are younger and smaller (in both height and diameter, at least for a while) than their canopy neighbors, conferring a high degree of physical complexity to the forest.

The species composition and size structure of Big Reed forests, then, are consistent with one another. Both the variety of tree sizes and the preponderance of shade-tolerant species is what one would expect to find in an area where disturbances are small gap-forming events that don't create enough light for intolerant species to survive. Remember, however, that we want to use information from this old-growth forest to make inferences about the presettlement forest. Thus, we want to know not just the composition and structure of today's old-growth forests, but whether they have changed much over time, in other words, whether what we observe today is indicative of presettlement conditions.

We can address this question with three different approaches: *dendrochronology*, presettlement survey records, and *paleoecology*. Dendrochronology is the use of tree ring analysis to study ecological patterns as well as questions in other disciplines such as archeology. Many of the studies in the Big Reed Forest Reserve have used the temporal patterns of tree ring widths (radial growth) and the distribution of tree ages to infer the history of disturbances. To use tree ring patterns as indicators of disturbance requires that such patterns are distinguishable from patterns caused by other factors, such as climate. A tree that survives a disturbance that kills one or more of its taller neighbors often exhibits a large, abrupt increase in its growth rate (a *release*) that is sustained for several years. Identification of such a pattern requires definition of the terms *large*, *abrupt*, and *sustained*. A commonly used definition of a disturbance-induced pattern is a twofold or greater increase in growth (averaged over ten years) that occurs within a three-year period and is sustained for ten or more years. Disturbances may also give rise to new trees, sometimes called *gap-origin trees*, which are characterized by wide growth rings around the pith when the tree was young. The definition we used to detect such trees was an average radial growth rate of 1 mm (about 0.04 inch) or more for the first five growth rings. The year of a release or the origin of a gap may be slightly variable among trees for any given disturbance event, so the disturbance is often assigned to a decade rather than to a year. The number of trees, or their crown areas, are summed for the decade and expressed as a percentage of the total in the stand. Thus, disturbance rate is the percentage of stand area disturbed in each decade.[46]

Using the dates of growth releases and gap origins, we have found that the average stand disturbance rates in the Reserve are 9–12 percent per decade. The rates are remarkably similar even among forest types with quite different mixes of species (except for cedar in wet areas, which experience about half the average). Indeed, disturbance rates vary more from decade to decade than the overall averages vary among stands. For example, one stand may experience virtually no disturbances in one decade, but have 25 percent of its area disturbed during the next. During many decades, the amount of disturbance among stands appears unrelated; for example, one stand might have a 5 percent disturbance rate, while another might have a 20 percent disturbance rate in that same decade. However, there are some

Spruce Budworm and Maine Presettlement Forests

Disturbances are multidimensional phenomena that can be difficult to describe in simple terms. For example, the term "small gaps" implies rather insignificant disturbances, ones that cause mortality of a few trees here and there. However, the term by itself can leave out important information and perhaps be misleading. For example, the agent of mortality may be specific to just one or a few species and thus be more significant for those species. Also, these agents of mortality may occur episodically and will appear more important at some times than others; the longer the time between these disturbances, the more difficult it is to correctly ascertain their role in the ecosystem.

Spruce budworm is an important disturbance in northern Maine, but outbreaks occur only every six to seven decades. To the untrained eye, the effect of spruce budworm may not be very apparent if the length of time since the last outbreak is considerable. On the other hand, it may appear quite dramatic if it has only recently occurred (such as in the 1970s; chapter 4). Analysis of tree rings can help put the role of spruce budworm over time in perspective. For example, when the radial growth rates of red spruce sampled across the Big Reed Forest are averaged and plotted over time, periods of dramatically reduced growth associated with spruce budworm are readily apparent (figure 3.13). The growth of a non-host species, in this case northern white cedar, can be used to distinguish the effects of other factors like climate, which might be common to most species, from factors like spruce budworm that are specific to only a subset of species.

At the Big Reed Forest Reserve, there have been at least five spruce budworm outbreaks since the 1700s, detected using analyses of growth over time (figure 3.13).[1] Paleoecology has given us an even longer perspective: head capsules of spruce budworms have been found in association with spruces and firs in peatland sediments on Anticosti Island in the Gulf of St. Lawrence dating back more than 3,000 years ago.[2] Twenty-five years have now passed since the end of the last outbreak in northern Maine. Despite the intense nature of that outbreak, its effects are diminishing visually as the skeletons of dead balsam firs (far more susceptible to mortality from spruce budworm than is red spruce) decompose and the gaps fill in with new or released stems.

Spruce budworm illustrates the importance of using multiple lines of evidence. In this case, tree ring analysis provides a clear indication of the disturbance over the last three centuries; other data, like the survey notes, did not reveal the disturbance because of the timing of the surveys relative to the timing of the outbreaks. Other episodic disturbances, like fire, may be more readily discerned by other lines of evidence, such as charcoal fragments in sediment cores. Disturbances have both spatial and temporal dimensions that require investigation by multiple research techniques if their role in forest ecosystems is to be revealed.

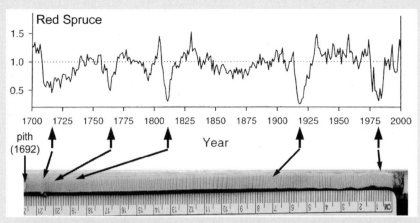

Figure 3.13 Index of diameter growth of red spruce in Big Reed Forest Reserve from 1700 to 2000. Upper graph is a chronology based on cores from 772 trees. Growth comparisons with non-host tree species suggest that the periods of low growth (arrows) were caused by outbreaks of spruce budworm. Lower photo is a core from a single 20"-diameter red spruce from Big Reed (T8 R10 WELS). The tree's pith (dated as 1692) is at 21.7 cm. Growth declines corresponding to those found in the composite growth chronology are shown with arrows.

NOTES

1. Shawn Fraver, Robert S. Seymour, James H. Speer, and Alan S. White., "Dendrochronological reconstruction of spruce budworm outbreaks in northern Maine, USA," *Canadian Journal of Forest Research* 37 (2007): 523–529.

2. Martin Lavoie, Louise Filion, and Elisabeth C. Robert, "Boreal peatland margins as repository sites of long-term natural disturbances of balsam fir/spruce forests," *Quaternary Research* 71 (2009): 295–306; see also Hubert Morin, Yves Jardon, and Réjean Gagnon, "Relationship between spruce budworm outbreaks and forest dynamics in eastern North America," in Edward A. Johnson and Kiyoko Miyanishi, eds., *Plant Disturbance Ecology: The Process and the Response* (Amsterdam, Netherlands: Elsevier, 2007), 555–577.

decades during which many stands show higher disturbance rates as indicated by many trees exhibiting accelerated growth rates and/or many trees of gap origin (a cohort of trees of similar age). These decades reflect disturbances, such as strong windstorms or insect outbreaks that are extensive across the landscape (sidebar 3.3; figure 3.13). Even during these periods of extensive disturbance, however, the area disturbed in any one plot seldom exceeded 50 percent; in fact, for the Reserve as a whole, any given plot would exceed 50 percent disturbance only once every 1,100 years or more. This is entirely consistent with the 1,012-year return time of catastrophic disturbance across northern Maine derived from presettlement surveys (see "Land Surveys . . ." earlier in the chapter). These data, then, tell us that at least over the past 150 years or more, the old-growth forests of Big Reed Pond Reserve have been characterized by a modest level of natural disturbance.

In addition to percent area disturbed, it is important to know the absolute sizes of disturbances in Big Reed Forest Reserve, because absolute size, not percent, is what determines how resources such as light are altered by the disturbance. For example, if 25 percent of a 0.5 ha (approximately 1¼ acre) plot is disturbed, we don't know if that 25 percent occurred as one large gap or as many small gaps; the difference is great enough to influence whether shade-intolerant species (or only tolerant ones) are capable of regenerating in the stand. To address this question, the locations of all trees in several plots in the Reserve were mapped. The resulting map of tree ages showed us which groups of trees reflected the disturbance in any given past decade. The disturbed areas indicated by these groups were typically small, about 1,000 square feet (the size of a small bedroom), only rarely reaching 3,000 square feet. Thus, the current gap structure of small treefall gaps in the Reserve, assessed using size structure, is well within the bounds of what has existed in past decades.

The evidence given above supports the hypothesis that the current forest in the Reserve is similar to what existed in presettlement times. The tree ring information discussed above, however, provides a very detailed look only for the past 10 to 15 decades, but then becomes less reliable as one goes back in time because of the scarcity of very old trees from which tree rings can be derived. The data can go back no further than the oldest trees.

As we demonstrated earlier in the chapter, land survey records provide a powerful lens on presettlement forests, and we can use surveys carried out in the mid-1800s to extend our examination of the Big Reed Forest Reserve back to those times.[47] In 1833, the outlines of T8 R10 had eighteen witness trees blazed at mile intervals and included 28 percent fir, 22 percent sugar maple, 17 percent yellow birch, 17 percent cedar, and 6 percent spruce. The middle two miles of the west line of T8 R10, over Big Reed Mountain within the Reserve, was an alternation of 11 forest stands containing 60 percent hardwood, 27 percent mixed woods, 10 percent softwoods, and 3 percent cedar swamp forest types.[48] These 1830s patches cleanly match, at the identical position and size, those on the 1976 forest type map of the same line. There is some increase in the proportions of hardwood and decrease in softwood types in presettlement forests, but this is evidently due to the differences in criteria used by the early surveyors.[49] These eyewitness observations confirm the remarkable similarity and stability of composition and

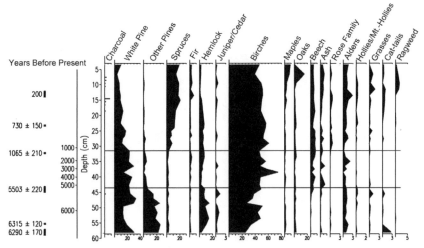

Figure 3.14 Pollen and charcoal profile for more than 6,000 years at Big Reed Pond Reserve. Thickness of the bars is proportional to abundance of pollen. Abundance of charcoal is proportionate to height of the bars. Depth is vertical distance from ground surface in the forest hollow.

vegetation types of the presettlement and modern old growth vegetation at Big Reed Forest Reserve.

Paleoecology provides another estimate of presetttlement conditions and can extend the time scale even farther back into the past. These techniques can also sometimes be applied at a scale comparable to a stand. The pollen, charcoal, and macrofossils that are deposited in lakes and bogs may originate from a large area, as we described earlier in this chapter. While important in their own right, the connection between these data and what is going on in a stand can be difficult to discern. However, there are sometimes features within a stand that collect pollen and other plant remains from a radius of only 200 feet or so. These features, which we collectively refer to as *small hollows*, are depressions in the forest floor where organic matter does not decompose rapidly, perhaps due to water accumulation and restricted aeration. Cores extracted from such hollows provide a local look at past species composition and disturbances. Such cores from the Reserve, from the dissertation research of Erika Rowland and Molly Schauffler at the University of Maine, reveal that the same general group of species has occupied the Big Reed area for the past 1,000 years, although relative abundances have shifted somewhat (e.g., increase in spruces, decrease in hemlock; figure 3.14). For that same time period, there is no charcoal evidence of local fires, suggesting that fires were not important in the sampled stands

Table 3.3 Key attributes of five old-growth areas in Maine

	Megunti-cook	Cobbossee-contee	Wizard Pond	Elephant Mountain	Big Reed
Forest type	Oak	Northern hardwoods	Spruce	Spruce-fir	Variety
Major species	Red Oak	Sugar Maple Beech White Pine	Red Spruce	Fir Red Spruce	Red Spruce Sugar Maple Cedar
Number of tree species[a]	7	8	4	3	5.7
Percent herbaceous cover[b]	3	30	12	95	5
Percent moss/lichen cover[b]	<1	2	95	30	87
Number of saplings[c]	170	117	235	186	607
Number of small trees[c]	108	154	257	146	104
Number of larger trees[c]	38	39	39	251	102
Basal area[d]	140	143	165	185	131
Largest tree[e]	19	51	26	22	36
Oldest tree (years)	170	213	287	319	436
Median age (years)[f]	104	148	203	164	104
Canopy height (feet)	67	70	56	53	63
Recent dead[d]	27	47	24	70	41

[a]Number of tree species per ~¼ acre sample plot.
[b]Cover is percent of ground covered by this group.
[c]Saplings are 4.5–10 cm (about 2–4") dbh, small stems 10–20 cm (about 4–8") dbh, and larger stems are > 20 cm (about 8") dbh—all per acre.
[d]Basal area is square feet of trunk cross section at tree breast height per acre.
[e]Tree diameter at breast height in inches.
[f]Median (the midpoint) of values for all trees > 20 cm dbh.

Note: See Figure 3.1 for locations.

in the Reserve. In summary, paleoecological research in small hollows yielded no information that refutes the picture of the presettlement forest in the Reserve as one of shade-tolerant species with small gaps. This does not mean that there were no large, severe disturbances in the presettlement forests; in fact, we know there were. When they occurred, they undoubtedly had a significant effect on forest composition and structure. However, they were relatively infrequent on any given piece of ground.

Remnant Old-Growth Stands: Museums of the Past

Even the description of Big Reed, the largest remnant and one of the most studied forests in the state, cannot do justice to the diversity or regional characteristics of the Maine Woods. Maine possesses old-growth areas other than Big Reed, but they are mostly tiny stands (or stunted vegetation on high mountains) that have survived because of an idiosyncratic history of ownership, condition, or location.[50] They lack the heterogeneity and mosaic of forest types and developmental stages of Big Reed. Although these stands are often uncharacteristic of the general forest landscape, they can provide illustrations of the likely composition and structure of presettlement forest stands. Indeed, old-growth remnants are exceptional examples of continuity with the past, conserved in a natural museum. In Maine, they have been considered sufficiently important to warrant a serious effort at discovery, cataloguing, and protection, for which one of this book's authors, Charles V. Cogbill, played an important role.[51]

The known remnant stands cover a total area of perhaps the same as Big Reed (5,000 acres), with a combined area of only 0.01 percent of the land in the state. There are hundreds more of these small scraps scattered mostly in the northern part of the state, unrecognized by their lack of outward distinguishing character.[52] We have chosen four remnant stands as examples of old growth, one in each of the four regions identified for the

---

Table 3.3 source: Megunticook data from Ecological Reserves Inventory, Maine Natural Areas Program, and Charles Cogbill sampled in 1995; Cobbosseecontee data from Alan White, Maine Critical Areas Program, and Charles Cogbill plot sampled in 1982 and whole stand sampled in 1992; Elephant Mt. data from Laura Conkey, Charles Cogbill, and Maine Critical Areas Program, sampled in 1985; Wizard Pond from Laura Conkey, Ed Cook, and Charles Cogbill, sampled in 1993; Big Reed plot cover data from Laura Conkey and Charles Cogbill sampled in 1991 and reserve wide from Lissa Widoff, Barbara Vickery, Shawn Fraver, Erika Rowland, Unna Chokkalingam, Alan White, and Charles Cogbill sampled in 1991–2001. Compiled by Charles V. Cogbill.

presettlement forest (figure 3.1, table 3.3). The southern oak-pine region is represented by the ten-acre Megunticook stand in Camden Hills State Park (Knox Co.), northern hardwoods by eleven acres of private land by Cobbosseecontee Lake in Monmouth (Kennebec Co.; figure 3.15),[53] the coastal conifers by a twenty-five-acre stand at Wizard Pond T10 SD (Hancock Co.) on Maine Public Reserve Land, and the northern conifers by twenty acres on Elephant Mountain in D Town (Franklin Co.) on the U.S. Park Service Appalachian Trail (figure 3.16).

The composition of these four stands follows the presettlement transition of plant communities across the state from oak dominance at Megunticook, through beech–sugar maple–white pine at Cobbosseecontee, to red spruce at Wizard's Pond, and finally to fir–red spruce on Elephant Mountain (table 3.3). Although old-growth landscapes often maintain a higher diversity of species than less developed forests, this is not the case in these stands, all of which support a moderately low diversity of plant species, especially the conifer stands. This might be explained in part by the lack of heterogeneity of forest types and developmental stages usually found in a full-blown old-growth landscape such as Big Reed. Basal area varies among the four stands, but all exhibit high values, which is typical of old-growth compared to younger forests (table 3.3). These seem to have reached a predictable ceiling, with hardwood levels (143 ft^2 per acre) lower than in softwood stands (186 ft^2 per acre). The basal area of Big Reed is slightly lower than these values, not surprising given that the Reserve supports much mixed deciduous forest and includes canopy gap areas with younger vegetation, which is generally not found in these tiny remnants.

Some of the trees at Cobbosseecontee, especially the supercanopy white pines, are very large (up to 4½ feet in diameter; figure 3.12); those at the other sites are modest in size (three feet or less; table 3.3). But size does not necessarily reflect age, for all four old-growth stands contain some very old trees, with a maximum of 319 years for a red spruce at Elephant Mountain, compared to the oldest tree in Big Reed, a 436-year-old hemlock (figure 3.17). Typical tree ages are higher in the softwood (150–200 years) than hardwood stands (100–150 years), with Big Reed more similar to the remnant hardwood stands.[54] Sidebar 3.4 combines the tree ages of these four stands, thirty-three other old-growth remnants, and those of Big Reed, and develops a reasonable estimate that an average of about 0.6 percent to 1 percent of Maine's old-growth and presettlement forests were disturbed in any given year, creating new patches with vigorous young trees. These

Figure 3.15 Drawing of the Cobbosseecontee old-growth stand, looking across stone wall into small uncut lot.

Figure 3.16 Elephant Mountain old-growth stand. Note the large fallen dead tree trunk.

Figure 3.17 Oldest cored red spruce in Maine (18.2" dbh in 1990), located at North Turner Brook, T4 R9 WELS, Baxter State Park. Note the snags and very ragged canopy in this stand of older trees.

four remnant stands show little evidence of past large disturbances,[55] consistent with the conclusion for Big Reed Forest Reserve that disturbances were mainly small, single-tree mortality events, due to wind or insects, rather than stand-replacing events.

Conclusion

For untold ages Maine had been one unbroken forest, and it was so still.
—Francis Parkman

The early 1700s was an important break point in the history of the Maine Woods. Thus began the fading of French influence, allowing the rise of English settlement and the profound disruption of the forest. Francis Parkman, the great historian of the French regime in North America, rightfully uses the metaphor of the undisturbed forest ("one unbroken forest") to introduce this theme.[56] Moreover, *"le grand dérangement"* (the great upheaval) of the French culture of *Acadie* was equally true for the Acadian forest of the region in the centuries to follow. Presettlement was the old regime, soon to be replaced by another, the postsettlement forest (chapter 4). Although 90 percent of Maine is covered with forest today, the 0.05 percent not altered by human land use is a vague shadow of the presettlement past. This chapter has attempted to reconstruct a reasonable picture of that long-past landscape. Historical and modern ecological evidence reveals a surprisingly fine-grained view of the species composition, heterogeneity, structure, and role of disturbance in the presettlement Maine Woods.

The species composition of the presettlement Acadian forest was not homogenous across Maine. At the broad scale, the forested landscape varied gradually among seven distinct clusters of similar composition over the state. These clusters form four presettlement geographic regions — south, central, coastal, and north — that were distinctive in flora and forest dominants. At the local scale, within each region, stands varied among widely different composition from hardwood to conifer to swamps, and these small patches alternated across the landscape. There was tremendous variation even among each of these forest types within each region, so that at the fine scale each stand was a unique mix of species, distinct from other, even nearby, forest communities. With the exception of the tension zone between oak-pine forests in southern Maine and the mixed northern hardwood forests in north-central Maine, these units from the finest to the broadest scales graded gradually into one another.

The presettlement forests of Maine were not especially tall, dense,[57] high in biomass, or composed of lots of gigantic trees.[58] There were some very large white pines and hemlock, such as those remaining in some old-growth remnants and reported by early loggers (chapter 4: "Before 1781...."), but most of the larger trees in presettlement stands appear to have been only two to three feet in diameter, considerably smaller than in old growth stands in places such as the Pacific Northwest or southern Appalachians. The canopy height, density, basal area, and tree sizes were variable depending on the setting and composition, but on average similar to modern older second-growth forests. The distinguishing *structural* characteristics of these old-growth, and presumably presettlement forests, are subtle.

The age structure of the presettlement forest was also not remarkably old (when compared to the quintessential ancient forests of the West), although obviously much older than typical managed woods today. Only three individual Maine trees of the more than nine thousand studied using dendrochronology are known to have exceeded 400 years old when cored.

Disturbance Return Time in Old Growth

Studies at Big Reed Forest Reserve reveal that presettlement disturbances were predominantly fine-scale tree-fall gaps rather than broad-scale catastrophic events. The age structure of old-growth remnants gives us a method for extending and quantifying these estimates of the disturbance regime across a wide area. Like at Big Reed, age structure of the small remnant stands displays the pattern of tree establishment over the life of the stand. If the trees regenerate after a single disturbance event, all would be nearly the same age (*even-aged*). The contrasting pattern of numerous smaller disturbances would result in a variety of ages (*uneven-aged* or *all-aged*). If we assume that the disturbances occur at random intervals, a fair assumption when averaged over a long time and large territory, the average of the trees' ages is a mathematical predictor of the "return time" for that disturbance (i.e., average time between disturbances at a point).[1] This "return time" is equivalent to the total time it would take a disturbance either to remove all trees at once from the area or for small disturbances to disturb an equivalent of the entire area. For example, the median age of 6,909 canopy trees cored at Big Reed is 104 years,[2] which translates to an average time between disturbances (point mortality) of 104 years. The reciprocal of this time (1 disturbance per 104 years) is the average disturbance rate or 0.96 percent of the total area per year. This estimate from tree ages is identical to the rate derived from studies of radial growth release histories of fifty stands at Big Reed, described earlier, which also give an average of 0.96 percent per year rate for the same return time of 104 years.[3]

Expanding these estimates across the state, we find that the median age of trees in the four old-growth remnants ranges from 104 to 203 years (table 3.2). Supplementing these data even further with aged trees from thirty-three other old-growth remnants across Maine gives an average return time of disturbance at 150 years for hardwoods and 171 years for softwoods.[4] Consistent with this age structure, Austin Cary, the first industrial forester in Maine, in a comprehensive study of uncut forests in northern Maine, calculated a median age of 180 years old in spruce logs removed from a first cut forest in the 1890s.[5] Significantly, none of the remnant old growth or Big Reed stands was even-aged. Many softwood stands have several age structure peaks (multiple cohort structure), indicating some role for moderate-scale disturbances, such as spruce-budworm outbreaks (sidebar 3.3), consistent with the episodic small disturbance regime found at modern Big Reed. All trees in the presettlement Maine Woods were subject to multiple disturbances.[6] The overall average odds of any tree being disturbed in any year varied from 1:104 for fine-scale sources of tree mortality determined at Big Reed to 1:1,500 for catastrophic fire disturbances and 1:3,250 for catastrophic wind disturbance determined from presettlement surveys.

NOTES

1. C.E. Van Wagner, "Age-class distribution and the forest fire cycle," *Canadian Journal of Forest Research* 2 (1978): 220–227. Although this calculation was first used with catastrophic disturbances, there is no difference when scaled down to within-stand dynamics. It is very robust to mixtures of scale such as episodic disturbances of different extent, and the averages represent the long-term landscape regime.

2. Ages are technically ring count or oldest cross-dated ring at breast height for larger trees. This gives a fixed minimum estimate of age for canopy trees. Median age of trees is a robust estimate of stand average age, as it effectively corrects for uncertainties due to missed centers, rotten cores, and counting indistinct or missing rings. Similarly, only trees greater than 10 cm (about 4') diameter and more than fifty years old are included in the calculation to accommodate the bias from consistently un-aged small trees. Although median age is theoretically an underestimate of mean age, given the truncation of cores less than fifty years old and the few very old trees in the tail of the distribution, median age is found to be a good approximation of mean age. Here these median ages are termed "average" ages referring to the documented age of canopy trees at breast height. Tree cores taken and analyzed by Shawn Fraver, Unna Chokkalingam, Erika Rowland, Laura Conkey, Charles Cogbill, Lissa Widoff, and Morton Mossewilde. Compiled by Charles Cogbill.

3. The canopy disturbance rates vary: 1.18 percent per year for hardwoods, 0.94 percent for mixed woods, 0.86 percent for conifers, and 0.51 percent for cedar swamps. This translates to a return time of 85 to 116 years in common forest types at Big Red. Shawn Fraver, "Spatial and temporal patterns of natural disturbance in old-growth forests of northern Maine, USA." (PhD dissertation, University of Maine, 2004); Unna Chokkalingam, "Spatial and temporal patterns and dynamics in old-growth northern hardwood and mixed forests of northern Maine" (PhD dissertation, University of Maine, 1998); Erika Rowland, "Disturbance pattern of multiple temporal scales in old-growth conifer-northern hardwood stands in northern Maine, USA" (PhD dissertation, University of Maine, 2006). This gap disturbance rate is higher than for most old-growth northern hardwood forests in Eastern North America, for which an average of 0.55–0.75 percent has been determined. See James R. Runkle, "Patterns of disturbance in some old-growth mesic forests of eastern North America," *Ecology* 63 (1982): 1533–1546; Lee E. Frelich, *Forest Dynamics and Disturbance Regimes* (New York: Cambridge University Press, 2002).

4. Tree cores taken and analyzed by Ron Davis, Ed Cook, Maine Critical Areas Program, Laura Conkey, Maine Natural Areas Program, and Charles Cogbill. Compiled by Charles Cogbill.

5. A. Cary, "On the Growth of Spruce" in *Second Annual Report, Maine Forest Commissioner* (Augusta: Maine Forest Commission, 1894), 20–36.

6. Perhaps the most significant disturbance affecting tree growth and establishment of new trees was experienced throughout Maine in the 1810s. The exact cause of this event(s) is unclear since this time was a concurrence of great disturbances including large fires, widespread budworm infestation (1805–1810), the destructive hurricane of 1815, and cold growing seasons culminating in "the year with no summer" (1816), thought to have been partly caused by the 1815 eruptions of Mt. Tambora in Indonesia.

Only these hemlock (436 years old), red spruce (423 years old), and northern white cedar trees (403 years old) were alive when the first English colonists landed at Popham in 1607.[59] There are certainly other old, undiscovered trees, but few trees extant in Maine today are actual survivors from the forest of that time.[60] As seen from the age structures of the old-growth forests (figure 3.10), even undisturbed stands exhibit trees of all ages, with an average between 100 and 200 years. Obviously most trees die well before their maximum biological longevity. Even in old growth, then, there is a constant turnover with individual tree mortality and regeneration of replacements. This small-scale dynamic, and its continuity, is the essence of both presettlement and old growth forests.

This small scale of disturbances is strongly reflected in the life history of the dominant trees of the presettlement forest: long-lived, shade-tolerant species, such as beech, spruce, hemlock, and sugar maple. Light-demanding, early-successional species, such as aspen, cherries, and white birch, were much less common, largely restricted to sites where specific large disturbances had occurred in the past. Observations of land surveyors, including in the area just outside of Big Reed Forest Reserve,[61] and early settlement era fires in southern (1761, 1762) and northern (1795, 1825) parts of the state, leave little doubt that Maine forests burned in presettlement times. Even after settlement, occasional large wildfires burned in Maine, such as the 1947 fires on Mt. Desert Island and in southwestern Maine and the 1977 fire in Baxter State Park. However, fires and wind storms were apparently not sufficiently frequent or widespread to be the dominant process in the establishment of trees across the state. This is consistent with the limited abundance of white pine in the presettlement Maine Woods, their economic importance bias notwithstanding. In northern Maine, the restricted charcoal in soils and sediment cores, the rarity of early successional species, the abundance of fire-intolerant northern hardwoods, uneven-age structure, and the old maximum age of trees in most stands all argue for the limited influence of presettlement fire. This is in striking contrast to the boreal forests to the north in Canada, which burn much more often and largely support relatively young, even-aged forests of fire-adapted species.[62]

Even for the southern oak-pine region, where fire would be expected to have a higher impact, the evidence for an overriding influence of fire in presettlement times is limited. We know that fire occurred, indicated by charcoal in presettlement portions of sediment cores from Phippsburg (The Basin study) and Waterboro Barrens in York County.[63] Frequent fire and open, fire-prone vegetation, however, appear to have developed largely after settlement at this latter site.[64] Despite the early explorers' coastal and river descriptions of oak forests that were obviously kept open, perhaps by Indians, there is no mention of actual wildfires or burned timber in any coastal Maine narrative.[65] Traditionally, scattered large trees, open understories, patchy thickets of sprouts, and oak or pine dominance are associated with frequent low-intensity wildfires and regeneration of these fire-adapted species. This is certainly well documented for typical oak-pine forests in southern New England, which burned commonly in the past.[66] For southern Maine, however, the evidence is inconclusive. We know that some oak and pine forests in Maine today can regenerate without fire. It is unclear, however, for presettlement times, if low-intensity ground fires frequently burned in these coastal areas, if the oak was regenerating on dry or rocky sites without fire, or if large but undocumented fires actually occurred.

We have analyzed four lines of evidence — early explorers, land surveys, paleoecology, and modern old growth — to reconstruct a picture of the presettlement Maine Woods. None of these types of data alone has been sufficient for that purpose; each has advantages but also weaknesses. Only the witness tree data provide a nearly comprehensive geographic view; only the paleoecology studies reveal changes over relatively long time periods; only the Big Reed old-growth landscape offers details about forest dynamics; and only the old-growth sites allow us to see in person and measure a forest that resembles those of presettlement times. Because of their complementarity, these multiple lines of evidence together bestow a remarkable picture of what the presettlement forest looked like, how it operated, and how it varied across the landscape — in some ways less obscured than our image of the modern forest, which has been shaped by so many ecological and cultural factors (chapters 4 and 6). It is important to emphasize that the presettlement view provided in this chapter is of roughly a single point in time, a snapshot, for a forest that in actuality was always changing, a sort of moving target. In chapter 2, we showed, for example, that not much more than 1,000 years ago, spruce was barely detectable in pollen cores from inland sites. By presettlement times, this species was the most abundant tree in Maine, accounting for 20 percent of the trees.[67] As we will see in the next chapter, in the almost four centuries since settlement, spruce and the entire, unbroken Maine Woods have been profoundly transformed by human culture.

APPENDIX 3.1

Observations by Early Explorers of the Forests along the Maine Coast

Bartholomew Gosnold 1602

Kittery (York Co.) trees "very high and straight"

Cape Porpoise (York Co.) "full of faire trees."

Martin Pring 1603

Fox Islands (Knox Co.) "Ilands we fond very pleasant to behold, adorned with goodly grasse and sundry sorts of trees, as Cedars, Spruce, Pine and Firre-trees."

Cape Neddick (York Co.) "beheld very goodly Groves and Woods replenished with tall Okes, Beeches, Pine-trees, Firre-trees, Hasels, Wich-hasels [ironwood] and maples."

Samuel de Champlain 1604

Ste. Croix Island (Washington Co.) "The island is covered with firs, birches, maples & oaks. . . . All the rest of the country is very thick forests."

Machais Bay (Washington Co.) coast "covered with pines, firs, & other inferior woods."

Mt. Desert Island (Hancock Co.) "The tops of most of [the mountains] are bare of trees, because there is nothing there, but rocks. The woods consist only of pines, firs, & birches. I named it Mount Desert island."

Bangor (Penobscot Co.) "I landed to see the country. . . . The part I visited most pleasant and agreeable. One would think the oaks there had been planted designedly. I saw few firs, but on one side of the river were some pines, while on the other were all oaks, together with underwood which extends far inland."

Samuel de Champlain 1605

Lower Kennebec (Sagahadoc Co.) "There are quantities of small oaks but very little cultivable land."

Saco River (York Co.) "The forests inland are very open, but nevertheless abound in oaks, beeches, ashes & elms, and in wet places there are numbers of willows."

Richmond Island (Cumberland Co.) "an island which is very beautiful . . . having fine oaks and nut-trees, with cleared land and abundance of vines . . ."

John Rosier (Waymouth) 1605

Monhegan Island (Lincoln Co.) "we laded our boat with dry wood of olde trees upon the shore side. . . . This iland is woody, grouen with Firre, Birch, Oke and Beech, as far as we saw along the shore; and so likely to be within."

Muscongus Bay (Lincoln/Knox Co.) "within the Illands growe wood of sundry sorts, some very great, and all tall: Birch, Beech, Ash, maple, Spruce, Cherry-tree, Yew, Oke very great and good, Firre-tree."

Allen's Island (Knox Co.) "spruce trees of excellent timber and height, able to mast ships of great burden."

Penobscot Bay (Knox Co.) "the wood she beareth is not shrubbish fit only for fewell, but goodly tall Firre, Spruce, Birch, Beech, Oke which in many places is not so thicke."

Camden Hills inland (Knox Co.) "passed over very good ground, pleasant and fertile, fit for pasture, for the space of some three miles having but little wood, that Oke like stands left in our pastures in England, good and great, fit timber for any use. Some small birch, Hazle and Brake. . . . In many places are lowe Thicks like our Copoisses of small yoong wood. And surely it did all resemble a stately Parke, wherein appeare some old trees with high withered tops, and other flourishing with living greene boughs. Upon the hills grow notable high timber trees, masts for ships of 400 tun."

Up Penobscot River (Penobscot Co.) "The wood in most places, especially on the East side, very thinne, chiefly oke and some small young birch, bordering low upon the river."

In general, "Oke, of an excellent graine, strait, and great timber, Elme, Beech, Birch, Wich-Hazell, Hazell, Alder, Cherry-tree, Ash, Maple, Yew, Spruce, Aspe, Firre, many fruit trees, which we knew not."

Popham & Gilbert 1607

Nova Scotia "gallant Illands full of heigh and mighty trees of sundry sorts."

Small Point (Sagadahoc Co.) "this Illand ys full of pyne trees and ocke and abundance of whorts [blueberry genus] of (l)ower Sorts of them."

Richmond Island (Cumberland Co.) "trees growing thear doth exceed for goodness and Length being the most p^t of them ocke and walnut growing a great space assoonder on from the other as our parks in Ingland and no thicket growing under them. . . . hear we also found a gallant place . . . whom Nattuer ytt Selfe hath already formed with out the hand of man."

Casco Bay (Cumberland Co.) "these Illands ar all over growen wth very thicke ocks walnut pyne trees and many other things growing as Sassaperilla, hassell nuts and whorts in aboundance."

Near Augusta (Kennebec Co.) upon the land "found aboundance of spruce trees such as are able to mast the greatest ship his majestie hath, and many other trees, oake, walnut, pine-aple."

Kennebec River at Augusta "found a gallant Champion Land and exceeddinge fertill. . . . by both syds of this river the grapes grow in aboundance and also very good Hoppes and also Chebolls [wild onions] and garleck."

John Smith 1614

Muscongus Bay region (Lincoln/Knox Cos.) "this coast is all mountains and

Iles of huge rocks, but overgrowen with all sorts of excellent good woods for building houses, boats, barks, of shippes."

Casco to Pennobscot "nothing but such high craggy Cliffy Rockes and stony Iles that I wondered such great trees could growe upon so hard foundations."

Southwest coast (York Co) "oke is the chiefe wood; of which there is a great difference in regard to the soyle where it growth, firre, pyne, walnut, chesnut, birch, ash, elme, cypresse, ceder, mulberrie, plum tree, hazel, saxefrage and many other sorts . . . with many faire high grouse of mulberrie trees gardens, also Okes, Pines and other woods."

John Pory 1622

Damerill's Cove (Damariscove Island, Lincoln Co.) "they have fortified themselves with a strong palisade of spruce trees of some ten foot high." Note that the island was cleared early, and by 1985 there was not a single spruce on the island.

Christopher Levett 1623

Near Kittery (York Co.) "I saw much good timber, but the ground it seemed to me not to be good, being very rocky and full of trees and brushwood."

Cape Newagen (Southport, Lincoln Co.) "I could see little good timber and less good ground."

Thomas Gorges 1640

Southwest coast (York Co.) "the woods well treed with stately cedar, lofty pines, sturdy oaks and walnut trees."

SOURCES

Gosnold: M. John Brereton (for Bartholomew Gosnold), "Briefe and true relation of the discoverie of the North Part of Virginia, 1602," in H.S. Burrage, *Early English and French Voyages, Chiefly from Hakluyt, 1534–1608* (New York: Charles Scribner's Sons, 1906), 330.

Pring: "A voyage set out from the citie of Bristoll, 1603," in Burrage, *Early English and French Voyages,* 345, 346.

Champlain: Biggar, H.P., ed., W.F. Ganong, translator, *The Works of Samuel de Champlain, Volume 1* (and *Voyages du Sievr de Champlain*), 270–342 (Toronto: The Champlain Society, 1922), 271, 274–281, 282–283, 291, 321, 323, 329.

Rosier (Waymouth): "Rosier's Relation of Waymouth's Voyage of 1605," in Burrage, *Early English and French Voyages,* 355–398, 363, 384–385, 386–387, 393.

Popham and Gilbert: [James Davies?], "Relation of a voyage to Sagadahoc, 1607–1608," in Burrage, *Early English and French,* 397–419, 410, 412, 413, 417.

Smith: Captain John Smith, *A Description of New England, 1614–15* (Boston: William Veazie, 1865), 42–43, 47.

Pory: John Pory to the Governor of Virginia, in Sydney V. James Jr., ed., *Three Visitors to Early Plymouth* (Bedford, MA: Applewood Books, 1963), 14–18.

Levett: Christopher Levett, *A Voyage into New England — Begun in 1623, and Ended in 1624 Performed by Christopher Levett* (London: William James, 1628), 80, 86.

Gorges: R.E. Moody, ed., *Letters of Thomas Gorges* (Portland: Maine Historical Society, 1978), 7.

APPENDIX 3.2
Observations of Forests by Travelers and Surveyors in the Interior of Maine before Division into Towns

Joseph Chadwick 1764

Near Orono (Penobscot Co.) "Trees large high Maples Black & Gray Oaks Black Birches, Little or no Under brush At about 4 or 6 furlongs from the River is a good growth of white pine Tember & Masts."

At Lake Onawa ... SE of Greenville (Piscataquis Co.) "Trees large Elems Maples & large O(A)lder bushes ... rises bearing a good growth of Large Black Birch & Bech &c."

Chesuncook Lake (Piscataquis Co.) "at Some distance backwards Riseing with easy asent Grows a thick Growth of young Trees ... Only sum small tracts of Good Land."

Paul Coffin 1768

Cornish to Kezar Falls (York Co.) "oak land and pitch pine... fine pitch pine plain ... A great deal of the road was pitch pine land, like a house floor."

John Pierce 1775

Great Falls on Dead River [to modern Flagstaff Lake] (T3R4 BKP-WKR, Somerset Co.) "The river is very Dead and with all very Pleasant — The Land on the West Side the river is not very Good and Descends from ye river and on the East Side is something better — Timber White Burch Black Do[t] Hemlock Spruce Pine Farr & maypole &C. and back from ye river is boggy and bad Travelling ... The Course of ye river is very Crooked — Timber on the river is Spruce Birch Pine Beech maypole Elm White Ceader Far &C. the Banks of the river are covered with Joint Grass ... Land poor and Timber small ..."

Thomas Pownall 1776

Land between Passamaquoddy and Penobscot Bays (Washington & Hancock Cos.) "... is White-pine Land, a strong moist Soil, with some Mixture of Oaks, White Ash, Birch, and other Trees, and in the upper Inland Parts has generally Beech Ridges."

W. M Morris 1791

William Bingham's Kennebec Million Acres near Bingham, (Somerset Co.):
"growth of timber on it hemlock, spruce pine, ceder, fir, etc.; [and better land]
beech, maple, elm, ash, linden, and some oak."

Joseph Pierpont & William Albee 1792

William Bingham's Penobscot Million Acres, Twp #23, Centerville (Washington
Co.) "Over high land — growth large birch, beech, rock maple, pine, spruce,
& hemlock . . . , Cedar swamp . ., low land Growth spruce, fir, & large white
pine, Growth beech . . ."

Levi Perham & Ephraim Ballard 1794

Eustis T1 R4 WBKP (Franklin Co.) "Timber spruce, fir, hemlock, pine, cedar,
beech, maple poplar"
Coplin Plt. T1R3 WBKP (Franklin Co.) "Fir and Spruce on 'Black Mountain,'
marked Spruce stub [on Crocker Mt.]"

Park Holland 1794

Yankeetuladi (T19 R11 WELS, Aroostook Co.) "[Mile] 31 on a birch on a high
mountain, 32d on a spruce in flat land, on top of a mountain, 33rd on a spruce
in a swamp at the foot of the mountain we came over. We mark our 34th on
a cedar in a swamp, the last mile good level land. We mark our 34th mile and
strike a stream running southerly. Mark our 35th mile on a birch at the foot
of a high mountain. We run 32 rods over our 36th mile and a yellow birch for
the corner 'H.M.W.B.N.W.C.B. 1794,' Holland Maynard, William Bingham's
Northwest corner Bounds, the year etc. This corner is part way down on the
west side of a mountain, land thickly wooded, and falling to the northward."

Solomon Adams & Lemuel Perham 1806

Bingham's Kennebec Purchase northeast of Kingfield (Somerset & Franklin
Cos.) "Rock, White and river maple, yellow & white birch, Black, brown or
swamp ash, Beech, Poplar, willow and alder, evergreens spruce and fir predominantly some Cedar, Hemlock and larch small white pine interspersed
on streams and lakes, small Norway Pine and elm on margins in vicinity of
Dead River."

Ezekiel Chase & John Towle 1825

Seboeis River (T4 R7 WELS, Penobscot Co.) "both sides had fell and hedged the
stream, which was occasioned by fires. The land especially on the west side
of the ponds was good with considerable pine timber growing on the shores.
Travelled about four miles through burnt land, not a living tree, the ashes
deep is slippery not a green bough. Land low cedar and spruce without any
pine."

John Johnson 1817

Eastern Boundary of Maine run North, "1st mile: Monument cedar swamp, hardwood, spruce & cedar, beech & hemlock, hardwood, 1 mile stone drove stakes at intervals, . . . stake 98 mile, birch maple fir & beech, birch, spruce fir cedar, cedar swamp, spruce tree marked 99 mile, Portage Path N54°W . . ."

W. F Odell 1818

North of the St. John River, New Brunswick "It was well wooded with a luxuriant growth of tall thrifty timber, a mixture of hardwood, fir, and spruce, with some pine, the mountain ash is abundant and there are a very few wild cherry trees; country however to what is usual met with the softwood grows mostly in the valleys, and the hardwoods on the tops of the hills."

W.G. Hunter 1819

Umbasucksus Portage T6 R13 WELS (Piscataquis Co.) "the timber immediately on the border of the Lake is ceder, spruce, poplar and white birch interspersed with inferior pine. (then) . . . is covered with a some growth of hardwood."

W.G. Hunter 1820 West Br. Penobscot & St. John at Baker Pond T7 R17 WELS (Somerset Co.) "the trees are of a very inferior growth and thinly scattered over the face of the lands, great portion of which are quite barren."

W.F. Odell 1820

N. Br. Penobscot to St. John T6 R17 WELS (Somerset Co.) "This portage is all bog and swamp; covered with small Yamaraes, except for about half a mile, which ground is a little more elevated, and the timber is chiefly spruce . . ."

Joseph Treat 1820

Islands in Penobscot River at Greenbush (Penobscot Co.) " . . . covered with a fine growth of hard wood, such as Grey oak, Rock, Red and White maple, yellow Birch, Beach and Elm."

Pemadumcook Lake (T4 Indian Purchase Penobscot Co.) "Burnt land round the south side . . . growth principally pine and mixed, now a small growth of hard wood. On the west . . . is a fine ridge hard wood land — On the East side . . . the land appears to be poor — the growth pine, spruce and hemlock &c. some Bogg."

Between Chessuncook and Umbazooksus Lakes (T6 R13 WELS, Piscataquis Co.) " . . . poor low juniper land near stream some mixed ridges appear at a distance — some small pine ridges also . . . the land around (Umbazooksus) generally poor — juniper and black growth . . ."

Round Pond (T13 R12 WELS, Aroostook Co.) " . . . around which are many hardwood ridges . . . — Balm of Gilead round the shores . . . leaving the Pond there appear high ridges and hills of mixed growth — and good pine on the shores — mixed with spruce, fir and hemlock."

Aroostook River between Fort Fairfield and Caribou (Aroostook Co.) " . . . the
growth near the shores and adjoining intervals mixed, pine hemlock, spruce,
fir, cedar, elm, ash, birch and maple on the shores grow very large Balm of
Gilead trees — also on the Islands — There are fine hard wood ridges in the
back ground — excellent land . . ."

Joseph C. Norris 1825

Monument line on Katahdin's northeast slope (T4 R9 WELS, Piscataquis Co.)
"There has been a growth of fir and spruce somewhat taller, but it is dead and
[has been] succeeded by an entire growth of spruce a little higher than one's
head."

Ezekiel Holmes 1839

Fish River vicinity (T13 R7 WELS, Aroostook Co.) "continually encumbered by
windfalls . . . Growth on the banks known by the name of black growth, that
is to say pine, cedar, spruce, fir &c little or no larch or hackmatack."

George William Featherstonhaugh 1839

Millinocket Lake (T7 R9 WELS, Piscataquis Co.) "This is a most beautiful lake . . .
studded with islands covered with cedar, spruce. &c, the shores thickly studded with trees, and lofty hills in the distance, covered to the top with trees. . . .
A great deal of pine."

Munsungan Stream (T8 R9 WELS Piscataquis Co.) "The timber is spruce and
cedar principally, with some small birch. On the ridges back hardwoods
grows, such as maple, large birch, beech &c."

Norway Bluff (T9 R9 WELS Piscataquis Co.) "The maples, birches and beeches
continued to within 300 feet of the top; here the spruces begin." Note summit
2250 feet elevation thus spruce line just below 2000 feet.

Richard Mudge 1839

Baker Lake (T7 R17 WELS, Somerset Co.) "a rising hill in the wood, covered
with silver and spruce firs. (then) . . . great difficulty to get through them, as
trees upon trees lie in every direction one upon another, which have fallen
unnoted from the earliest ages. The cedar forms the great majority of these;
it is a wood which takes many years to rot; it may be called the wood of the
country."

Three Maine Yellow Birch Witness Trees 1789–1820

Three large yellow birch trees were initially marked as survey monuments on
the three corners of Maine's border with British Canada. The "Crown Monument" at the New Hampshire-Canada-Maine corner is at the northwest corner of the state. In 1789, Joseph Crane and Jer[h] Eames marked a "large (yellow) birch tree" with stones piled around it at the northern end of the New
Hampshire-Massachusetts state line. It is on the height of land (2568 ft. elev.),

was 282 feet south of current international monument 475 and was reduced to a "stubb" before 1858. The "Eastern Monument" was 5¾ miles north of Schoodic Lake and was the starting point for the eastern boundary of the state and the Monument Line of the northern Maine surveys. This corner was a "stake near a yellow birch tree hooped with iron" and marked by Samuel Titcomb and John Harris in 1797. Joseph Bouchette and John Johnson replaced it with a cedar post in 1817. The "Mile Tree" is the western end of the "Highlands," the northern line of Maine ambiguously defined in the Treaty of Ghent in 1814. It is exactly 50 miles northeast of the New Hampshire border along the divide between the Chaudiere-Kennebec drainage. In 1820, Colin Campbell found "a large Birch Tree standing on the road [now Rte 201 at the Quebec frontier, in Coburn Gore] . . . marked by the Canadian and American Surveyors" for the International Boundary.

REFERENCES

Chadwick: Fannie H. Eckstrom, "History of the Chadwick Survey from Fort Pownal in the District of Maine to the Province of Quebec in Canada in1764," *Sprague's Journal of Maine History* 14 (1926): 63–89, quotations pages 78, 80, 85.

Coffin: "Journal of a tour from Buxton to Piggwacket," *Coll. Maine Historical Soc.* 4 (1856): 275–293.

Pierce: "Journal of John Pierce surveyor for Arnold's Expedition," in K. Roberts, compiler, *March to Quebec, Journals of the Members of Arnold's Expedition* (Camden, ME: Down East Books, 1980), 653–675.

Pownall: Thomas Pownall, *A Topographical Description of the Dominions of the United States of America* (Pittsburg, PA: University of Pittsburg Press, 1949), 69.

Pierpont & Albee: "A Journal over 1,000,000 Acres of Land in the Counties of Hancock & Washington sold by Government to General Jackson and others containing a description of the Soil, Growth, Rivers, Lakes, Ponds, Brooks, Mountains, Plains, Swamps, Meadows &c commencing Monday 15 October 1792 . . . ," Manuscript, Bingham Papers (Philadelphia: Historical Society of Pennsylvania).

Morris: F.S. Allis, ed., *William Bingham's Maine Lands, 1790–1820*, I. Observations on William Bingham's Kennebec Million Acres along Kennebec River (Boston: Publication of the Colonial Society of Massachusetts, 1954), Volume 36: 200–216.

Perham & Ballard: Field Notes Ballard and Perham survey of land west of Bingham's Kennebec Purchase, Land Office Records, Maine State Archives.

Holland: F.S. Allis, ed., *William Bingham's Maine Lands, 1790–1820*, Holland

Autobiography III (Boston: Publication of the Colonial Society of Massachusetts, 1954), Volume 36: 217–232: quotation page 223.

Adams & Perham: Adams & Perham Field Notes, "Survey of three Ranges of Townships on the West side of the Kennebec River part of Bingham Purchase," Manuscript in Bingham Papers (Philadelphia: Historic Society of Pennsylvania, 1806).

Chase & Towle: "Report of Ezekiel Chase and John Towle of the tract they travelled from Fish River at its junction with the river St. John to Penobscot River in the autumn of the year 1825," Manuscript, Land Office Records, Maine State Archives.

Johnson: John Johnson Papers, Wilbur Room, University of Vermont, carton 7, volume 22, "Record of U.S. Canada Border Survey 1817 John Johnson U.S. Surveyor 1817"; Maine Historical Society, *Documentary History of the State of Maine,* vol. 8: Joseph Bouchette and John Johnson surveyors (Portland, ME: The Lafavor-Tower Co., 1902).

Hunter: "Northeastern Boundary arbitration Statement on the part of the United States of the case referred to the King of the Netherlands," appx. 56, Extracts of the reports of the surveyors, 404–426, quotations 414, 421–422.

Odell: "Northeastern Boundary arbitration," appx. 56, 405–406, 416–417.

Treat: Joseph Treat, "Journal and Plans of Survey by Joseph Treat," manuscript in Maine State Archives, Maine Land Office, Field Notes, transcribed in Micah A. Pawling, *Wabanaki Homeland and the New State of Maine, The 1820 Journal and Plans of Survey of Joseph Treat* (Amherst: University of Massachusetts Press, 2007), 17, 45, 73, 93, 143.

Norris: Myron H. Avery, "The monument line surveyors on Katahdin 1928," *Appalachia* 17: 33–43.

Holmes: Ezekial Holmes, *Report of an Exploration and Survey of the Territory on the Aroostook River during the Spring and Autumn 1839* (Augusta, ME: Smith & Robinson, 1839).

Featherstonhaugh: Alec McEwen, *In Search of the Highlands, Mapping the Canada-Maine Boundary* (Fredericton, NB: Acadiensis Press, 1839), 43–47.

Mudge: Alec McEwen, *In Search of the Highlands, Mapping the Canada-Maine Boundary* (Fredericton, New Brunswick: Acadiensis Press, 1839), 96–97.

Yellow Birch Boundaries: William D. Williamson, *The History of the State of Maine* (Hallowell, ME: Glazer Masters, & Co., 1832); Maine Historical Society, "Report on New Hampshire Boundary Line by the Committee on Waste Land, June 6, 1790," in *Documentary History of the State of Maine,* vol. 8: 64 (Portland, ME: The Lafavor-Tower Co., 1902); "Northeastern Boundary arbitration . . . ," appx. 56: 411.

Equivalent Vernacular Names for Maine Trees in Old Records

Old Vernacular Name	First Maine Use	Current Name	First Maine Use
wild pear	1667	black gum	not known
walnut[b]	1668	hickory generally[c]	1734
white maple (in part)	1669	red maple	1820
black birch[b]	1679	yellow birch	1742
black oak[b] (in part)	1686	red oak	1687
red ash	1687	black ash?	1686
spruce pine	1687	spruce	1658
rock maple[a]	1720	sugar maple	1796
popple[a]	1720	aspen	1703
yellow ash	1720	black ash	1686
red birch	1720	yellow birch	1742
poplar[b]	1729	aspen	1703
hornbeam[a]	1730	ironwood	1795
juniper[b]	1730	tamarack	1822
Norway pine	1732	red pine	1723
witch hazel[b]	1732	ironwood	1795
plum	1733	shadbush (or plum)	not known
hazel[b]	1734	ironwood?	1795
grey oak	1738	red oak	1687
burch cherry	1766	yellow birch	1742
leverwood	1767	ironwood	1795
stinking wood tree	1769	black gum?	not known
yellow pine	1781	pitch pine	1686
oilnut	1798	butternut	1817
river maple	1805	silver maple	not known
hackmetack	1812	tamarack	1822
brown ash[a]	1818	black ash	1686
hornbine	1823	ironwood	1795
bald spruce	1829	tamarack	1822
silver fir	1829	balsam fir	1720
whitewood	1829	basswood	1720
white maple (in part)	1832	silver maple	not known
boxwood[b]	1832	shad tree	1832
shad-blossom	1832	shad tree	1832
spruce fir	1839	spruce	1658
roundwood	1845	mountain ash	1831
grey birch[b] (in part)	1850	white birch	1673

[a]Still in general use
[b]Current use of this name not the same tree as old use.
[c]Except for very early butternuts.

4 | From European Settlement to Modern Times

How Did Human Culture Transform Maine Forests?

Ecology becomes a more complex but far more interesting science when human aspirations are regarded as an integral part of the landscape.—Rene Dubos

The Perham Farm, Sandy River Valley, Maine

In the spring of 1787, seventeen-year-old Silas Perham traveled alone from Dunstable, Massachusetts, to backlot 24 on the east side of the Sandy River, in what is now Farmington, Maine (figure 4.1).[1] The first Euro-Americans had settled in the valley only six years earlier, lured by the wide, fertile intervale spanning the river's flanks. Twenty thousand years earlier, a continental glacier churned over this land, grinding down exposed edges, transporting mud, sand, and stones. After the ice melted by about 15,000 years ago, an estuary filled the valley, as far inland as Farmington.[2] Later, unburdened by the weight of the ice sheet, the land rebounded, moving the coastline to near its present position. Indians settled the area, making a living through hunting and gathering. By first contact with Europeans, these Indians had lived in the Sandy River valley for many generations. But by the late eighteenth century, they had been decimated by war and foreign diseases. When the first Euro-American settlers arrived in 1781, two Indian

Figure 4.1 Part of the Sandy River Valley today, taken from Mt. Blue High School.

families lived along the Sandy, although there was evidence that far more had occupied the area in the not too distant past.[3]

Silas Perham and his family were part of a wave of pioneers penetrating the interior of Maine after the French and Indian Wars and the American Revolution. That first year, Silas cleared land, planted and harvested crops, and returned to his family in Dunstable in the fall. The next spring, the entire family made "a slow and painful journey of twenty-three days"[4] to their new home on backlot 24, on a rolling ridge, with a generous allocation of deep soil, looking out toward the prominent rounded peak of Mt. Blue and the western mountains. Less than a mile away, Ephraim Ballard in 1789 documented the presence of hemlock, beech, cedar, birch, maple, oak, and basswood, as described in chapter 3 ("Land Surveys . . .").[5] It didn't take long for the Perhams to clear five acres of this forest and build a barn and log home. They prospered, as did the community, incorporated as the town of Farmington in 1794. Silas was a carpenter. His older brother, Lemuel, well educated before moving to the farm, opened the first school in town and be-came a surveyor, helping Ephraim Ballard run the lines for the towns west of the Bingham Purchase, over Mt. Abraham, Sugarloaf, and the Bigelows.[6] Mainly, the family farmed what became known as Perham Hill, clearing

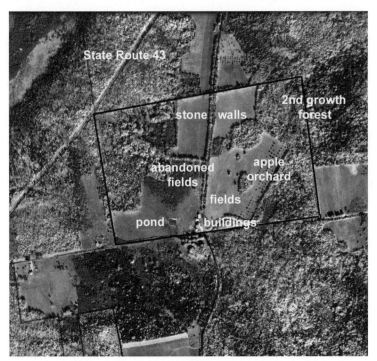

Figure 4.2 Aerial photograph of the Perham Farm, Farmington, Maine, in 1951. Approximate current ownership boundaries shown by polygon outline.

increasing amounts of land. By 1850, Silas and Hannah Perham's son, Silas Decatur Perham, and his wife Mary Ann Hobbs Perham, had taken over the farm,[7] which consisted of thirty acres of woodlots and seventy acres of land cleared for crops, hay, and pasture, separated into sections by stone walls that persist today. They raised a variety of vegetables, including Indian corn, potatoes, beans, peas, wheat, and oats, maintained a large flock of sheep and other animals, planted a large apple orchard on top of the highest hill and, in 1865, completed an impressive Greek Revival house.[8] Two town roads crossed in front of the home, forming Perham Corner.[9]

Like the Perhams, the town of Farmington was enjoying great success by midcentury: a rapidly growing population, sawmills, tanneries, grist mills, a brick-making operation, a railroad connection, an agricultural fair, a noted preparatory school, and a new college.[10] Forests around Farmington, in the Kennebec River drainage, had become an important source for a thriving lumber and later pulp and paper industry. By the 1860s, however, external forces began to fray the fabric of rural Maine — and that of the Perham family as well. People left their Maine farms for opportunities in the Midwest and in urban areas. Silas Decatur's first wife, Mary Ann, died in 1874, and his son, Silas French who lived on the farm, died in 1885 at the age of thirty-five. By 1892, Silas Decatur was seventy-seven years old with no obvious heir to the farm, and he and his second wife, Electa, sold to Marinda Holt, ending more than a century of Perhams on backlot 24. For the next eighty years, the land changed hands frequently, and agriculture ceased being the chief occupation of its owners. Fields were abandoned and reverted to forest. A 1951 aerial photograph shows about 50 acres in forest, 20 acres in old fields reverting to forest, and 50 acres still in fields, which were hayed by local dairy farmers (figure 4.2).[11] In 1973, a psychiatrist and his family purchased the land and home, which they carefully renovated. In 1999, two biology professors, including Andrew Barton, one of the authors of this book, took over the Perham Farm. Today, 100 acres are used as a woodlot, which consists mainly of ash, maples, beech, white birch, and popple. About 20 acres provide hay for a neighboring organic dairy farm. Falling stone walls abound on the land; a few of the original apple trees linger on.[12]

Changes in the Land, Changes in the Forest

The complex patterns of land use in Maine over the past four centuries cannot be captured in a narrative about a single piece of land. Not all land has gone from forest to farmland and back to forest. Large areas of forest were never cleared; some properties have supported farming continuously since settlement; and some land has been overtaken by residences for an expanding and sprawling population. Nevertheless, the Perham Farm and the Sandy River Valley provide a touchstone for changes in the relationships of humans with forests across the state that have reshaped the extent, composition, and structure of forests.

In this chapter, we will step back to examine this chronology, from Euro-American settlement in the early 1600s to today (see Appendix 4.1 for a chronology of key events). There is no single plotline that applies to the entire state, for each region followed its own trajectory, dependent on geography (e.g., coastal vs. interior), ecology (e.g., northern hardwoods vs. spruce-fir), and human enterprise. This complicates the tale, but it is a necessary complication that reveals much about the ultimate anthropogenic drivers and impacts on forests. Plainly put, our goal is to understand how fundamental aspects of the forests — how much exists, the species, the physical structure — have responded over the past four centuries to human culture.

Although we discuss people and economics and politics, we do this in the service of understanding how the forest changed. As Phillip Coolidge wrote at the beginning of his classic, *History of the Maine Woods*, "Our subject is the woods, and we leave the usual subject of historical writing — the details of exploration, Indian wars, politics and land grants — largely to other writers."[13] Of course, as our narrative shows, sometimes politics and changes in the woods are inextricable.

David C. Smith, one of the most prominent writers about the timber industry in Maine, divided up the state's history into periods of long logging (up to 1880), forest products industry consolidation (1880–1914), and mechanization (1914–1965).[14] We more finely partition these years, using turning points that transformed forests: (1) before 1781, during which settlement was erratic and harvesting minimal; (2) 1781–1840, which includes the first wave of land clearing and modest harvesting, mainly of white pine; (3) 1840–1880, the height of land clearing and a period of rapid migration of the harvesting frontier into northern Maine, which entailed removal of white pine and then spruce; (4) 1880–1920, when huge areas of farmland were abandoned, the pulp and paper industry took control of large forestland areas, key wildlife were extirpated, and forest conservation emerged; (5) 1920–1970, a relatively stable period during which technology increased the intensity of harvesting, but also a time marked by the resurgence of wildlife; and (6) 1970–2011, a volatile era of shocks to the state's land traditions, changing harvesting methods, a historic spruce budworm attack, land speculation, turnover in land ownership, and reinvigorated conservation.

Before 1781: Colonial Maine and Its Forests

The colonial period abounds with stories of bravery, treachery, and the collision of cultures. From the point of view of the forests of Maine, however, the impacts were modest, largely because of a small population, relatively primitive technologies, and rudimentary transportation networks. In the decades after abandonment of the Popham colony in 1608, the English established year-round coastal settlements, focused on fishing and fur trading, including at Pemaquid, Damariscove Island, Monhegan Island, Richmond Island, and Machias (figure 4.3). The first sawmill was built in 1631 in South Berwick on the Salmon Falls River.[15] The initial ecological impacts of Europeans occurred well before the establishment of these outposts, however. A thriving business since the mid-1500s, fur trading must have strongly affected local animal populations, especially beavers. The intro-

Figure 4.3 Incorporation dates for select Maine towns. Dates before the Revolutionary War are shown in light italics. In some cases, the indicated town was part of another jurisdiction, and that earlier date is used.

duction of trade goods, such as guns and iron products, created new Indian seasonal movements and land use activities.[16] Europeans also brought with them the pathogens of the Old World, to which Native Americans had not been previously exposed, leading to epidemics that decimated many Indian villages. The most severe of these, between 1616 and 1619, is thought to have killed up to 75 percent of the estimated 20,000 Maine Indians.[17] Such depopulation must have also temporarily mitigated local hunting and gathering pressure, and, in southern areas of the state, where maize and squash farming occurred, led to reversion of some farmland to forest.[18] This modest lull in the human ecological footprint would, of course, eventually be swamped by the tide of Euro-American settlement. By the 1640s, new land grant settlements sprang up on the mainland, occupied by families who drew their sustenance from farming rather than fish and furs.

By 1670, about 3,500 Anglo inhabitants lived in a thin strip of settlements stretching along the coast and in tidal areas of major rivers of present-day Maine. At the same time, the French had established outposts of the colony of Acadia south and west to the Penobscot River (figure 4.3).[19]

The years from 1630 to 1670 represent the humble beginnings of forest clearing for settlement and commercial logging in Maine. Although oaks were harvested for staves for casks used in trade with the West Indies, mainly for rum and molasses, the chief target was large white pines. Every Maine schoolchild knows this story: the massive trunks crashing to the ground, extricated from the woods by oxen, floated down rivers (figure 4.4), and eventually employed as masts for British naval ships, a role so crucial that larger pines were reserved for the Crown by the Broad Arrow policy of 1691. Some of these old-growth pines were impressive indeed — up to 6 feet in diameter and 150 feet tall. In the early years, only the most accessible trees were taken, but over time loggers were forced to search beyond the areas near major rivers. Likewise, the mast trade frontier moved slowly east along the coast, from the Piscataqua River in the 1600s to the Casco Bay Region and eventually to the Androscoggin and Kennebec Rivers in the 1700s. Nevertheless, estimates suggest that only a slim portion of the old-growth pine in Maine, nearly all of that along the coast and tidal parts of major rivers, was taken by the time of the American Revolution.[20]

The timelines of Maine and southern New England, namely the Massachusetts Bay Colony, diverge in the late 1600s. Starting in 1675, war between most of the Indian population and Euro-American settlers (with some Indian allies) engulfed New England. By war's end in 1676, the Massachusetts Bay Colony had largely subjugated Indians in that region and settlement resumed. In contrast, for nearly the next century, Maine was a theater of periodic warfare among the English, the French, and various coalitions of native peoples. These insecure conditions quelled the drive for settlement and exploitation of Maine forests. Even places with good soil and navigable rivers, like the Sandy River Valley, remained the domain of Indians. By the end of the second major period of war in 1697, nearly all farmhouses east of Wells had been destroyed, and by 1713, the number of white settlers was lower than it had been at any time since the 1660s. This period of warfare in Maine ended in 1758, followed the next year by the fall of the French in Quebec to the English. Because this period was quickly followed by the American Revolution, the gates of settlement of Maine did not fully open again until the defeat of the British and end of hostilities in 1781.[21]

The colonial period up to 1781 was important for the forests of Maine

Figure 4.4 White pine logs were driven down large rivers, such as the Andro- scoggin, shown at the top, and even streams, such as Temple Stream, shown below, from the seventeenth well into the twentieth century.

not because of direct impacts, for those had been quite minimal, especially beyond a very thin strip along the coast and the lower parts of major rivers. Instead, the colonial property system taught settlers and especially the new class of proprietors to view land as a means to profit by utilization, improvement, and speculation rather than subsistence, a value framework that has driven the relationship of humans with forests up to this day.[22]

1781–1840: Settlers Come, White Pine Goes

The last two decades of the eighteenth century brought a flood of settlement and timber harvesting to Maine, then a district of the Commonwealth of Massachusetts, in the new United States of America. This sharp break

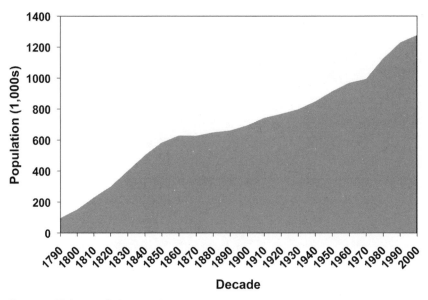

Figure 4.5 Maine population growth, 1790–2000.

with the past was partly stimulated by the end of warfare, but also by attempts to spur settlement through land warrants for Revolutionary War veterans and the sale of cheap land. From 1783 to 1820, Massachusetts sold 5.5 million acres of Maine land, including 2 million acres to Senator William Bingham and partners in 1792, part of which was surveyed by Ephraim Ballard and Lemuel Perham, a recent immigrant to the district. Immigration led to a rapid increase in the population (figure 4.5). By 1790, the population of 100,000 was five times that estimated for Indians during the precontact period. At statehood in 1820, this number had tripled to nearly 300,000, and by 1840 there were more than 500,000 Mainers, a rate of increase slightly faster than that for the entire United States, which had grown from 3.9 to 17.1 million.[23] Settlement spread geographically eastward along the coast and northward up the major river valleys, a pattern clearly revealed by the establishment dates of towns in figure 4.3.[24]

Settlement was accompanied by rapid clearing of land, as most of the pioneers drew their sustenance from farming. The best estimates suggest that by 1760 little more than 10,000 of the 19.8 million acres of land area had been cleared, rising to about 650,000 by 1820, and a million by 1840.[25] Land was cleared for crops, hay, and pasture, a land use that became increasingly common by the 1830s, with the rise of the wool industry in New

THE CHANGING NATURE OF THE MAINE WOODS

Figure 4.6 Stone wall and adjacent field at the Perham farm.

England. The first half of the nineteenth century was also the major period of construction of stone walls in Maine, which, like on the Perham farm today, still demarcate the boundaries of ancient fields that have reverted to forest (figure 4.6). A variety of purposes drove the construction of these walls, including boundaries, fences, field clearance, and the Puritan ethic of improving the landscape. Settlers quickly discovered that rocks continue to appear after clearance of open land, as a result of freeze-thawing and erosion. The large size of rocks in stone walls at the Perham Farm suggests boundary walls and the clearance of stones from hay fields (whereas many small rocks implies continual removal of rocks in tilled ground). Why include a discussion of stone walls in a book about forests? At least for the more settled part of Maine, these stone walls are very much part of today's human-created forested landscape.[26]

The settlement of Maine also brought with it unprecedented exploitation of its vast forestlands, both by settlers and by the new class of New England entrepreneurs living in the growing cities of southern New England. White pine continued to be far and away the main target. Although pine ship masts were exported as late as the 1850s, the majority of the harvest was for lumber during this period, cut in large mills on rivers near the coast and then exported. The industry was centered in Saco around 1800, but slowly moved east and upriver as supplies of large trees were depleted, first to the Presumpscot River in the 1820s, then to mills in Brunswick, Augusta, and Fairfield along the Androscoggin and Kennebec, and eventually to the Penobscot and St. John. As an indicator of the level of cutting in target areas, the Saco and Presumpscot Rivers were largely out of large pines by 1832.[27]

Large quantities of wood in addition to exported white pine were also harvested. By 1820, there were 248 tanneries, mainly in the central part of the state, which created a large demand for the tannin-rich bark of hemlocks. Oaks were used extensively for barrels and shipbuilding (sidebar 4.1). Hardwoods must have been cut in prodigious quantities for fuel. The typical New England farmhouse during that time required ten to twenty cords of wood per year for heating. By the turn of the century, coastal Maine had also become part of the fuel-shed for Boston, which had far outstripped its local supplies. Finally, the construction of commercial buildings and settler homes exerted strong demand for white pine, spruce, and other species. This largely unrecorded harvesting for local uses probably rivaled the export market at least until the latter parts of the nineteenth century.[28]

Having meticulously detailed the sharp break between the pre- and postcolonial periods in terms of land clearing and tree harvesting, we want to end with the seemingly paradoxical statement that Maine in 1840 was still a largely unexplored and unexploited place, with abundant old forest and wildlife populations. The post-American Revolution decades for Maine were a turning point in its history, but these changes in the land were to be dwarfed by what was to come. Annual harvests of pine rose from several million board feet per year before 1781 to 225 million by 1839. By this time, a total of 5–7.5 billion board feet of wood had been harvested (with perhaps twice that much wasted) since settlement and a million acres cleared. But that left more than 90 percent of Maine in forests, much of it untouched by the axe.[29] In 1846, Thoreau wrote, "Twelve miles in the rear, twelve miles of railroad, are Orono and the Indian Island, the home of the Penobscot tribe, and then commence the batteau and the canoe, and the military road; and sixty miles above, the country is virtually unmapped and unexplored, and there still waves the virgin forest of the New World."[30]

The Coast Bears the Brunt

The impacts of early settlement and forest exploitation were not felt equally across the state. The coast received the brunt of development between the American Revolution and 1840. By 1820, nearly every island larger than twenty-five acres supported people, livestock, or both, and most of the large towns were located along the coastal mainland—a result of their accessibility by water. Combined with aggressive hunting for birds for meat, feathers, and eggs, this settlement pattern probably accounts for Audubon's observations of few birds along the Maine coast in 1832. Around the same time, large areas were cleared to accommodate high numbers of sheep, a history reflected in island names such as Ram Island and Sheep Island. Overgrazing and repeated burning to keep land cleared led to erosion, loss of soil and nitrogen, and increased rockiness.

Easy access to water encouraged enormous harvesting pressure. Large white pines were cut for masts and lumber. So much beech and red oak were taken for barrels that the supply of these species often ran short locally.[1] In the mid-coast area, large quantities of wood fed the lime industry. By 1835, 750,000 casks of lime were being produced annually by 150 kilns, each of which demanded about thirty cords of wood for each seven- to ten-day firing. In *Islands in Time*, Phillip Conkling argues that, from colonial times to the advent of steel ships in the 1850s, the state's world-renowned shipbuilding industry had a larger impact on islands than any other activity: "Much of the original forest growth of Maine's islands could at one time be found sailing one of the world's seven seas." Ships required prodigious amounts of white and red oak, beech, ash, sugar maple, larch, hophornbeam, spruce, and white pine. Two thousand oak trees went into the building of a single ship for the British Admiralty.[2] By 1850, then, most of the islands and coast of Maine had been settled, cleared, and cut over, whereas the deep interior of the state was still frontier.[3]

NOTES

1. Phillip W. Conkling, *Islands in Time: A Natural and Cultural History of the Islands of the Gulf of Maine* (Camden, ME: Down East Books, 1999), 65, 68, 101, 112.

2. Conkling, *Islands in Time*, 97–100

3. Conkling, *Islands in Time*, 68, 97–100, 102, 105, 112; Lawrence C. Allin and Richard W. Judd, "Creating Maine's resource economy, 1783–1861," in Judd et al., eds., *Maine: The Pine Tree State from Prehistory to the Present*, 280–282.

1840–1880: Into the Interior — Forest Clearing, Tree Harvesting

The landscape of Maine changed more and faster from 1840 to 1880 than for any comparable period since deglaciation. By 1840, it had taken Euro-American settlers more than two centuries to clear the first million acres of Maine forests. Over the next decade, another million were cleared, and by 1880 nearly 3.5 million acres, or 15 percent, of Maine were open farmland[31] (figure 4.7[32]). On the Perham farm, cleared land grew from 5 acres to encompass most of the more than 200 acres. This remarkable change was the result of two pressures. First, the number of Mainers grew rapidly from 1800 to 1860, adding nearly 500,000 new people. The number of farms increased similarly — about 8 percent per year — exceeding 64,000 by 1880. The advent of railroads, better roads, new scientific agricultural practices, sheep pasturing, and vigorous regional markets drove farming to new levels of success by the mid-1800s, stimulating many farmers to clear additional land, reflected in increased farm size, from an average of 43.6 to 54.2 acres between 1850 and 1880. As a result of land clearing and the establishment of towns, the areal coverage of forest in Maine declined rapidly during this period (figure 4.8).

The wave of settlement that spread across Maine's southern border during the first half of the nineteenth century stalled by the 1860s and 1870s, reaching a frontier little more than one-fourth the way up the state and fifty to eighty miles inland, as shown in figure 4.9. More modest settlement pushed south from Canada, especially into the St. John and Aroostook River valleys. Beyond these lines, settlement and farms were scattered and few large towns established. Figure 4.7 reveals this pattern in numbers: clearing amounted to 35 percent of the land in southern counties, but less than 10 percent in northern and Aroostook counties.[33]

The still forested, largely unsettled interior of Maine formed the core of a burgeoning lumber industry (figures 4.10, 4.11). In the 1840s alone the harvest level doubled. A portion of this increase was to meet the demands of new settlers, but most was pine and later spruce lumber for the export market. More than just demand propelled Maine's lumber industry forward. Incremental but important advances in woods work — the replacement of the axe with the cross-cut saw, oxen with draft horses, squalid conditions with modest amenities — bolstered the efficiency of tree harvesting. The development of river driving, boom companies, dams, and redirection of waterways[34] improved transport and sorting on the way to mills. Upgraded

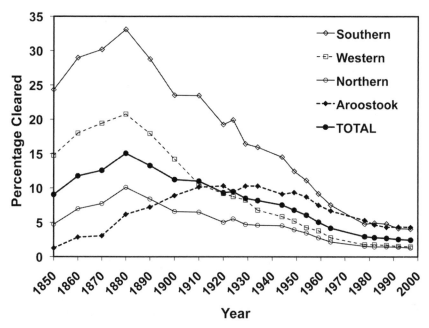

Figure 4.7 Land cleared for agriculture in Maine, for the entire state and by region, as a percentage of total land area of the state. Regions are as follows. Southern: Androscoggin, Cumberland, Kennebec, Knox, Lincoln, Sagadahoc, Waldo, and York counties; western: Franklin and Oxford counties; and northern: Penobscot, Piscatiquis, Somerset, Hancock, and Washington counties. The percentages of cleared land in this figure are lower than those reported elsewhere because we include only land that was actually cleared and open, used as croplands, hayfields, pasture, orchards, and such. Most articles quote total farmland, which also includes farm woodlands, which make up a substantial portion of total farm area. For example, in 1880, the peak year of clearance, woodland made up over 40 percent of the total 6.55 million acres in Maine farms. Although the general trends revealed here are robust, the specifics must be viewed with caution because of changing definitions of farmland categories over time, a point made forcefully in recent articles: Navin Ramankutty, Elizabeth Heller, and Jeanine Rhemtulla, "Prevailing myths about agricultural abandonment and forest regrowth in the United States," *Annals of the Association of American Geographers* 100 (2010): 502–512; SoEun Ahn et al., *Agricultural Land Changes in Maine: A Compilation and Brief Analysis of Census of Agriculture Data, 1850–1997*, Technical Bulletin 182 (Orono: Maine Agricultural and Forest Experiment Station, 2002).

sawmills, some of them of gargantuan size, such as the mile-long F.W. Ayer mill in South Brewer, increased the rate of lumber production. Finally, the introduction of railroads opened up previously inaccessible areas to logging and facilitated access to outside markets, including from Farmington into the upper reaches of the Sandy and Carrabassett Rivers.[35]

Large white pines were still common and accessible at the beginning of this period. White pine was not an abundant species across Maine (chapter 3; table 3.1), however, and eventually the frantic rate of harvesting exhausted

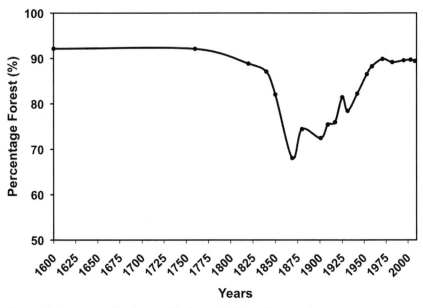

Figure 4.8 Percentage of land covered by forests in Maine 1600–2008.

Figure 4.9 Approximate frontier in Maine of dense settlement and land clearing by 1880.

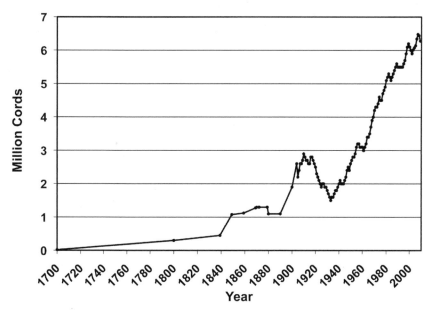

Figure 4.10 Total wood harvest in Maine 1700–2009.

supplies until large pines were common only in remote places such as the St. John, Aroostook, and Allagash watersheds.[36] Over the short term, this had the effect of pushing the harvesting frontier into the farthest reaches of the state, but this was only a temporary remedy. The end of abundant large pine is dramatically demonstrated by the rapid decline in the size of trees floated down the Penobscot to the Old Town Boom, just above Bangor, the logging capital of Maine during that time (figure 4.12). In 1851, a state report declared prophetically that, "In a few years our pine timber will become exhausted, and the spruce and hemlock must take its place."[37] In 1850, the first spruce trees were driven down the Kennebec River. By 1861, spruce had supplanted white pine as the major lumber species in Maine. By 1880, spruce constituted 80 percent of the lumber harvest (figure 4.10).

This period from 1840 to 1880 was a key break in the history of Maine forests. In portions of the state, the landscape became agricultural, with forests relegated to woodlots. In the two centuries prior to 1840, five to seven billion board feet of trees had been harvested; in the following four decades, about twenty billion board feet were extracted from the forest (figure 4.10).[38] Plant species indicative of settlement and clearing — ragweed, English plantain, sheep sorrel, and introduced European grasses — show up as a major component of pollen chronologies in paleoecological studies.[39]

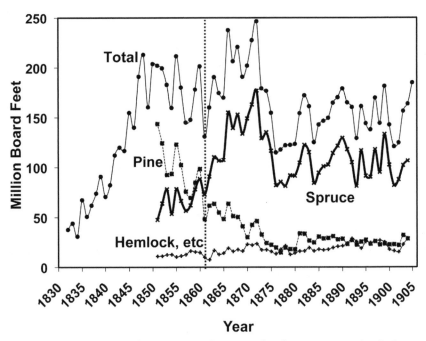

Figure 4.11 Total board feet of lumber surveyed in Bangor for white pine, spruce, hemlock (and other species), and total. Dashed vertical line shows the year (1861) in which spruce surpassed pine as the predominant lumber species in Maine.

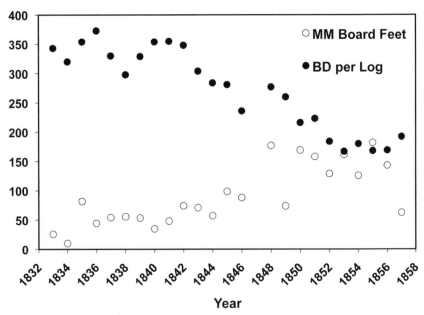

Figure 4.12 Total board feet of lumber processed and board feet per log at the Old Town Boom. As the harvest increased, the size of trees declined.

Wildlife such as beaver and the gray wolf were largely extirpated from the state by human activities (sidebar 4.2). For the first time, the abundance of a tree species, especially of large specimens, was substantially driven down across large portions of the state over a short period of time.[40] These alterations in the presence of previously abundant plant and animal species must have had enormous cascading effects on other ecological components of forested ecosystems, undocumented by foresters and ecologists.

1880–1920: The Remaking of the Maine Forests

After a century of accelerating expansion of farming and lumbering, the trajectory of Maine forests changed abruptly from 1880 to 1920 because of farm abandonment, the shift to pulp and paper, and the emergence of tourism and conservation. The seeds for these transformations were planted in the political, economic, and technological upheavals of the previous decades, which left a clear imprint on Maine's population. From 1800 to 1860, the population grew briskly, but then, for the first time since the warfare of 1675–1759, declined slightly in the 1860s (figure 4.5). The 1870s eked out only a 4 percent increase, similar to that for the next four decades. Outmigration resulted in considerable downsizing and disruption, especially for small towns, such as Buckfield in Oxford County, which peaked at 1,705 in 1860 but was reduced to 957 by 1920, and did not reach its former population until the year 2000. Many Maine towns are still well under their nineteenth-century peak populations.[41]

This abrupt change of fortunes for Maine had multiple causes. The Civil War claimed the lives of more than 9,000 Maine soldiers, and it revealed opportunities outside of the state to the more than 60,000 survivors.[42] The Homestead Act of 1862 opened up free, fertile land west of the Mississippi. Industrialization, particularly textile mills, created new opportunities in larger cities, some in Maine, such as Lewiston, but most south of the border. The railroad allowed young people a glimpse of the vibrant urban life they were missing. At the same time, these new rail networks allowed products from the Midwest — sheep, wheat, and lumber, for example — to compete with those of Maine. These were major fractures in the history of the United States that affected the country at multiple levels — political, social, and economic. For Maine, these forces aligned to discourage immigration and encourage outmigration, with repercussions for Mainers and their forests.[43]

Extirpation of Wildlife

Writing about Farmington in 1865, Francis Butler observed,

> The fiercer animals, as the bear and the wolf, have long since disappeared, although a gray wolf was killed in the northern section of the town as late as February, 1844. The otter, too, has sought more retired surroundings, and some sixty years have passed since the last beaver rewarded the hunter's toil. In former times the river and streams of the town teemed with fish. Salmon and alewives were taken in great quantities by the early settlers. But with the building of dams they disappeared. Few salmon have been taken since 1795; probably none since 1820.[1]

By 1880 or soon thereafter, the eastern gray wolf, mountain lion, beaver, American marten, Canada lynx, caribou, otter, wild turkey, puffins, great auk, Labrador duck, and passenger pigeon were largely extirpated from Maine—these last three driven to extinction across their entire range. Other species, including the black bear, white-tailed deer, moose, salmon, alewife, bald eagle, osprey, common raven, pileated woodpecker, and great blue heron had also experienced serious declines. Habitat loss, lack of large prey, and especially unrestrained hunting were the main culprits. For wolves, mountain lions, bears, bobcats, and lynx, the state made concerted efforts, using bounties, to entirely eliminate these predators, which were seen as a threat to human livelihood. Not until 1967 was the bounty on lynx, today a federally threatened species, ended.[2]

Many of these species played key ecological roles: top predators (wolves, mountain lions, lynx, bear), major herbivores (deer, moose, caribou), habitat engineers (beaver), and seed dispersers and prey (passenger pigeons, wild turkeys, salmon). Their decline must have had deep ecological impacts: loss of wetlands, increased

The effects on farming, land clearing, and forests were not instantaneous. In fact, the zenith for cleared land did not show up until the 1880 agricultural census (figure 4.7), the rationale for our designation of that year as the beginning of this period. Starting then, the number of farmers and the amount of open farmland declined abruptly. In the 1880s, more than 400,000 acres reverted from fields to regenerating forest, followed by another 600,000 in the 1890s. It's during this time that the 100-plus-year tenure of the Perhams on backlot 24 in Farmington came to an end, apparently of a common populations of small mammals, reduced browsing pressure on plants, to name a few. One study speculates that the loss of beavers might have even intensified the Little Ice Age in the nineteenth century by curtailing wetland emissions of methane and carbon dioxide.[3]

NOTES

1. Francis G. Butler, *A History of Farmington, Franklin County, 1776–1885* (Farmington, ME: Press of Knowlton, McLeary, and Co., 1885), 16.

2. Dean B. Bennett, *The Forgotten Nature of New England* (Camden, ME: Down East Books, 1996), 301–330; David R. Foster et al., "Wildlife dynamics in the changing New England landscape," *Journal of Biogeography* 29 (2002): 1337–1357; R.M. DeGraaf and M. Yamasaki, *New England Wildlife, Habitats, Natural History and Distribution* (Hanover, NH: University Press of New England, 2001), 8–12; Kevin S. McKelvey, Keith B. Aubry, and Yvette K. Ortega, "History and distribution of lynx in the contiguous United States," in Leonard F. Ruggiero et al., eds., *Ecology and Conservation of Lynx in the United States* (Fort Collins, CO: U.S. Department of Agriculture, Forest Service, Rocky Mountain Research Station, General Technical Report RMRS-GTR-30WWW, 1999), chapter 8; Philip T. Coolidge, *History of the Maine Woods* (Bangor, ME: Furbish-Roberts, 1963), 680–705; D. Köster, J. Lichter, P.D. Lea, and A. Nurse, "Historical eutrophication in a river-estuary complex in mid-coast Maine," *Ecological Applications* 17 (2007): 765–778; David. C. Smith, *A History of Lumbering in Maine, 1861–1960*, Maine Studies No. 93 (Orono: University of Maine Press, 1972), 339; Richard W. Judd, *Common Lands, Common People: The Origins of Conservation in Northern New England* (Cambridge, MA: Harvard University Press, 1997), 43; William B. Krohn and Christopher L. Hovin, *Early Maine Wildlife: Historical Accounts of Canada Lynx, Moose, Mountain Lion, White-Tailed Deer, Wolverine, Wolves, and Woodland Caribou, 1603–1930* (Orono: University of Maine Press, 2010), 15; Craig R. McLaughlin, *Black Bear Assessment and Strategic Plan 1999* (Bangor: Maine Department of Inland Fisheries and Wildlife, 1999); Neil Rolde, *The Interrupted Forest: A History of Maine's Wildlands* (Gardiner, ME: Tilbury House, 2001), 295; Arthur Spiess, "Comings and goings: Maine's prehistoric wildlife," *Habitat: Journal of the Maine Audubon Society* 5 (1988): 30–33.

3. Johan C. Varekamp, "The historic fur trade and climate change," *Eos* 87 (2006): 593, 596–597.

cause: no descendents left to take over the farm. By 1920, nearly half of the previously cleared land had lapsed into forest.[44] It had become clear that Maine would remain, for the foreseeable future, largely a forested landscape with a relatively sparse population.[45] It's worth emphasizing the brevity and intensity of land use change in New England during the 100 years from 1820 to 1920. As Richard DeGraaf and Mariko Yamasaki remark in *New England Wildlife*, "Nowhere else in the world was so large an area cleared and rapidly abandoned as in New England."[46]

Figure 4.13 Pulp and paper mill in Rumford, Maine. The original plant was established by Hugh Chisholm in 1893 as Oxford Paper Company and shortly after as International Paper Company. The mill is now owned by the New Page Corporation and employs 1,100 people.

Population stabilization and land abandonment set the stage for Maine to expand its role as a forest products supplier. Lumber output (i.e., trees cut for long boards, box boards, match sticks, and short-board items) increased and, by 1910, reached levels unsurpassed until the 1990s. This belied the poor internal health of the industry, however, and many companies failed in the face of competition from the Midwest and exhaustion of "old-growth spruce suitable for dimension lumber south of the St. John."[47] It's hardly an exaggeration to say that the emergence of the pulp and paper industry in the 1880s saved Maine's forest products economy — and, of course, set it on a path that continues to this day.[48]

The pulp and paper industry arose in Maine and elsewhere in the 1860s in response to increasing demand for paper and the high cost of rags, the raw ingredient in paper making up to that time, and the development of technology to grind or dissolve wood fiber to make pulp. In the 1890s, production ramped up tremendously,[49] so much so that overproduction and falling prices led to consolidation, most notably by the formation in 1897 of the huge International Paper Company (figure 4.13). The crowning achievement, however, was the creation of the Great Northern Paper Company in 1899, which constructed a gigantic mill and associated town in Millinocket

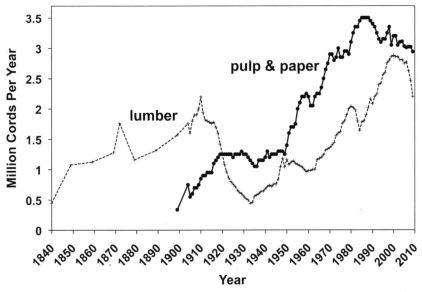

Figure 4.14 Harvest of wood for lumber and pulp from 1839 to 2009.

and eventually owned about one-tenth of Maine's timberlands. In the next several decades, the pulp and paper industry rose as rapidly as the lumber industry fell (figure 4.14).[50]

The impacts on forests were considerable. Harvesting levels and intensity rose sharply (figure 4.10). At first, only aspen was used for papermaking, but, by the late 1880s, this shifted to spruce, which produces long, strong wood fibers. Much of the original old-growth spruce had been cut out by that time, but smaller trees worked fine for pulp production. This meant that areas that had in the past been logged selectively for big trees were now being harvested again, this time more intensively.[51] In 1896, Maine forester Austin Cary described a pulpwood harvest as "the hardest cutting ever seen by the writer."[52] This change in the forest did not sit well with everyone.

"Injudicious destruction of the woods," "improvident waste," "want of foresight in the economy of the forest . . . in many parts of New England": these words were spoken in 1847 by then congressman from Vermont George Perkins Marsh.[53] Henry David Thoreau's *Walden*, published in 1854, and Marsh's *Man and Nature*, published in 1864,[54] inspired a generation of writers, artists, scientists, and politicians to call for a new relationship between humans and nature. This movement was led in the late 1800s by two powerful advocates with differing points of view: John Muir, writer, scientist, and founder of the Sierra Club, argued for *preservation* of forests as a remedy for the harm waged by humans on nature, whereas Gifford Pinchot, eventually first chief of the Forest Service and governor of Pennsylvania, advocated for *conservation*, that is, sustainable use of forests. This dichotomy has run through the history of debates surrounding the forests of Maine and the entire country ever since that time.[55]

These issues resonated with Mainers, who had witnessed unprecedented levels of harvesting on already cutover lands, as well as the decline of wildlife such as moose and deer. In *Common Lands, Common People*, Richard Judd takes this further, arguing that "The key ideas [nationally] that guided this movement gained vitality at a grass-roots level, in consonance with the real experiences, dreams, and desires of people and classes in conflict in local situations."[56] There were at least three main concerns about forests: the effects of heavy cutting on soils and climate, the possibility of a timber famine, and the threat of fire — all integrated into "a cosmic view of the importance of trees in the balance of nature."[57] The rapid rise of tourism in Maine in the last three decades of the nineteenth century also created new economic considerations. Well-off summer visitors came to the state for "remote wilderness or pastoral landscape, and plenty of fishing and hunting opportunities nearby."[58] The proprietors of these resorts became strong advocates for reining in what they viewed as the destruction of the scenic backdrop of their business.[59]

Conservation issues had been discussed, especially among farmers, in Maine since the middle 1800s, but it took two forest men, George Talbot and Austin Cary, both from East Machias and graduates of Bowdoin (although fifty years apart), to turn concern into practical action. Talbot was responsible for the first statewide forest conservation legislation, leading to the establishment of a state forest commission in 1891. Cary brought scientific forestry ideas and practices to Maine, first as an early employee of the commission and then as the first private forester hired in New England.

In the end, Maine's early conservation actions were modest, focused mainly on state game laws, cooperative prevention of fire and insect outbreaks, and the introduction by paper companies of inventories to assess the "growing power" of the woods — the first attempt at sustained-yield forestry. Some agitated for forest preservation and state regulation of harvesting, but they were strongly opposed by others, including the powerful forest industry.[60] An outcome of this debate was to play a central role in the future of Maine and its forests: timber companies agreed to public recreational access to their vast forests, a *quid pro quo* for the lack of forest regulation. The first larger public lands were also established during this time, including White Mountain National Forest (1918), Acadia National Park (1919), and the state forest system (1921), as were the first state agencies responsible for assessing and informing management of forests, inland fisheries, and wildlife.[61]

Maine forests experienced enormous transformation from 1880 to 1920. Forest area increased substantially for the first time in recorded history, but harvesting on these lands was intense: about 100 million cords over 40 years, five times the total from the previous 250 years. Four significant patterns regarding forests emerged: (1) the rise of the pulp and paper industry and the fall of lumbering (logging for boards, slats, veneer, etc.); (2) the rapid consolidation of this new enterprise into several large, powerful companies that controlled mills, timberlands, and water; (3) the first stirrings of conservation, forestry, and the idea of preservation; and (4) a largely private rather than public approach to forest management and conservation, in contrast to some other states, such as New Hampshire, Vermont, and New York.[62] These paths were to become even more entrenched over the next half-century.

1920–1970: Well-Worn Patterns

In 1860 a traveler from the 1820's would not have noticed much difference in the life of the Maine woods. Indeed, a traveler from the 1860's would not have found that much difference in the 1890's. However a traveler from the nineties, put down in the Maine woods in 1930, would have recognized the trees but not much else.
—David C. Smith, in *A History of Lumbering in Maine, 1861–1960* (383)

During the half century from 1920 to 1970, harvesting systems went through a major transition in mechanization, planning, transportation, and intensity. In the nineteenth century, saws cut, horses yarded, and rivers transported logs to the mills. In the 1920s and 1930s, motorized vehicles were first introduced, and railroads extended the reach into previously inaccessible sites. Foresters, using maps and inventories, planned out harvests for the big companies. After World War II, tractors were regularly used to drag logs out of the woods and trucks to haul them to the mill, slowly replacing horses and river driving. In the 1950s, the chainsaw began to replace the cross-cut saw, and by the 1960s skidders largely ended the work of horses in the woods. Equally important as these mechanizations, the development in the early 1950s of a process for using hardwoods for pulp greatly expanded the range of forest types and species harvested (figure 4.14).[63]

By midcentury, a well-worn pattern of forest harvesting had developed: trees of nearly all sizes and species were cut, "the huge companies controlled nearly all the cut, and the cut was to their specifications."[64] According to Lloyd Irland, the transition from the harvesting of large pine and spruce with men, horses, and rivers to this new more mechanized system greatly intensified the environmental impact of logging, from "an ecologically benign system of exploiting the forest, in which the forest's floor and its long-term productivity were scarcely disturbed . . . to one in which the forest floor was disturbed by large machines, most trees were harvested, and little residual timber was left." At the same time, the contribution of the paper industry to Maine's economy and forest access was prodigious.[65]

Most of the forest statistical trends established between 1880 and 1920 continued in the following five decades, but not without some deviations. Total harvest levels increased about 1½-fold, despite a severe decline during the Great Depression (figure 4.10). Paper production and pulpwood harvesting surged far ahead of lumbering, especially in the 1950s and 1960s (figure 4.14).[66] Tourism continued to grow, transformed by the automobile, which broadened the tourist profile to include visitors of lesser means who came for shorter periods of time. The depression also hit the tourism industry hard, and not until after World War II did it fully recover and then gradually increase its share of Maine's economy. Over this fifty-year period, about another million acres of farmland was abandoned and reverted to forest (figure 4.7), with the economic woes of the depression contributing to this process. By the 1970s, most of the former agricultural landscape of even the southern quarter of Maine had reverted to a largely forested one. This landscape transformation is actually recorded in a decline in weed pollen in lake sediments dated to that time.[67]

Human culture also reshaped the Maine forests in indirect ways during this period through the introduction of nonnative organisms. Species from the Old World had been brought into the state, on purpose or accidentally, since the earliest days of European settlement,[68] evident from fossil pollen in lake sediments dating to the 1700s.[69] Although most of these were benign, the introduction of disease-causing fungi had rapid and profound effects on forests. One of the earliest was the virulent chestnut blight, which, across the eastern United States, "killed three to four billion trees . . . enough trees to fill nine million acres."[70] Originating in Asia, *Cryphonectria parasitica* appeared in New York City in 1904 and southern Maine by 1916. Spores enter wounds in the bark, and the growing fungus kills cambial tissues, constricting the flow of food. Although chestnuts occurred only in scattered populations in southern and central Maine, its near total disappearance must have had local effects on wildlife utilizing its abundant nut crop. Dutch elm disease is caused by at least two pathogenic fungi, which are dispersed by the European elm bark beetle, accidentally introduced in the early 1900s. Not as virulent as chestnut blight, it has nevertheless killed a large portion of the population of American elms, one of the most beloved shade trees and an important lowland species in central and southern Maine.[71] Beech bark–*Neonectria* disease, which was first documented in the state in the late 1920s, has led to the most damage to Maine forests, because of the abundance and widespread distribution of American beech. This is a complex disease caused by a native and an alien fungus (*Neonectria* species), but also an introduced European scale insect that weakens the tree and wounds the bark, allowing fungal penetration of spores. A second killing front moved through northern Maine from 2000 to 2003, killing up to 50 percent of trees in some stands, apparently in response to warm winters and dry summers that stressed trees and allowed the scale to increase. In general, however, mortality in "aftermath" forests is on the order of 1–2 percent annually.[72] Unlike harvesting, these epidemic diseases can permanently alter the mix of tree species in forests because of their continued pathogenicity.

The period 1920–1970 was relatively quiescent regarding the conservation of forests and wildlife, but there were stirrings of the modern environmental era to come. Earlier conservation principles were further developed in state regulatory frameworks and natural resource agencies, and some of the first modern environmental groups in the state, notably the Maine chapter of The Nature Conservancy (1956) and the Natural Resources Council of Maine (1959) emerged. One of the most prominent conservation successes in Maine's history — Baxter State Park — was founded (1931) and expanded during this time, largely the work of one person, former Governor Percival P. Baxter. At the end of this period, the state's legislature and voters also approved the establishment of the Allagash Wilderness Waterway. Finally, expanding forests and enlightened conservation policies, enacted earlier in the century, had begun to work their magic on wildlife populations, which began a remarkable recovery on the Maine landscape (Sidebar 4.3).[73]

1970–2010: Conflict and Change

After a half-century in which "New England forests were of little concern to legislators and the public . . . during the 1970s a series of dramatic changes

Wildlife Returns—At Least Some of Them

Maurice Morin watched Tuesday as Lewiston Public Works crews cleaned out a culvert on the Stetson Road—again. Beavers did what beavers do, built a dam. But their dam has trapped water and at times flooded the road. Despite strong hints for them to leave, like running a telephone pole through their dam, the beavers have stayed and kept rebuilding.
—*Lewiston Sun Journal*, October 1, 2010

One of the most remarkable stories of resilience in the Maine Woods is the resurgence of wildlife, whose populations were drastically reduced or extirpated in the nineteenth century (see sidebar 4.2). Beavers are so common today that they interfere with human activities even in the state's second-largest city.[1] The black bear has recolonized most areas of the state, and their population has reached about 23,000. Moose have increased from about 2,000 to 29,000 over the past century. River otter numbers have increased to about 20,000. As a result of reintroductions, the wild turkey now inhabits more of Maine than before European settlement. These and other forest-dwelling species have recovered in part because of the reversion of abandoned agricultural land to forest, but also because of the curtailment of indiscriminant hunting. At the same time, species requiring open lands, such as the New England cottontail and bobolink, have declined to levels probably more typical before European settlement. Populations of the gray wolf, the mountain lion, and the caribou have not been reestablished in Maine, although possible transient wolves from Canada have been reported in the state. Conditions may be marginal for caribou, given the failure of two past attempts at reintroduction, but Maine provides suitable habitat—large tracts of forests with large prey and low road densities—to accommodate wolves and cougars.[2]

NOTES

1. Bonnie Washuk, "Persistent beavers cause problems on Stetson Road," *Lewiston Sun Journal*, October 1, 2010.

2. R.M. DeGraaf and M. Yamasaki, *New England Wildlife, Habitats, Natural History and Distribution* (Hanover, NH: University Press of New England, 2001), 10, 341, 353; "Moose," Maine Department of Inland Fisheries and Wildlife (retrieved on April 10, 2011 at www.maine. gov/ifw/wildlife/species/moose/index.htm); Craig R. McLaughlin, *Black Bear Assessment and Strategic Plan 1999* (Bangor: Maine Department of Inland Fisheries and Wildlife, 1999); John H. Hunt, *River Otter Assessment* (Bangor: Maine Department of Inland Fisheries and Wildlife, 1986); Philip T. Coolidge, *History of the Maine Woods* (Bangor, ME: Furbish-Roberts, 1963), 504–507, 629–661, 680–705; Susan C. Gawler et al., *Biological Diversity in Maine* (Augusta: Maine Natural Areas Program, Department of Conservation, 1996), 47–50; Neil Rolde, *The Interrupted Forest: A History of Maine's Wildlands* (Gardiner, ME: Tilbury House, 2001), 313–347; David. C. Smith, *A History of Lumbering in Maine, 1861–1960*, Maine Studies No. 93 (Orono: University of Maine Press, 1972), 373–374.

swept over the region."[74] Long-held assumptions about land in Maine were suddenly in play, as the state's forested region was hit by the first of several waves of surging land prices, land speculation, and second-home divisions. New investments in paper and lumber mills to meet rising demand led to an increase in harvesting (figure 4.10). Fuelwood cutting swelled in response to a spike in fuel oil prices. From 1975 to 1984, Maine was struck with the worst spruce budworm outbreak on record. The timber industry responded with aerial insecticide spraying — 3.5 million acres in 1976[75] — and salvage harvesting of spruce-fir forests, mainly with *clear-cutting*. This harvest method was rarely used until the 1960s, when modern chainsaws, powerful skidders, and demand for small trees made it economical. The advent of giant mechanical harvesters, which rapidly cleared areas in northern Maine, "introduced new elements of controversy in logging practices."[76] By 1988, 60 percent of all harvesting by industrial landowners was carried out with clear-cutting, often followed by herbicide spraying to suppress hardwoods, which compete with the desired spruce and fir.[77] This "perfect storm" of forces affecting land use occurred at a time when the Maine public was already sensitized to environmental issues such as acid rain and the pollution of the Androscoggin and Kennebec Rivers.[78]

Some of the concerns emerging from these events were reminiscent of the late 1800s. Anxiety about running out of trees. Worries about the destruction of the expansive green backdrop treasured by tourists. Alarm about the potential environmental consequences of heavy cutting. Nancy Allen of the Maine Green Party expressed these fears in stark terms: "The forest is in deep trouble. Anyone familiar with the woods sees the destruction everywhere. Huge clear-cuts are visible from nearly every high spot . . . an ecosystem obviously stressed to the max."[79] Echoing the views of some Maine farmers a century earlier, objections emerged to the fundamental way the timber industry viewed forests: "From the industrial perspective, the forest is not a biological community to which we belong and which we must maintain. It is a resource to exploit," wrote Mitch Lansky in *Beyond the Beauty Strip*.[80]

Other concerns were new. After nearly a century of reforestation, the Maine Woods was suddenly being pecked away at for other uses, raising doubts about the future land base for forestry and public access for recreation. Harvesting increased, but the number of forest industry jobs declined, making those in rural areas wonder about their futures.[81] These worries were not allayed in the coming years. In 1986, Great Northern-Nekoosa Corporation announced plans for laying off up to 1,400 mill workers in its Millinocket and East Millinocket plants.[82] Then, in 1988, Diamond International Corporation, headed by corporate raider Sir James Goldsmith, sold 790,000 acres in Maine[83] (and 186,000 elsewhere in the northeast), not to another timber company but to a French communications giant, which promptly hired LandVest to sell off parcels to the highest bidders. This event was so momentous that Congress created the Northern Forest Lands Study, which led to a set of recommendations in 1994, aimed primarily at land ownership but also at improving forest stewardship and rural economies.[84] In retrospect, the Diamond International sale was the beginning of a process of divestiture of timberlands by Maine paper companies that has continued to this day. After nearly a century of profiting from vertical integration, new global realities changed the game: now the parts of a paper company (timberlands, mills, power-generating dams), sold separately, were worth more than the whole. New tax structures had also begun to favor forestland ownership by Timber Investment Management Organizations (TIMOs) and Real Estate Investment Trusts (REITs) rather than timber corporations. This has led to a shift in timberland ownership from industrial toward investment landowners (see chapter 6: "Changing Disturbance").[85]

By illuminating the ecological consequences of land use at the landscape scale, the emerging field of *conservation biology* also raised compelling new considerations.[86] Subdivisions and high levels of harvesting meant less mature forest — and also less habitat for species requiring conditions supported by those areas. As regard by scientists for the ecological importance of old-growth rose, many pointed out the almost complete lack of such forests in Maine as a problem in and of itself.[87] Research revealed further that intensive land use was not just a problem of *habitat loss*, but also of *habitat fragmentation*, a process that can splinter species into isolated populations, increasing their vulnerability to local extinction, especially for species requiring large areas of undisturbed forest. Conservation biology instilled a greater appreciation for the entire range of *biological diversity* in Maine, which became an important topic even in some timber management plans. According to *Biodiversity in the Forests of Maine: Guidelines for Land Management*, a practical guide published in 1999, "A primary goal for biodiversity in Maine's managed forest is to ensure that adequate habitat is present over time across the landscape to maintain viable populations of all native plant and animal species currently occurring in Maine."[88]

The massive spruce budworm attack in the 1970s gave forest scientists a unique opportunity to consider the entire landscape in another way, in assessing the relationships between natural and human-caused disturbance (harvesting) in Maine forests. Bob Seymour, a professor at the University of Maine, in particular, provided key insights into how the combination of "large-scale, episodic" logging and budworm attacks "transformed Maine's spruce-fir forest from one dominated by mixed-age, old-growth stands, to a forest dominated by younger, more uniform stands" — which he called the "new forest" because of its unprecedented nature. According to Seymour, this new forest is vulnerable to insufficient regeneration[89] of desirable species (e.g., spruce) after harvest, especially clear-cutting, which can lead to shortfalls in wood supply or dominance by intolerant hardwoods, usually controlled with herbicide spraying. This was a lucid explanation of the problems and outcomes encountered by paper companies after the spruce budworm attacks, which were part and parcel of the controversial atmosphere surrounding Maine forests during this period.[90]

The momentous events, perceived threats, and new perspectives in the Maine Woods led to profound changes, not seen since the early 1900s — with important consequences for forests.[91] First, Maine's forest practices laws and regulations were modified several times (1989, 1999, 2003) to reduce impacts on streams and lakes, to mandate adequate postharvest regeneration, to discourage liquidation harvesting (land purchase followed by removal of all commercial timber and resale within five years), and to limit clear-cut size (less than 250 acres) and proximity (no closer than 250 feet of another).[92] Clear-cutting declined dramatically from about 40 percent of all harvests in the late 1980s to less than 5 percent in the 2000s,[93] largely attributable to the new practice of leaving just enough cover to avoid the clear-cutting limit.

Second, Maine was in the vanguard of the *forest certification* movement, "a process whereby independent third parties review forest practices to determine if they meet the standards of a well-managed forest."[94] The two most widespread third-party certification systems in the world — Forest Stewardship Council and the Sustainable Forestry Initiative — are both present in Maine, which as of 2005 had nearly seven million certified acres, 40 percent of all timberlands, one of the highest percentages in the country.[95]

The third response was heralded in 1998, when The Nature Conservancy purchased 185,000 acres for $35.1 million along the St. John River, an unprecedented acquisition — for its scale, for its allowance of traditional uses, and for the pursuit of sustainable forestry on a large portion of the land. Although important conservation acquisitions had occurred earlier in the "modern" era (e.g., Bigelow Preserve, Allagash Wilderness Waterway), the

Table 4.1 Conservation purchases (acres) in Maine from 1990 to 2010, separated into fee purchase land and conservation easements

Year	Fee Purchase	Conservation Easement
1990	25,887	2,065
1991	1,958	4,220
1992	4,467	498
1993	4,308	4,465
1994	4,390	1,076
1995	3,561	952
1996	3,674	9,339
1997	15,069	3,357
1998	176,372	2,264
1999	9,259	7,604
2000	7,438	26,055
2001	13,678	764,428
2002	60,355	197,756
2003	51,503	321,036
2004	73,261	23,670
2005	33,501	313,990
2006	13,319	9,099
2007	10,153	15,704
2008	15,736	11,237
2009	57,835	13,382
2010	5,611	1,413
TOTAL	591,335	1,733,702
1990–1997 (ac/yr)	7,914	3,247
1998–2010 (ac/yr)	40,617	131,357

Source: Daniel Coker, The Nature Conservancy in Maine, Brunswick, Maine.

period from 1998 to 2010 was unparalleled, with purchases of more than 2.2 million acres for conservation in Maine (table 4.1[96]). Most of this land — 1.7 million acres — was in conservation easements, an agreement whereby a landowner retains ownership but relinquishes development rights and often agrees to additional covenants such as sustainable harvesting and public recreational access. The rest — 528,000 acres — was fee purchase land, added to Maine's public-private portfolio of lands dedicated to conservation goals. These acquisitions were funded by a combination of private, federal, and state funds, which have declined in the past few years.[97]

How effective have changes in forest practices, certification, and acquisition been in addressing the concerns raised during this time period? Have they helped mesh dominant land uses with other forest values important to Mainers? Some compare 1970 and 2010 and see important steps toward minimizing the negative impacts of harvesting, insuring a forestry land base and public access, and enhancing the ecological integrity of the landscape. Others argue that conservation easements are poor investments, forest practice regulations merely "legitimized what they were already doing,"[98] and land acquisition has been insufficient. Still others argue against the changes as an invasion of private property rights.[99] As this book is going to press, a proposal has been floated in Maine that is sparking further debate over these issues. Long-time conservationist Roxanne Quimby has offered land for a national park (70,000 acres), restricting many traditional uses, and a state park (30,000 acres), where they would be maintained.[100] Whatever one's views on these changes in policies and practices, they are having and will continue to have impacts on the extent, composition, and structure of the Maine Woods. They point to the germination of a new trend in Maine, a sort of bifurcation of land management. One path is leading to a modestly increasing portfolio of preserved lands where ecological goals reign, land that will support older, climax forests. The other path, on the bulk of forests, continues the tradition of a high level of wood production, where frequently disturbed, younger forests will predominate.

Maine Forests Today

How can we make sense out of these four centuries of forest history? What persistent forces and attributes made the Maine Woods what they are today? Despite its location at the edge of the United States, seemingly out of the mainstream, Maine has never truly been isolated from global currents. From that world came European settlers, an expanding demand for a series of wood products, new technologies and corporate structures to meet those demands, an ever-growing appreciation for the "wildness" offered by Maine, new ways to access and own a piece of that remoteness, waves of forest-altering tree pests, and atmospheric pollutants from thousands of miles upwind. That external imprint is obvious in Maine's forest history and its current forest. But equally crucial are the state's intrinsic elements. Its long attraction as a source of timber is a direct consequence of biogeography: interior Maine is an elevated land mass where merchantable trees

The Besse Farm:
Long-Term Legacies of Land Use

Areas that have never been harvested, harvested but never cleared, and cleared and then reforested tend to support different kinds of forests. The name "old field pines," for example, acknowledges that white pine readily invades abandoned fields and is more common on these sites than on continuously forested lands—a pattern recognized throughout New England since the early 1900s.[1] One of the best case studies of such patterns in Maine is Theresa Kerchner's work at the Besse Farm in Wayne. Nearby paleoecological studies reveal a presettlement forest of beech, birch, and hemlock, with smaller amounts of spruce, pine, oak, and maple, which corresponds well to the observations of land surveyor Ephraim Ballard in the late 1700s.[2] By 1860, 80 of the 130 acres of the Besse Farm were cleared, leaving some acreage as woodlots. By the late 1990s, all but 4 acres had reverted to forest. White pine predominates on former cultivated land, hayfields, and wooded pasture, where it is joined by eastern hemlock. White pine and red maple are most common in areas previously used as pasture. In former woodlots, which have been hardly logged since the early 1900s, red oak is most abundant, with some white pine, white ash, eastern hemlock, red maple, and sugar maple. Thus, not only are distinct forests associated with different past land uses,[3] but the modern forest as a whole differs from the original presettlement forest. Clearly, the legacies of land use linger long in the forests of central Maine.[4]

grow well (and crops not as well); where a long snow season allowed easy skidding of logs over the ground; where large rivers provided cheap transportation of logs to, and electrical and mechanical power for, mills; and where remoteness discouraged competing land uses until recently. At the same time, the people of Maine developed "a New England town-meeting culture and a way of life that depends to a large degree on the illusion of isolation."[101] This encouraged an abiding suspicion of government and outside control, the eschewing of public approaches to land regulation (until recently), and adoption of a conservation ethic, influenced by national conversations certainly, but still by and large homegrown. Although the complexities are manifold, in the simplest sense, the history of the Maine Woods over the past four centuries is the story of the evolving interactions of these unique internal elements — the land and the people — with the powerful current of external forces.

NOTES

1. For example, see R.T. Fisher, "Second-growth white pine as related to the former uses of the land," *Journal of Forestry* 16 (1918): 253–254.

2. Ballard also found white oak, which in that area is at the northern limit of its distribution: "5th mile a white oak tree white oak land." (Ephraim Ballard's survey notes taken in Wayne; November 9–10, 1789, Maine State Archives, from Theresa Kerchner, *The Improved Acre: The Besse Farm, Maine, as a Case Study in Land Clearing, Farm Abandonment, and Reforestation in Northern New England*, unpublished report, 2009, p. 5.)

3. One could argue that this variation may simply reflect differences in the potential of the original sites selected for crops, hayfield, and woodlots, but Kerchner has found that these patterns are expressed even across adjacent areas that exhibit no obvious environmental differences except for land use history and are also repeated on other abandoned farms in the area.

4. Unless indicated otherwise, information is from Theresa Kerchner, "The improved acre: the Besse Farm as a case study in land-clearing, abandonment, and reforestation," *Maine History* 44 (2008): 77–102, as well as her unpublished report, Kerchner, *The Improved Acre*. Pollen data from K. Gajewski, "Late Holocene pollen stratigraphy in four northeastern United States lakes," *Geographie Physique et Quaternaire* 41 (1987): 377–386; pollen data retrieved on May 26, 2011 at "Pollen Search," NOAA Paleoclimatology, www.ncdc.noaa.gov/paleo/paleo.html. Witness tree data from database of Charles Cogbill. For other case histories of farmland clearing, abandonment, and forest reversion, see Kristen Hoffman, "Farms to forests in Blue Hill Bay: Long Island, Maine, as a case study in reforestation," *Maine History* 44 (2008): 50–76 and Elizabeth H. Moore and Jack W. Witham, "From forest to farm and back again: land use history as a dimension of ecological research in coastal Maine," *Environmental History* 1 (1996): 50–69. For a comprehensive ongoing study of the relationships between presettlement forests, land use, and current forests in central Massachusetts, see D.R. Foster and J.D. Aber, eds., *Forests in Time: The Environmental Consequences of 1,000 Years of Change in New England* (New Haven, CT: Yale University Press, 2004).

One of the most remarkable outcomes of this history is that Maine's forest area today is hardly smaller than it was four centuries ago. The state covers 21.3 million acres, 19.8 million of that in land. Ninety percent or 17.7 million acres is forested today compared to an estimated 18.2 million in the year 1600. This is the highest percent forest coverage in the nation. The trajectory of forest area has been more complicated than that, of course, declining rapidly during the 1800s to a low point around 1880 and then recovering since as a result of farmland abandonment.[102] Following this forest recovery, populations of many wildlife species — deer, moose, beaver, woodpeckers, and others — made an equally vigorous return. Perhaps not as dramatic as that of southern New England, where most of the landscape was cleared at one time, Maine's forest history is still a story of resilience and recovery — part of what Bill McKibben has called "the great environmental story of the United States."[103]

This modern Maine forest can be divided into three parts, according to its land use history. Not much more than 10,000 acres are remnant old growth, little altered by harvesting or clearing, including the 5,000-acre Big Reed Forest Reserve, the state's only old-growth landscape (see chapter 3).[104] About 14–15 million acres have never been cleared for agriculture, but have been harvested, in most cases multiple times, and contain a vast network of logging roads. About 3 million acres were cleared for farmland, mainly in the 1800s, and reverted to forest, almost all of which has been cut subsequently. Together, these three forest types cover 17.7 million acres, about 97 percent of which is open for harvesting; the remaining 3 percent is protected from logging. The best estimates suggest that in the past decade these timberlands have experienced the highest rate of wood harvesting in the state's history, on the margin of sustainability (sidebar 4.5).[105]

How does the forested landscape of Maine today compare to that of the presettlement forest described in chapter 3? In terms of species composition, a visitor from the seventeenth century would find many familiar organisms, but key species, such as the gray wolf and caribou, are missing. Many nonnative species have joined the Maine fauna and flora, and predominate in some sites (see chapter 6). A recent assessment of Maine's biodiversity found that about one-third of the 2,107 plant species in the state were exotics. There is also circumstantial evidence that the rarity of some forest-dependent herbs, mosses, and lichens is the result of land clearing and harvesting.[106] Whether other presettlement species, especially inconspicuous ones, were extirpated by the past four centuries of land use is an open question. Our knowledge even of Maine's current biota is woefully incomplete.

Table 4.3 compares the relative abundance of tree species in Maine for the modern versus the presettlement forests. The largest declines have been for beech, yellow birch, hemlock, oaks, and spruces; the largest increases have been for red maple, poplars, balsam fir, and white pine. Analysis of pollen cores throughout the Northeast by Emily Russell and her colleagues corroborate these findings: modern forests in the region have less spruce, hemlock, and beech, and more birch, fir, and red maple. These transformations were the result of complex direct and indirect forces exerted on forests by humans, starting in the early colonial period. Increased fire incidence during settlement favored birches and white pine and acted against fire-sensitive species such as red spruce, hemlock, and beech. Beech bark–Neonectria disease has decimated beech. Frequent harvesting has favored species that respond well to disturbance, such as poplars (aspens), paper and

Table 4.2 Average annual acreage harvested (2000–2009) in Maine for different types of harvests

Type of Harvest	Acreage Harvested	Percentage
Partial harvest[a]	282,723	55.1
Shelterwood harvest[b]	208,413	40.6
Clear-cut harvest[c]	16,482	3.2
Harvest to change to nonforest use	5,862	1.1
Total harvest	513,479	100.0

[a] Trees are removed individually or in < 5-acre patches.

[b] Harvest of mature trees in two or more stages. First stage removes a portion of trees to allow establishment of regeneration before remaining trees are removed in subsequent harvest.

[c] Harvest on > 5 acres that results in a residual basal area of acceptable growing stock trees > 4.5" dbh of less than 30 square feet per acre, unless after harvesting the site has a well-distributed stand of acceptable growing stock 3 feet tall for softwood and 5 feet for hardwoods.

Source: Maine Forest Service, *Silvicultural Activities Report* — 2002 and 2009 (Augusta: Maine Forest Service, Department of Conservation, 2003 and 2010).

Table 4.3 Tree species composition in presettlement versus modern forests in Maine

Tree Species	Presettlement (%)	2003 (%)
Spruces	20.2	17.4
American beech	12.2	4.8
Balsam fir	10.5	15.2
Yellow birch	9.3	4.7
Eastern hemlock	8.9	5.8
Northern white cedar	8.0	10.3
Sugar maple	5.6	5.3
Paper birch	4.9	6.1
Oaks	4.6	2.4
Eastern white pine	4.5	5.2
Red maple	3.6	13.4
Ashes	2.0	1.6
Poplars	1.4	4.0

Note: Percentages are from statewide numbers of survey witness trees for presettlement and for live trees greater than 5" dbh (table A8 in the U.S. Forest Inventory and Analysis) for 2003.

Source: Charles Cogbill, Maine Witness Tree Data Set; W.H. McWilliams et al., *The Forests of Maine: 2003*, Resource Bulletin NE-164 (Newtown Square, PA: U.S. Department of Agriculture, Forest Service, Northeastern Research Station, 2005).

Are Harvest Levels in Maine Sustainable?

Of the 17.7 million forested acres in Maine, 17.2 million are open for harvesting. Here are some statistics for these timberlands for the decade 2000 to 2009. Most stands were harvested using partial cutting or shelterwood, where some trees are left; only a small percentage was clear-cut or converted to nonforest uses such as housing (table 4.2). On average, 513,000 acres per year were subject to some sort of commercial cut. At that rate, the entire 17.2 million acres of timberlands would be harvested over a thirty-three-year period (again, mainly using methods that leave some trees). The average annual harvest during the decade for pulpwood, sawlogs, biomass, and firewood combined was 6.18 million cords—the highest reported level in the state's history.[1]

This raises an important question: Was this a sustainable rate of harvesting? At first glance, the concept of sustained-yield harvesting seems simple: the amount of wood harvested should not exceed the growth over the managed acreage.[2] The 2000–2009 average of 6.18 million cords over 17.2 million acres gives a harvest level of 0.36 cords per acre per year—a slim 0.01 cord less than Maine's average forest growth rate of 0.37 cords per acre.[3] It is important to point out that current harvest estimates are incomplete. Firewood harvesting, a new component of annual reports, was underestimated because only a subset of dealers could be included, raising the possibility that statewide cutting actually slightly outstripped growth during that ten-year period.[4]

There are deeper issues involved in sustained-yield harvesting, beyond the scope of this book, but worth pointing out. The statewide average, for example, conceals important details about the harvest-growth calculus: whether current wood removals are sustainable for each economic region of the state, for each individual important timber species, and for each type of wood (veneer, sawtimber, pulp, etc.). These are crucial assessments, for growth-harvest balance at the state- level might well be the result of compensation for overcutting in some regions (e.g., the north) or of some species (e.g., spruce) by low harvest rates in other regions (e.g., the south) or other species (e.g., hardwoods). Sustained-yield harvesting also bears only faint relevance to ecological sustainability—whether at current harvest rates, ecosystems can maintain viable populations of all species, sufficient nutrient stocks, and other ecological services over the long term. The crude statewide averages reported above do, nevertheless, provide a starting point and an absolute limit that cannot be exceeded by forests for very long. At the very least, they tell us that harvesting levels in Maine from 2000 to 2009 were high relative to that objective benchmark. A recent report predicts that, at least for spruce-fir forests, growth rates will increase over the next twenty years, as these stands continue to recover from the past spruce budworm outbreak, prompting a recommendation to increase harvesting on that forest type by as much 64 percent.[5]

NOTES

1. Maine Forest Service, *Silvicultural Activities Report*—2003 and 2009 (Augusta: Maine Forest Service, Department of Conservation, 2004 and 2010). *Partial harvest*: trees are removed individually or in small (< 5 acre) patches. *Shelterwood*: harvest of mature trees from a forest site in two or more stages. The first stage removes only a portion of the trees to allow establishment of regeneration before the remaining trees are removed in subsequent harvests. *Clear-cut*: harvest on a site larger than 5 acres that results in a residual basal area of acceptable growing stock trees > 4.5" dbh of less than thirty square feet per acre, unless after harvesting the site has a well-distributed stand of acceptable growing stock three feet tall for softwood and five feet for hardwoods (overstory removal). Refer to the latest copy of the Maine Forest Service Rules, chapter 20, for additional information. It can be found on the Maine Forest Service website (www.state.me.us/doc/mfs/rules_regs/index.htm). As mentioned previously, harvest estimates before the Forest Practices Act of 1989 are somewhat uncertain, but the best we have.

2. The idea of sustained-yield harvesting came to the United States from Germany in the late nineteenth century with foresters such as Bernhard Fernow, Carl Schenk, Austin Cary, and Gifford Pinchot: Char Miller, *Gifford Pinchot and the Making of Modern Environmentalism* (Washington, DC: Island Press, 2001); Gifford Pinchot, *Breaking New Ground* (Washington, DC: Island Press, 1998 from original 1947).

3. It has been estimated that intensive silvicultural treatments could push yields to 0.50 cords per acre across the state, and well over one cord per acre on the most productive sites; Maine Forest Service, Department of Conservation, *Maine State Forest Assessment and Strategies* (Augusta: Maine Forest Service, Department of Conservation, 2010), 4; Michael Greenwood, Robert S. Seymour, and Marvin W. Blumenstock, *Productivity of Maine's Forest Underestimated—More Intensive Approaches Are Needed*, Maine Agricultural Experimental Station Miscellaneous Report No. 328 (Orono: Maine Agricultural Experimental Station, 1988). Thus, at a state level, it would be possible to increase sustainable harvesting levels above 0.37 cords per acre per year.

4. In the 2009 report, seventy dealers were surveyed, adding up to 102,000 cords, a value lower than the 400,000 estimated in a 1990s State Planning Office report. If the higher figure is used, the average harvest rises to 0.377 cords, slightly higher than growth. Maine Forest Service, *Silvicultural Activities Report*—2003 and 2009 (Augusta: Maine Forest Service, Department of Conservation, 2004 and 2010).

5. The report is by James W. Sewall Corporation in Old Town, Maine, commissioned by the Maine Forest Service; Kevin Miller, "Report: Maine woods can support heavier logging," *Bangor Daily News*, August 19, 2011 (retrieved on August 21, 2011 at bangordailynews.com).

gray birch, white pine, black and pin cherries, and red maple. In the north, 150 years of harvesting has targeted spruce, to the benefit of balsam fir.[107]

The physical structure of Maine's forests has also changed from pre-settlement to today. Maine never did support giant trees, with the exception of some white pines. But because of high harvest levels, trees in Maine today are much more skewed toward small sizes than during presettlement times (or in old-growth remnants). The most recent statewide Forest Inventory and Analysis found that, for trees at least 5" dbh, only 7.2 percent are 13–21" dbh and 0.5 percent larger than 21".[108] For comparison, for trees at Big Reed Forest Reserve larger than 4" diameter, 20 percent are greater than 14" dbh and 5 percent greater than 20". Paul Frederic, a geographer and longtime observer of the Maine landscape, tells a story of a puzzled visitor from Namibia who upon touring the Maine Woods described them simply as "a forest of small trees." Today's forests are also young. Whereas 59 percent of presettlement forest stands were older than 150 years, this value had declined to 1 percent by 1995 (figure 4.15). Of the original old-growth forests of Maine, only about 0.05 percent remains today. Because of harvesting, modern Maine forests also tend to be lacking in large-diameter classes of dead organic matter, such as snags, cavity trees, and logs, which add structural complexity to forests, provide habitat for vertebrates, invertebrates, bryophytes, lichens, and fungi, and supply nutrients to plants and other organisms.[109]

In sum, the Maine Woods remains a vast territory that continues to support a wide range of species, resources, and values. The last four centuries of land use, however, have altered the original forest, favoring generalist and disturbance-associated species, reducing tree size and age, and decreasing structural complexity. The southern Maine forest has changed from an old-growth mixed hardwood forest to a network of suburban and rural forests, much of it established on formerly cleared land. The northern forest has changed from a structurally heterogeneous spruce-hardwood forest to a more homogeneous, younger "new" fir-spruce forest. The previous chapters of this book detailed the intimate relationships of Maine forests with underlying ecological factors. This chapter demonstrates that the Maine Woods of today are also very much a product of human culture.

What might these land use changes mean for the future of Maine forests? Returning to the presettlement forest is not ecologically possible, given the scale of forest and environmental change. On the other hand, elements of that landscape — mature forest, coarse organic matter, structural complexity —

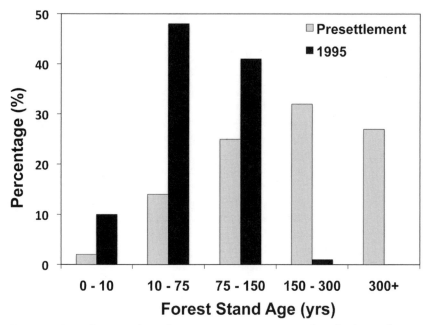

Figure 4.15 Presettlement and 1995 forest age structure in Maine. Technically, this graph shows the time since the last disturbance, from harvesting or natural causes, revealing that modern forests are much younger than forests before Euro-American settlement.

that are essential for the maintenance of biological diversity and other values can be supported. For land managers invested in those goals, an understanding of the trajectories of Maine forests is crucial. Past legacies clearly constrain the universe of future conditions, but Maine's forest history suggests manifold potential paths within those ecological boundaries. We will discuss those futures in more detail at the end of chapter 6.

Chronology of Important Events in the History of the Maine Woods from European Settlement to Today

Year	Event
1497	First purported visit of Europeans to Maine: John and Sebastian Cabot
1604–1605	Champlain explores mid-coast Maine
1607	First Euro-American attempt to establish permanent town in Maine
1620s	First permanent settlements
1631	First sawmill, Salmon Falls River, South Berwick
1645	First bounty on wolves
1675–1759	Warfare among British, French, and Indians
1763	End of hostilities between British, French, and Indians; Treaty of Paris
1783	Treaty of Paris; end of the American Revolution
1787	Settlement of Perham farm
1814	Treaty of Ghent between United States and Great Britain, ends War of 1812 and designates northern border of Maine as following "the highlands" south of the St. Lawrence River
1820	Statehood
1830	Hunting season restrictions on moose and white-tailed deer. Bounties on wolves, bobcats, Canada lynx, porcupines, crows, and black bears
1836	First railroad, the Bangor and Veazie Railroad
1840	Maine population surpasses 500,000
1842	Webster-Ashburton Treaty between the United States and Great Britain, ends "Aroostook War" and settles location of northern border of Maine
1846	Henry David Thoreau's first of three trips to Maine
1861	Spruce surpasses pine in lumber harvest
1864	*Man and Nature* published by George Perkins Marsh
1876	Poland Spring Hotel resort built
1880	Percentage of cleared land peaks and then begins to decline
1891	Establishment of Maine Forest Commission
1893	Construction of huge pulp and paper mill in Rumford, part of International Paper Company at its formation in 1897
1899	Construction of huge pulp and paper mill in Millinocket by Great Northern Paper Company
1899	Hunting season on caribou closed
1907	Last recorded shooting of passenger pigeon in Maine, which becomes extinct several years later
1915–1920	Harvest of pulpwood surpasses that for lumber
1917–1919	Major spruce-budworm outbreak
1918	White Mountain National Forest established
1919	Creation of Lafayette National Park (later Acadia)
1928	Last long log drive on the Penobscot

Appendix 4.1 *(continued)*

Year	Event
1931	Baxter State Park established after donation from Governor Baxter
1940s	Skidder and chainsaw introduced
1956	Formation of the Maine chapter of The Nature Conservancy
1959	Formation of The Natural Resources Council of Maine
1975	Formation of the Sportman's Alliance of Maine
1976	Last log (pulpwood) drive in Maine (on the Kennebec)
1976–1981	Major spruce-budworm outbreak
1988	Sale of 790,000 acres by Diamond Intl. Corporation in Maine
1989	Formation of the Northern Forest Lands Study
1989	Forest Practices Act passed
1990s	The beginnings of forest certification in Maine
1998	The Nature Conservancy purchases 185,000 acres along the St. John
1999	762,192-acre Pingree lands conservation easement
2003	Liquidation harvesting law passed
2006	Katahdin Forest 241,000-acre conservation deal
2008–2011	Closure of former Great Northern paper mills in Millinocket area
2011	Plum Creek Plan passed by LURC, but tied up in courts

Note: Sources can be found through endnotes in the text.

5 | The Length and Breadth and Height of Maine

How and Why Do Forests Vary across the State?

Maine. How many images that name brings to mind! And how many of those images relate to the state's outstanding natural assets: mountains, forests, coast, wildlife, its unspoiled land and free-flowing rivers. Maine's is indeed a rich natural legacy, thanks to its geographic location where temperate and northern climates merge, its varied topography, and relatively sparse human population.

—Dean Bennett, *Maine's Natural Heritage* (1988)

Even from his coastal perspective in 1614, Captain John Smith suspected that Maine was a land "of diverse sorts of wood."[1] He was right. Owing to broad gradients of latitude, altitude, landform, and proximity to the coast, the state indeed expresses all forms of nature in great variety. Twenty-one hundred plant species, 226 bird species, 158 dragonflies and damselflies, 34 reptiles and amphibians, and much more. But these numbers are only a small part of the story. What's truly impressive about the ecology of Maine is the degree to which nature differs from one place to another, from the southwest to the far north, from the coast to the interior, from the flatlands to the mountains. Just as they have varied over time (chapters 2–4), Maine's ecosystems are highly heterogeneous over space.[2]

The scientific study of the diversity of the natural world began long ago with piecemeal descriptions of species and natural communities. This produced a bewildering panoply of named natural elements with little apparent order, at least initially. Slowly, natural historians began to glimpse patterns amid the chaos and to construct organizing frameworks based on these patterns. In the eighteenth century, Carl Linnaeus devised the first comprehensive naming system and classification that situated each species with respect to its similarity to others. In the nineteenth century, the first attempts were made to identify and classify the Earth's distinct ecological regions, expressed as follows by the German naturalist and explorer Alexander Von Humboldt: "As in all other phenomena of the physical universe, so in the distribution of organic beings: amidst the apparent disorder which seems to result from the influence of a multitude of local causes, the unchanging law of nature become evident as soon as one surveys an extensive territory."[3] By the twentieth century, detailed descriptions of natural communities and ecoregions in hand, ecologists took on the serious investigation of the underlying environmental and geographic factors responsible for such remarkable diversity in species and communities. Finally, the last three decades have witnessed the application of these three centuries of scientific study to the conservation of biological diversity, threatened by human culture during that same period.

In this chapter, we will explore the ecological diversity of Maine by reprising this narrative. We will begin with six ecosystem snapshots, descriptive and impressionistic, that illustrate the complex dimensions and textural range of ecodiversity in the state (figure 5.1). Then, we will bring order to this complexity by discussing the idea of ecological classification, the Maine Woods' place in the natural geography of North America, and the prevailing frameworks that divide up Maine into a diverse array of communities and ecoregions. Along the way, we will describe representative species in the context of their natural place in the state. The tremendous diversity revealed in these discussions will lead us to address the question, "Why is there so much ecological variety in Maine?" We will conclude the chapter by assessing the extent to which Maine's network of conservation lands adequately represents the contrasting ecoregions and natural communities across the state.

Six Maine Nature Snapshots

Kennebunk Plains: Grassland and Fire

Nancy Sferra, five feet four inches tall, is decked out in flame-retardant khaki pants and bright yellow shirt, a backpack full of fire-fighting gear, a bulky helmet, and a chattering radio attached to her chest. In her right hand is a drip torch spewing licks of flames from its narrow tip. Sferra is the *fire boss* on one of the frequent prescribed fires at Kennebunk Plains (figures 5.1 and 5.2), part of a preserve complex in southwestern Maine owned and managed by The Nature Conservancy and the Maine Department of Inland Fisheries and Wildlife. Situated on a dry outwash sandplain deposited during the last glaciation (chapter 2: "How Glaciation Shaped . . ."), 600 acres of grassland and nearly 2,000 acres of open pitch pine–scrub oak woodland support a suite of warm-region species occurring in few other places in Maine: prairie grasses, state endangered grasshopper sparrows, horned larks, upland sandpipers, cobweb skipper butterflies, state endangered black racer snakes, and many more. The largest population in the

Figure 5.1 Map of Maine showing location of each of the six nature snapshot sites.

Figure 5.2 Kennebunk Plains, showing sandplains grassland with northern blazing star. Inset: prescribed fire on the preserve.

world of the rare northern blazing star flourishes on the plains, producing a spectacle of purple blossoms in late summer (see sidebar 5.2). This species and the Sandplain Grassland are critically imperiled in the state, "s1" (see sidebar 5.1). The open structure of the grassland is intimately connected to its long interaction with humans — first, Native Americans, who might have burned the area long ago, and then twentieth-century Mainers who removed trees to create blueberry barrens and then maintained them with regular burning. Without the current prescribed fire program, trees would invade, forming woodland unsuitable for species dependent on open conditions.[4]

Sferra barks commands into her walkie-talkie and then drips a line of liquid fire along one of the preset fire boundaries. A low wall of flames works its way slowly into the prescription area, leaving behind smoking, blackened stubble. In a few hours, Sferra and her large crew will be mopping up, expunging the last sparks, soaking down the boundaries. By next summer, northern blazing star and the rest of the grassland ecosystem will resprout from vigorous root systems and return to full bloom.

Katahdin: Alpine Communities

In some ways, the top of Katahdin couldn't be more different from the Kennebunk Plains: frigid, high mountain pitches, rocks, patches with the barest

The Maine Rarity Ranking System

Conservation efforts should be focused on native species vulnerable to global extinction or regional extirpation. But how do scientists and managers keep track of the conservation status of species? In 1978, NatureServe, a nonprofit organization that addresses "the scientific basis for effective conservation action," developed a straightforward system for designating the potential risk to any given species or community type.[1] The most recent protocol uses documented ecological studies and expert advice to score ten conservation factors, grouped into three categories: *rarity* (e.g., population size, range extent, number of occurrences), *trends* (changes in populations), and *threats* (to persistence). Scores from these criteria are pooled to calculate a conservation status rank, the five levels of which are shown below for Maine. The Maine Natural Areas Program (Department of Conservation) applies the NatureServe conservation rank system to plants and natural communities, whereas the Maine Department of Inland Fisheries and Wildlife applies it to animals. Global conservation status rank for species is determined by NatureServe. In addition to these guidelines, the state has two legal categories for species rarity: "endangered," which means a high level of risk of the species being lost from the state, and "threatened," for species that with continuing decline could become endangered. Such designation for native wild animals, which technically belong to the state, mandates protective regulation (Maine statutes 12 M.R.S.A. § 12801—12810), whereas for plants, which technically belong to the landowner, designation is merely advisory and educational (5 M.R.S.A. § 13076–13079).

Maine Species and Natural Community Conservation Status Ranking[2]

Rank	Criteria
S1	Critically imperiled because of extreme rarity (5 or fewer occurrences or very few remaining individuals or acres) or because some aspect of its biology makes it especially vulnerable to extirpation from Maine.
S2	Imperiled in Maine because of rarity (6–20 occurrences or few remaining individuals or acres) or because other factors make it vulnerable to further decline.
S3	Rare in Maine (20–100 occurrences).
S4	Apparently secure in Maine.
S5	Demonstrably secure in Maine.
SU	Possibly imperiled in Maine, but status uncertain; need more information.

NOTES

1. This description of conservation status ranking is from D. Faber-Langendoen et al., *NatureServe Conservation Status Assessments: Methodology for Assigning Ranks* (Arlington, VA: NatureServe, April 2009); L. Master et al., *NatureServe Conservation Status Assessments: Factors for Assessing Extinction Risk* (Arlington, VA: NatureServe, April 2009); Susan C. Gawler, and Andrew R. Cutko, *Natural Landscapes of Maine: A Classification of Vegetated Natural Communities and Ecosystems* (Augusta: Maine Natural Areas Program, Department of Conservation, 2010), 11 and 279.

2. Faber-Langendoen et al., *NatureServe Conservation Status Assessments*; Master et al., *NatureServe Conservation Status Assessments*; Gawler and Cutko, *Natural Landscapes of Maine*, 11 and 279.

skim of soil (figures 5.1 and 5.4). But like the Kennebunk Plains, this is a treeless place that supports rare creatures, species found more commonly elsewhere — in this case, far to the north. Like other prominent isolated mountains in Maine, at the core of Mt. Katahdin is a granitic pluton, a mass of rock formed from underground magma, exposed through eons of erosion of the overlying, less resistant rock.[5] This pluton raises the mountain into alpine climate zones. Forget about growing tomatoes up here. The steep uppermost slopes, peaks, and tablelands support at least thirty alpine species and nine distinct alpine communities, many of these typically found in the arctic tundra.[6]

Lapland rosebay is a prostrate rhododendron with small bell-shaped, purple flowers that occurs in arctic and subarctic regions, is absent over the vast boreal forest, and then reappears in the highest mountains of the Northeast. In Maine, it occurs only on Katahdin and is listed as endangered in the state (and as an S1 species).[7] Growing with Lapland rosebay and sharing a similar disjunct geography is alpine bearberry (another S1), a squat shrub with small leaves that turn crimson in the fall. These two join other low-growing shrubs, herbs, and grasslike species to make up the Windswept Alpine Ridge community (an S1 community type), which occurs above tree line in New England on exposed areas of gravel among fractured rocks. Accompanying the plants is a group of northern animals, including the ground-dwelling American pipit (state endangered), with a breeding range restricted to the arctic and high mountain ranges in North America, and the Katahdin arctic butterfly (state endangered), which occurs nowhere else in the world.[8]

Northern Blazing Star
(*Liatris scariosa* var. *novae-angliae*)

In a famous essay, Stephen Jay Gould defended the use of the popular name *Brontosaurus*, rather than the technically correct *Apatosaurus*, for the giant herbivorous dinosaur that flourished during the Jurassic period, as well as in the imaginations of countless children (and adults) in modern times.[1] One could make a similar argument for the Maine native, northern blazing star, which at one time was given the moniker *Liatris borealis* (figure 5.3). *Borealis* means northern. The derivation of *Liatris* is unknown, but it's the genus for thirty-seven perennial herbaceous plants, known as blazing star or gay feather, in the composite family (Asteraceae), with grasslike leaves and tall, brilliant flowering spikes.[2] *Liatris borealis,* the northern blazing star—how appropriate! But, alas, botanists are a meticulous and law-abiding lot, and for technical reasons, there is a new name: *Liatris scariosa* var. *novae-angliae*. *Scariosa* means shriveled, not green, or membranous; *novae-angliae* is Latin for New England. So, the shriveled blazing star of New England.

Northern blazing star is rare, occurring only from New Jersey west to Pennsylvania and north to southern Maine (S1 in the state; see sidebar 5.1)—in other words, an endemic to that region.[3] There are fewer than 100 known populations in New England, six of which were extirpated in the 1980s and 1990s. It grows in open, dry woodlands, barrens, and grasslands, most commonly on coastal sand plains. Fire maintains these open conditions, but for blazing star, dependence on fire runs even deeper. Ecologist Peter Vickery, who with his wife, Barbara (director of conservation programs for The Nature Conservancy in Maine), were early promoters of protection of Kennebunk Plains, found that burning keeps in check a larval moth seed predator that otherwise would destroy most of the plant's seeds.[4] Seeds of northern blazing star in sites unburned for about two years suffered seed predation on about 90 percent of their seeds, whereas seed predation in recently burned sites was only about 17 percent. The key to maintaining this rare species is the protection of scarce coastal sand plain habitat, along with active management by fire to maintain open conditions and relatively low seed predator loads. Where such conservation occurs—such as in Kennebunk Plains—the plant returns the favor with spectacular purple displays each midsummer.

Figure 5.3 Northern blazing star (*Liatris scariosa* var. *novae-angliae*) at Kennebunk Plains.

NOTES

1. Stephen Jay Gould, *Bully for Brontosaurus* (New York: W.W. Norton, 1991), 79–93.

2. Information on northern blazing star from *Flora of North America*, vol. 21: 512, 517, 532, 534 (retrieved on February 10, 2010, at www.efloras.org/florataxon.aspx?flora_id=1&taxon_id=242416765). Information on plant names from David Gledhill, *The Names of Plants*, 3rd edition (Cambridge: Cambridge University Press, 2002), 85, 180, 207, 259.

3. Distribution and status from Ailene Kane and Johanna Schmitt, *New England Plant Conservation Program Conservation and Research Plan:* Liatris borealis *Nuttall ex MacNab (Northern Blazing Star)* (Framingham, MA: New England Wildflower Society, 2001).

4. Peter Vickery, "Effects of prescribed fire on the reproductive ecology of northern blazing star *Liatris scariosa* var. *novae-angliae*," *American Midland Naturalist* 148 (2002): 20–27.

Figure 5.4 Alpine zone on Pamola, Mount Katahdin.

The Orono Bog: Walkable Wetlands

Dan Grenier, then a student, now a preserve manager for The Nature Conservancy in Maine, was stuck in muck. We shouldn't have laughed. After all, he was carrying the $15,000 global positioning system. But there he was, teetering, mired mid-thigh, in the middle of a large bog on Great Wass Island, yelling for help, the satellite dish above his head ominously swaying. We had our laugh, Dan eventually slogged to shore, equipment intact, and our research continued.

Given that about one-quarter of the state's land area is wetlands — more than 5 million acres — getting wet and muddy is a natural hazard in Maine. Which makes the boardwalk on the 616-acre Orono Bog near Bangor very special indeed (figures 5.1 and 5.5).[9] The brainchild of Dr. Ron Davis, professor emeritus at University of Maine and a pioneering wetland ecologist, the one-mile-long boardwalk was built in 2002, forming a loop that highlights the unique natural features of these ecosystems. Bogs make up one of several types of wetlands where peat accumulates as a result of slow decomposition, stemming from acidic conditions and oxygen depletion. Because only species adapted to these stresses are able to persist, bog communities differ in structure and species composition from better-drained sites.

Figure 5.5 Orono Bog and boardwalk.

Peatlands across Maine support a wide diversity of natural communities as a result of variation in peat depth, hydrology, nutrient availability, microclimate, and other factors. At the Orono Bog, for example, Conifer Wooded Fens, Mixed Wooded Fen, Wooded Shrub Heath, and Bog Moss Lawn all prosper in different sites. Some of the species, such as red maple, also occur in drier habitats, but others are wetland dependents, such as tussock and tall cotton grasses; rose pogonia and white-fringed orchids; and the carnivorous round-leaved sundew, naked bladderwort, and pitcher plant.

Coplin Plantation Center Public Lot: Northern Hardwood Forest

Four hundred thousand acres is a lot of land to misplace, but that's what happened. In 1972, Bob Cummings, a Portland newspaper reporter, discovered that the state of Maine had forgotten that it owned hundreds of public lots, originally established in each township in 1786 by the Massachusetts General Court. After extensive legislative wrangling and land swaps, these lots eventually formed the core of the now extensive state public reserved land system, totaling more than a half million acres today.[10]

This is the backstory to a late fall hike through a spectacular northern hardwoods forest on one of the original public lots, Coplin Plantation Center in western Maine (figures 5.1 and 5.6).[11] Leading the way is Bill Haslam, forester for the Bureau of Parks and Lands, the state agency that manages these lands. Tagging along are two writers, Bob Kimber and Bill Roorbach, and one of your authors, Andrew Barton. The 562-acre public lot is located at about 1,500 feet elevation, in a tier of foothills just above soggy spruce-fir flats, in an area sometimes called the High Peaks region, boasting some of the state's most rugged mountains — Bigelow, Saddleback, East Kennebago, and others. Surrounded by heavily cut forest, this lot is carefully managed with selection harvesting to promote large timber, old-growth structure, and wildlife, but also to make money to support the bureau's operations. One can hardly tell that only three years before scattered trees were cut across the entire lot.

As we hike through the stand, we hear the soft call of chickadees, flush some partridges (ruffed grouse), and observe bear scratches on trees, but mainly we marvel at this northern hardwoods stand. Widely spaced sugar maples, some as much as three feet in diameter, stems with huge, vigorous crowns, trees well into their second century. Big yellow birches, their bark forming the gnarly plates of old timers. Solid-looking red spruces and hemlocks. Even a few robust beeches, gun-metal gray, smooth with no sign of beech bark–*Neonectria* disease. Clinging to the lower trunks of several of the largest maples are mosses and leafy lungwort, a leafy lichen common on old trees.[12] The forest floor is chock full of vegetation, especially sugar maple seedlings, which will provide the next generation of big canopy trees in this Northern Hardwood Forest community.

The St. John River Forest Preserve: Conifers and Lynx

The breakfast bell clangs at 4:30 AM at The Nature Conservancy (TNC) logging camp deep in the spruce-fir woods of far northwestern Maine, near the Quebec border. Nature Conservancy, logging camp? Yes, indeed, for this is TNC's St. John River Preserve: 47,000 acres with no timber harvesting, where natural conditions hold sway, and 133,000 acres dedicated to sustainable forestry (figures 5.1 and 5.7). Here are miles and miles of dense coniferous, hardwood, and mixed forest stands, for most of the last two centuries part of the industrial forest landscape of northern Maine. In terms of species diversity, the forests are relatively depauperate, composed mainly of balsam fir and red and black spruce on water-logged flats and hardwoods such as sugar maple, yellow birch, and beech on drier ridges.

Figure 5.6 Coplin Plantation northern hardwood forest.

Temperate tree species, common to the south and east of this region, can't tolerate the harsh climate.[13]

The river is another matter. The longest free-flowing river in the northeastern United States, the St. John supports more rare species than any other environment in Maine except Mt. Katahdin. The Riverside Seep community (s2 and globally imperiled) and five other rare river-associated communities provide vital habitat for species few Mainers will ever see: the state endangered dwarf rattlesnake root (s2), auricled twayblade (s1), Mistassini primrose (s3), the pygmy snaketail dragonfly, and the wood turtle. Furbish's lousewort, first recognized by skilled amateur botanist Kate Furbish in the 1880s and rediscovered by Professor Charles Richards of the University of Maine in the 1970s,[14] occurs along the banks of this river — and nowhere else in the world. The biological richness of these sites is the product of calcareous parent rock, which buffers the usually acidic Maine soil, as well as dramatic spring ice floes, which scour the riverbank. These disturbance-dependent species live a precarious life in which they

Figure 5.7 Spruce-fir forests in St. John River Forest Preserve.

must move (by legs or wings for animals, by seeds for plants) from sites overrun by shrubs and trees to open ones, newly created by the river.

Mt. Agamenticus: Temperate Hardwood Forests

"Check this out! Chestnut oak! Sassafrass!" I was a little embarrassed by my geeky enthusiasm. I continued anyway. "I can't believe it! Shagbark hickory! Sweet birch! Taste those twigs!" My students were thinking, "Sort of cool, but these are just trees." I tried to explain. "Here we are in Maine, surrounded by a hardwood forest from southern New England, even Virginia. Completely different from the rest of the state. It's as if someone picked up Mt. Agamenticus from hundreds of miles south and plopped it onto southern Maine." Polite smiles. Enthusiastic nodding.

Mt. A is located only about twenty miles from New Hampshire and ten miles from the Gulf of Maine (figures 5.1 and 5.8). Its 691 feet elevation is hardly impressive, but it's the only hill around, providing unhindered views of the ocean and the White Mountains, and those at sea have a similarly clear sight of it. Mt. A is so prominent that Captain John Smith featured it in his famous 1614 map of New England.[15] The mountain has undergone tremendous changes over time: a huge volcano 200 million years ago,

Figure 5.8 Mt. Agamenticus temperate hardwood forest.

scouring by repeated ice sheet advances over the past million years, life as a literal island when the sea moved inland briefly during the most recent glacial retreat.[16]

Today, Mt. Agamenticus, with its exposed slopes and thin soil warmed by southern geography, moderated by the ocean, provides sanctuary for "southern" species: chestnut oak (s1), shagbark hickory, sweet birch, flowering dogwood (s1), sassafras, mountain laurel (s2), whip-poor-will, Blanding's turtle, banded bog skimmer dragonfly, and red-winged sallow moth. At its base is one of the northernmost examples of an Atlantic White-cedar Swamp (s2), a community type found as far south as Florida. What's most amazing is to find all of these species, rare and scattered elsewhere in the state, together, constituting not just a fragment, but a fully intact temperate deciduous hardwood forest.[17]

Making Sense of the Diversity of Maine Forests

These six snapshots provide a glimpse of the remarkable natural variety of Maine: cold but also temperate, wet but also dry, cloaked in dense forests but also treeless, populated by both conifers and hardwoods, and controlled by snow and ice but in some places also by fire. The multiplicity of these snapshots also illustrates the challenge of systematically organizing this diversity, which hardly falls neatly along a single dimension such as north-south. We will use three types of ecological classifications–*biomes*, *ecoregions*, and *natural communities* — to situate Maine in the natural geography of North America and to divide up and order the ecological diversity within the state.

Biomes

The simplest type of ecological classification is the division of the biosphere into *biomes* (figure 5.9), which are huge areas characterized by a distinct climate and type of vegetation. In their ecology textbook, Begon et al.[18] identify nine of these: tundra, boreal or northern coniferous forest, temperate deciduous forest, temperate grassland, desert, chaparral (a Mediterranean-climate shrub vegetation), tropical grassland-scrub, tropical seasonal deciduous forest, and tropical rain forest.[19] As we'll see later, Maine supports a transitional forest between the boreal biome to the north and the temperate deciduous forest biome to the south.

Biome delineation relies mainly on the *physiognomy* (the physical structure) of the vegetation. Zones are defined by predominance of forests vs. grasslands vs. shrublands, coniferous vs. broadleaved forests, and evergreen vs. deciduous species. These physiognomic traits are not haphazard; they are in fact adaptations for dealing with the rigors of the particular environment in which they are found. For example, evergreen needles and pyramidal crowns of conifers in the boreal biome promote efficient use of energy in the face of short growing seasons and effective snow shedding during long, cold winters, respectively.

Perhaps the most revealing aspect of biomes is that they can be neatly ordered along global gradients of temperature and precipitation, as shown in figure 5.10. Moving both from cold to warm and from dry to moist biomes, canopy height, forest density, living biomass, leaf size, and ecosystem productivity all increase.[20] The number of species also increases along these axes for nearly all taxonomic groups.[21] The powerful implication of this graph, therefore, is that many key global ecological patterns can be understood with reference to just nine biomes and two fundamental aspects of climate.

Ecological Classifications Used within Maine

Two types of ecological classifications are applied at scales smaller than biomes, including within the borders of Maine. *Ecoregional* systems divide up

Figure 5.9 Biomes of Central and North America. Tropical grassland-scrub, mentioned in the text, is not included in this map.

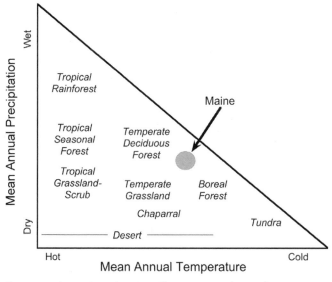

Figure 5.10 Approximate location of biomes on gradients of temperature and precipitation. Boundaries between adjacent biomes are indistinct and are not shown here. Maine occurs between the boreal forest biome and the temperature deciduous forest biome.

landscapes into blocks, each expressing a unique combination of climate, landform, soils, and predominant vegetation.[22] A common ecoregional system in Maine employs the national Bailey-United States Forest Service framework,[23] which parcels the United States into three extensive domains (Dry, Humid Temperate, Humid Tropical), each of which is divided into multiple divisions, each of which is divided into multiple provinces, and so on. The final result is a nested hierarchy of contiguous chunks of land with distinct boundaries, the smallest units of which could be traversed in a car in one or two hours or on foot in one to several days. Maine is divided into (or, rather, belongs to) three ecological provinces (see figure 5.11) and, at the lowest level, 19 "subsections," each unique in environment and biota. We'll explore these later.

Natural community classifications focus on assemblages of organisms rather than on the environmental settings that support them. These occur at smaller scales than ecoregions, and many different community types occur within each ecoregion. The *Forest Cover Type* and *Küchler Type* systems,[24] which identify forest types (e.g., red spruce, beech–sugar maple) and vegetation types (e.g., bluestem prairie, northern hardwoods), respectively, are commonly used community classifications. But these are coarse national frameworks that, for Maine, include only 20 forest cover types and 6 Küchler types, an insufficient level of detail to capture the entire spectrum of diversity. Maine has accordingly developed its own detailed natural community classification, with 104 different types (e.g., Spruce–Fir–Cinnamon Fern Forest).[25]

Classifications Are Subjective

It is important to recognize that ecological classifications are subjective human constructs of nature.[26] As such, systems developed by different scientists for the same landscape almost always differ. This is largely because classifications attempt to create discrete units for a natural world that is largely continuous, with one community or ecoregion or biome very gradually grading into another across the landscape. Although there are exceptions, the spatial or geographic distributions of different species in a community usually don't coincide; instead, they overlap broadly, even with those in neighboring communities.[27] How one divides up such continuous gradients into discontinuous parcels involves subjective judgments and a whiff of the arbitrary. In fact, the authors of the biome system described previously cautioned, "The number of biomes that are distinguished is a matter of taste." This process of identifying discrete modes for continuous

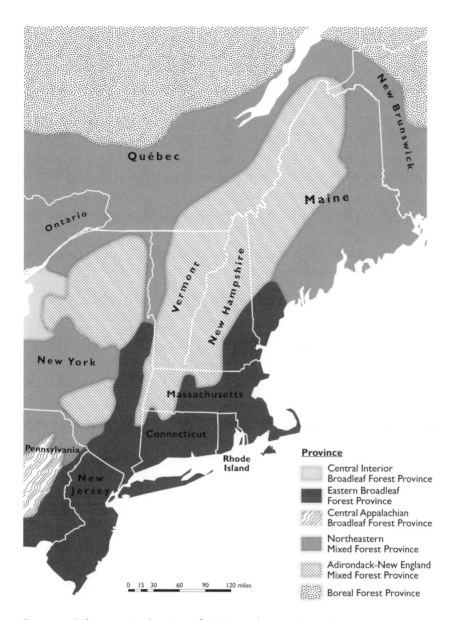

Central Interior
Broadleaf Forest Province

Eastern Broadleaf
Forest Province

Central Appalachian
Broadleaf Forest Province

Northeastern
Mixed Forest Province

Adirondack-New England
Mixed Forest Province

Boreal Forest Province

Figure 5.11 Bailey ecoregional provinces for Maine and surrounding regions.

phenomena is hardly confined to ecology — consider, for example, political philosophy, music genres, and ethnicity.

It's most useful to think of ecological classifications as tools that help humans simplify the complex — sometimes bewildering — spatial variation of nature. They have the power to reveal fundamental patterns of the natural world, as shown earlier for biomes, but they are equally crucial for protecting nature, as we will see in the final section of this chapter. According to Nancy Sferra, director of science and stewardship at The Nature Conservancy in Maine, "These classifications are essential for seeing what kinds of ecosystems are missing from our Maine and New England preserve systems, as well as for just agreeing on what we're talking about." As expressed in a recent article, the hope is that, classifications "can heighten awareness about the urgency of biodiversity loss and play an important role in conserving the extraordinary variety of life on Earth."[28]

Maine's Place in the Natural Geography of North America

How does Maine fit into the ecological diversity of North America? Is it part of the boreal biome, the temperate deciduous forest biome, both, neither? The Bailey ecoregional map,[29] shown in figure 5.11, proposes that the state's lands belong to three extensive ecological provinces. The eastern two-thirds of the state are part of a large transitional Northeastern Mixed Forest Province, north and south of the international border, stretching from the Canadian Maritimes to Lake Erie, sandwiched between the boreal and temperate forests.[30] The mountainous areas of north-central to western Maine, along with similar areas in the rest of northern New England, upstate New York, and Quebec, constitute a second, mountainous portion of this transition zone (the Adirondack–New England Mixed Forest Province). The eastern portion of these two provinces combined, from the Adirondacks to the Maritimes, is often referred to as the *Acadian Forest*,[31] in reference to the former French colonial empire of northeastern North America. Red spruce is a diagnostic species for this ecoregion; the geographic boundaries of the two are nearly coincident. Finally, southwestern Maine is mapped as a northern extension of the Eastern Broadleaf Forest Province that stretches west to Ohio and south to the Tennessee-Alabama border. The Bailey ecoregion map, then, characterizes Maine as part of a large transition zone between the two great biomes of eastern North America, except for the southwestern corner, which belongs to the temperate deciduous forest. This is not a new idea: most forest vegetation maps of Maine since Merriam's 1898 nationwide effort have emphasized the transitional nature of the state's forests.[32]

Despite the Bailey map's clarity about the transitional nature of Maine, confusion often creeps into discussions about the relationship between Maine and the true boreal forests of Canada. For example, in an excellent chapter on the boreal forest in the standard reference book, *North American Terrestrial Vegetation*, one map (the chapter cover page) includes northern Maine as part of the boreal forest, but another puts the state more than 100 miles (160 km) south of the biome.[33] This uncertainty reveals much about the essence of the Maine Woods. Forests in parts of Maine do indeed exhibit traits associated with boreal forests: cold winters with lots of snow and abundant evergreen, pyramidal-shaped, needle-bearing spruces and firs. Thus, physiognomic criteria might lead one to include these areas with northern zones. On the other hand, the most common spruce in Maine is red spruce rather than the typical boreal white or black spruce and temperate deciduous species are common even in northern Maine.[34] Furthermore, fire plays a modest role in the dynamics of the Maine forests, whereas "it is almost dogma that the boreal forest in North America is a fire-dependent forest."[35] Using floristics (species identity) and ecosystem function, therefore, one would conclude that Maine's forests are not truly boreal.[36] These contradictory attributes, of course, make sense when the state is viewed as a transition zone, possessing traits of both the temperate deciduous and boreal forests. This is not to say that the state is *merely* a mixing ground of these two great biomes. Instead, chapter 3 and this chapter reveal that the Maine Woods should be viewed as unique in its own right, part of the Acadian forest, with distinctive elements, such as red spruce, and a natural disturbance regime different from both the boreal and temperate hardwood forests.

The transitional nature of Maine's biogeography is also strongly reflected in the ranges of individual species[37] (figure 5.12). Southern Maine, for example, is chock full of deciduous hardwood species, such as chestnut and white oak, shagbark hickory, sweet birch, sassafras, tupelo, and flowering dogwood — southern species that reach their northern limits in Maine. Likewise, many tree species common in Canada, such as white spruce, balsam fir, jack pine (figure 5.13), balsam poplar, and showy mountain-ash, reach their southern boundary in the state. In fact, out of forty-seven common tree species whose distributions include Maine, a remarkable 49 percent (twenty-three) exhibit a range boundary in the state, compared to just 28 percent (twenty-four out of eighty-seven) for Kentucky and Tennessee

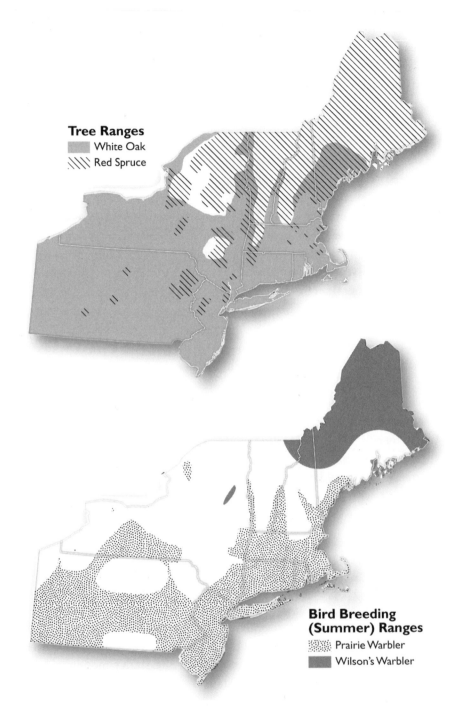

Figure 5.12 Range maps contrasting the distributions of species most common north versus south of Maine: red spruce vs. white oak and Wilson's warbler vs. prairie warbler. White color is for areas where neither species occurs.

Figure 5.13 Jack pine (top), a northern species, reaches its south-eastern range boundary in Maine. The black racer (below), a widely distributed snake, reaches part of its northern boundary in southern Maine. The ranges of the two species do not overlap.

combined, a comparable area in terms of latitudinal range.[38] This striking pattern of transition is also clearly revealed in the analysis of presettlement survey witness trees described in chapter 3 ("Land Surveys . . ."). Similar range limits occur for animal species in Maine: the eastern tiger swallow-tail, the black racer (figure 5.13), little blue heron, prairie warbler, cotton-tail rabbit, and eastern pipistrelle bat, among others, reach their northern range limit in the state; the bog fritillary, mink frog, spruce grouse, Wilson's warbler, American marten, and Canada lynx reach their southern limit in Maine.[39]

Dividing Up the State: The Ecoregions of Maine

Janet McMahon began her master's program at the University of Maine in the mid-1980s. Up to that time, conservation efforts tended to focus on rare species and communities — that is, the organisms themselves — as a way to figure out where to invest time and dollars. Find a rare species or community, raise money, buy the land, and protect it. By the 1980s, the ground had shifted: ecological research had revealed that communities were ephemeral over the long term, moving and reconfiguring in response to changing environments (chapter 2: "15,000 Years..."). Scientists argued, accordingly, that conservation efforts should also focus on protecting the variety of physical sites that support different types of communities, that is, the landscape settings that over the long term would allow communities to reform and species to evolve in response to a changing Earth.[40] Working for state conservation agencies before graduate school, McMahon experienced firsthand how the lack of such landscape concepts hindered statewide conservation efforts. She conceived her master's research immersed in these new ideas and practicalities, and her thesis, completed in 1990, became a watershed document for our understanding of biological diversity and conservation in Maine.[41] In fact, the Bailey ecoregional classification system for Maine described previously leans heavily on McMahon's work.

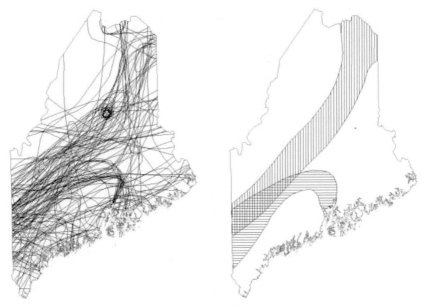

Figure 5.14 Left map shows the range boundaries of woody plant species in Maine. Right map shows two transition zones visually delineated from these boundaries, described in the text.

McMahon methodically plotted the distributions and range boundaries of nearly all of Maine's woody plant species (figure 5.14, left map). One of her key discoveries was that these boundaries are concentrated in two geographic zones (figure 5.14, right map). The most prominent zone runs along the edge of the foothills of western Maine, turns abruptly south near the Penobscot River, continuing to Penobscot Bay. Sixty-seven of the 240 (28 percent) woody plant species occurring in Maine exhibit either their northern- or southern-range boundary in this zone, which locates the transition from temperate hardwood to coniferous–northern hardwood forests. The ranges of many Maine bird and other species also correspond closely to this plant transition belt.[42] A second zone parallels the foothills of the western mountains and extends to the northern extreme of the state, running along an elevational contour of about 1,000 feet. Forty-four woody species reach their western range boundary here. These two transition zones, based apparently on species responses to environmental gradients, divide the state into three major ecological areas: southwestern temperate, mountainous, and a large area encompassing central and eastern Maine. This pattern should sound familiar, for it forms the ecological basis for and corresponds to the three provinces identified for Maine by the Bailey ecoregional system.

McMahon then subdivided these three provinces into fifteen *biophysical* regions, using differences in climate, landform, soils, and predominant vegetation. This number was increased to nineteen in the most recent (2007) Bailey classification, in part to achieve congruence with the ecoregions of other northeastern states (see appendix 5.1). The Bailey-McMahon system (or modifications of it) is the mostly widely used ecoregional framework in Maine for ecological, conservation, and management applications. For our treatment, we simplify this system by coalescing the units into seven major ecoregions (figure 5.15), following the lead of the Maine Natural Areas Program, with the goal of capturing the essence of the state's landscape diversity. Detailed descriptions are provided in sidebar 5.3 (and appendix 5.2). The contrasts among the seven ecoregions are considerable: a 2½-fold range in annual snowfall; a freeze-free span from about three months in the coldest to five months in the warmest ecoregion; prevailing physiography ranging from low-elevation flatlands to mountains; mainly acidic soils but nearly circumneutral substrates in one ecoregion; extensive peatlands in some regions but excessively drained outwash sand in one; and a twofold span in the number of woody plant species. The remarkable range of physical settings in Maine is exemplified by the contrast between the cold, snowy

Three Provinces, Seven Ecoregions of Maine

The *Northeastern Mixed Forest Province* exhibits broad climatic and ecological range, from the coast to the 1,000-foot contour in the interior, and includes four ecoregions (figure 5.15).[1] *The Aroostook Hills and Lowlands Ecoregion*, the most northern and coldest, consists of rolling hills, scattered small mountains, and lowland croplands. Northern hardwood and northern hardwood-spruce mixed forests predominate. Maine's most extensive calcareous formations occur in the lowlands, producing relatively alkaline soils, which support rare plants, such as cut-leaved anemone (s1), and communities, such as Boreal Circumneutral Open Outcrop (s2). The *Central Foothills and Eastern Lowlands* includes well-drained sites with hardwood and mixed forests, but also poorly drained soils underlain by marine clay, deposited when the ocean intruded during the most recent deglaciation. These support the state's largest and most diverse concentration of wetlands. Some, such as ribbed fens, reach their southern range limits here. The *East Coast and Interior Ecoregion* ranges from low relief to small coastal mountains (e.g., Cadillac Mountains). White pine used to be abundant on inland dry sandy outwash sites, most of which are managed blueberry barrens today. Mixed forests occur inland, whereas coastal spruce-fir forest and large peatlands predominate on the seaboard, a pattern attributed to a cool, moist growing season.[2] Because of its location between the temperate southwest, the mountains, and Down East, the *Central Maine Coastal and Interior* region is the nexus of ecological transition in Maine. Northern species (e.g., spruces) mix with temperate ones (e.g., white oak, pitch pine), which largely disappear beyond the northwestern boundary. Along the coast, the eastern coastal spruce-fir forest gives way to temperate communities, such as white pine–red oak forest. As a result of the combining of northern and southern elements, this coastal area sustains the most woody plant species (191) in Maine.

The *New England Mixed Forest Province*, bounded on the east by the 1,000-foot elevation contour, encompasses two high-elevation ecoregions (figure 5.15). The colder, snowier conditions beyond this key ecological boundary exclude many temperate species. The *Boundary Plateau and St. John Uplands Ecoregion* supports spruce-fir forest on wet lowlands and northern hardwood forest on well-drained uplands. This is the most depauperate ecoregion for woody plant species, with only about half the richness of the mid-coast. On the other hand, the riverbank along the St. John River provides conditions for rare herbaceous plant species and their associated communities. The *Central and Western Mountains Ecoregion* includes the highest peaks in the state. Red spruce and balsam fir predominate on poorly drained valley sites and at cold, higher elevations. Northern hardwoods prevail on better-drained valley locations and at middle elevations. Approaching the higher summits, the height of trees declines rapidly, and many take on a flagged appearance, the result of strong prevailing winds. Near the top, only dwarf, stunted conifers remain, a formation known as *krummholz*, from the German for "crooked wood." On the highest mountains, conditions become untenable for trees, and tundra communities of shrubs, herbs, mosses, and grasses predominate, as described for the Mt. Katahdin tablelands. Some of the rarest communities in Maine occur in these nonforested habitats above the tree line, which make up a tiny portion of the landscape.

In Maine, the *Eastern Broadleaf Forest Province* is represented by the *Southwest Interior and Coast Ecoregion* (figure 5.15), which has more affinity with areas to the south than to the rest of the state. The climate is relatively temperate, with some stations recording growing seasons twice as long and snowfall a third as much as in the mountain ecoregion. The coast is a northern extension of the broad coastal plain, characterized by sandy beaches, which runs along the Atlantic seaboard from Florida to Casco Bay. Temperate oak and pine forests, including oak-hickory and pitch pine–scrub oak, predominate. Some sites are even sufficiently warm and dry to support fires, which maintain open conditions, such as at Kennebunk Plains. Spruce-fir forests and raised bogs, which typify large areas of the rest of the state, are relatively rare here, a result of moisture stress imposed by excessively well-drained soils (e.g., sandy outwash plains) and higher temperatures, to which these northern species are poorly adapted.

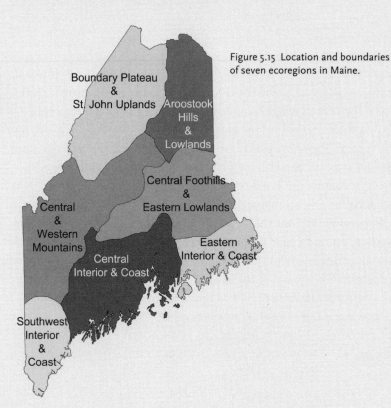

Figure 5.15 Location and boundaries of seven ecoregions in Maine.

1. Information is primarily from Janet S. McMahon, *The Biophysical Regions of Maine— Patterns in the Landscape and Vegetation* (master's thesis, University of Maine, 1990). Additional information from Dean Bennett, *Maine's Natural Heritage: Rare Species and Unique Natural Features* (Camden, ME: Down East Books, 1988); S.C. Gawler et al., *Biological Diversity of Maine: An Assessment of Status and Trends in the Terrestrial and Freshwater Landscape* (report prepared for the Maine Forest Biodiversity Project) (Augusta: Maine Natural Areas Program, Department of Conservation, 1996); William B. Krohn, Randall B. Boone, and Stephanie L. Painton, "Quantitative delineation and characterization of hierarchical biophysical regions of Maine," *Northeastern Naturalist* 6 (1999): 139–164; Ronald B. Davis and Dennis S. Anderson, "Classification and distribution of freshwater peatlands in Maine," *Northeastern Naturalist* 8 (2001): 1–50; Cleland et al., *Ecological Subregions: Sections and Subsections of the Conterminous United States*, Gen. Tech. Report WO-76 (Washington, DC: United States Department of Agriculture, Forest Service, 2007); W.H. McNab, D.T. Cleland, J.A. Freeouf, J.E. Keys, Jr., G.J. Nowacki, and C.A. Carpenter, compilers, *Description of Ecological Subregions: Sections of the Conterminous United States*, Gen. Tech. Report WO-76B (Washington, DC: U.S. Department of Agriculture, Forest Service, 2007); Susan C. Gawler, and Andrew R. Cutko, *Natural Landscapes of Maine: A Classification of Vegetated Natural Communities and Ecosystems* (Augusta: Maine Natural Areas Program, Department of Conservation, 2010).

2. Ronald B. Davis, "Spruce-fir forests of the coast of Maine," *Ecological Monographs* 36 (1966): 79–94; A.N.H. Damman, "Geographic changes in the vegetation pattern of raised bogs in the Bay of Fundy region in Maine and New Brunswick," *Vegetatio* 35 (1977): 137–151.

Boundary Plateau and St. John Uplands Ecoregion, supporting spruce-fir and northern hardwoods, and the climatically moderate Southwest Interior and Coast Ecoregion, where temperate hardwood forests predominate.

Reading through the descriptions in sidebar 5.3, one can't help but notice the resemblance between these seven modern ecoregional divisions and the four regional presettlement forests, ascertained from witness trees, described in chapter 3 ("Land Surveys . . ."). The southern oak and pine presettlement forest region maps directly onto the Southwest Interior and Coast Ecoregion; the northern presettlement mixed forest with abundant spruce, fir, and cedar corresponds to the two northern ecoregions; the coastal spruce and hemlock forests of presettlement times are the seaboard portions of the Central and Eastern Interior and Coastal ecoregions; and the central presettlement forest region dominated by beech, hemlock, and spruce maps onto the interior portions of these ecoregions and the entire Central and Western Mountains and Central Foothills and Eastern Lowlands.[43] The presettlement transition zones — between southern pine-oak and northern hardwoods in central Maine and between northern hardwoods and spruce-fir to the north — are easily recognized in the ecoregional classification. One of the main conclusions of our analyses of Maine's presettlement forests and modern ecoregions, in fact, is that, despite considerable change in climate, land use, and many other environmental factors, the distributions of species, the ecological regions, and the transition zones have all been remarkably stable from the seventeenth and eighteenth centuries to today.

The Natural Communities of Maine

The Maine natural community system identifies 104 distinct terrestrial community types, described in rich detail in the recently published 300-page guidebook, the *Natural Landscapes of Maine,* by Susan Gawler and Andy Cutko,[44] long-time ecologists at the Maine Natural Areas Program.[45] The communities are grouped into twenty-four "ecosystem types," which are combined into five "ecosystem groups": forests, freshwater shorelines, open uplands, peatlands, and tidal (see appendix 5.3).

The ecosystem snapshots at the beginning of this chapter described some of the natural communities found in Maine. Let's return to five of those locations as a way to elucidate the scale and nature of this framework. Kennebunk Plains is in the Southwestern Interior and Coast Ecoregion. The open grassland community upon which the snapshot focused is technically a Little Bluestem–Blueberry Sandplain Grassland. Little bluestem grass and blueberries are the most common species, grassland is the physiognomic type, and sandplain is the geological setting. Other community types, such as Pitch Pine–Scrub Oak Barrens (open habitat with short trees and shrubs) and Red Maple–Sensitive Fern Swamp, occur in this complex of conservation lands. The tablelands on Katahdin, in the Central and Western Mountains Ecoregion, support the Alpine Snowbank community described in the snapshot, but also many other alpine types, including Spruce-Fir Krummholz (evergreen shrubs and prostrate trees), Heath Alpine Ridge, Windswept Alpine Ridge, and Alpine Cliff. These community types, which often are only a few yards apart, differ mainly in their topographic locations and species composition.

The Coplin Center Plantation Northern Hardwood Forest, also in this mountain ecoregion, is a Beech-Birch-Maple community type. The private land surrounding this site would presumably support the same community

Figure 5.16 Four natural community types on Tumbledown Mountain. Communities and major species are as follows. *Crowberry-Bilberry Summit Bald:* red spruce, alpine bilberry, alpine blueberry, three-toothed cinquefoil; *Acidic Cliff–Gorge:* common hairgrass, Rand's goldenrod, marginal woodfern, rocktripe lichen; *Red Spruce–Mixed Conifer Woodlands:* balsam fir, red spruce, paper birch, black huckleberry, dicranum moss; *Spruce Rocky Woodland:* red spruce, reindeer lichen, mountain holly.

type, but because of years of heavy cutting, an early successional Aspen-Birch Woodland/Forest Complex community predominates. And just below the Coplin site, where drainage is poor, a mix of spruce-fir communities resides, including Spruce-Fir Wet Flat. The St. John River Forest, in the adjacent Boundary Plateau and St. John Uplands Ecoregion, shares many of these same forest communities. But it also supports rare communities associated with the river, described in the snapshot, as well as other forest types, such as Spruce-Heath Barren, a relatively dry fire-prone open woodland, and the water-logged Spruce-Larch Wooded Bog. Mt. Agamenticus takes us back to the Southwestern Ecoregion. The many temperate species found on Mt. Agamenticus combine in myriad ways, constituting a large set of community types that occur only in the southern part of the state, such as Chestnut Oak Woodland, White Oak–Red Oak Forest, Atlantic Cedar Swamp, Pitch Pine Bog, and Hemlock-Hardwood Swamp. These communities range from drier sites on the top and slopes of the mountain (oak forests) to poorly drained areas at the base (wetlands).

These descriptions and Appendix 5.3 reveal five key characteristics of the Maine natural community system. First, community types are described and delineated based chiefly on dominant species and vegetative physiognomy. Often, the difference between similar communities is a quantitative one. For example, a Beech-Birch-Maple community can have spruce but not lots of it; if spruce numbers reach 25 percent of the stand, then it's a Spruce–Northern Hardwoods community. Where possible, communities are also described according to their topographic or geological setting. Second, many communities are very small-scale, and a diverse mosaic of different community types can occupy a local area. An example from Tumbledown in western Maine is provided in figure 5.16. Third, some community types are very specific to a particular biophysical region or even a local site (for example, Kennebunk Plains and the St. John River), occurring nowhere else in Maine, whereas others (spruce-fir types, Beech-Birch-Maple) are quite common throughout the state. Fourth, wetlands are some of the most diverse types of natural areas in the state, represented by many community types: seven of the twenty-four ecosystem groups are peatlands, each including between six and thirteen communities (appendix 5.3). In fact, the original peatland natural community classification for Maine by Ron Davis and Dennis Anderson formed part of the foundation for the current larger classification.[46] Finally, whereas most of the differences from one community type to another appear to be driven by climate, landform, or soil factors, two sites identical for these can produce different communities as a result of contrasting disturbance history, either natural, or more commonly, anthropogenic, especially timber harvesting.

Why Is Maine Ecologically Diverse?

Seven ecoregions, four presettlement forest zones, nineteen biophysical areas, 104 natural communities, over 2,100 plant species, 226 bird species. To what can we attribute such ecological diversity? Why isn't Maine dominated by one type of community consisting of species optimally adapted to the state's environment? The simple answer, of course, is that conditions are far from homogeneous across the state, and this variety favors a correspondingly diverse set of species and communities, a fundamental thesis of this chapter. What is it about Maine that engenders such a diverse environmental template?[47]

The crux of the answer is Maine's geography: latitude, the ocean, elevation, and landform. First, average temperature declines with increasing latitude from the equator, and Maine spans 320 miles from its southern to northern point. This pattern results from the fact that the Sun's radiation is most concentrated at the equator and becomes increasingly diluted toward the poles. In Maine, this is accentuated by the tendency of frigid polar air masses to dip only into northern parts of the state during portions of the winter. Second, because water warms and cools more slowly than land, proximity to the ocean can buffer temperature, keeping it relatively warmer in the winter and cooler in the summer. This so-called *maritime effect* is most pronounced within about ten or twenty miles of the ocean. The famously cold Labrador Current, which flows southwesterly into the Gulf of Maine, creates a particularly steep maritime-inland temperature gradient in the state and also produces the fogs and drizzles that frequently bathe coastal lands. This coastal climate regime supports the strip of coniferous forests and unusual species described previously in this chapter and in chapter 3. A third major controller of climate is elevation, which in Maine ranges considerably, from sea level to almost a mile high (5,267 feet) at the top of Mt. Katahdin. Temperature declines 3.6 to 5.4°F per 1,000 feet, all others things being equal.[48] Elevated areas can also enhance precipitation. This is especially true in drier, hotter climates, but even in Maine, mountains in the western part of the state encourage moisture-laden warm air to rise in the summer, which can induce afternoon thunderstorms.

Latitude, maritime effects, and elevation act geographically in parallel in Maine, moving inland and northward, producing a very steep climate gradient. The coldest parts of the state exhibit about half the growing season length and three times the snowfall as the warmest (see appendix 5.2). Remarkably, the temperature range in Maine's 3° of latitude is comparable to that found over a 20° swath in Western Europe.[49] This is reflected in the remarkable range of plant hardiness zones in the state, from 3a east of Fort Kent to 6a along the southeast coast,[50] as well as in the delineation of seven distinct climate zones by former Maine State Climatologist, Gregory Zielinski.[51]

Because of varied topography and landforms, Maine climate can also vary substantially even locally. As Moses Greenleaf concluded in his 1829 survey of Maine, "The various configurations of mountains, plains, hills, and valleys, lakes and streams, which diversify the face of a country, have so important an influence on its climate, agriculture, nature."[52] Higher insolation on south facing slopes, for example, supports warmer, drier conditions than found on north slopes, and the sinking of heavy, cold air can cause considerable temperature differences over very short distances — a pattern illuminated during bragging sessions on frosty winter mornings about the lowest overnight minimum temperature.

Janet McMahon's master's thesis suggests that these four physical drivers — latitude, maritime effects, elevation, and landform — distill down to one master environment gradient, *total available energy*, that accounts for most of Maine's ecological diversity. At the "high-energy" end of this axis are southern, low-elevation, flat-land sites, warm with relatively long growing seasons and little snow, supporting temperate deciduous hardwood species and communities. At the "low-energy" end are northern, higher-elevation sites distant from the coast, cold with relatively short growing seasons and lots of snow, supporting boreal-type conifers and northern hardwoods. In between are intermediate environments, supporting transitional associations of species (such as those near the two main transition zones). Imagine multiple, parallel southeast to northwest lines crosshatching the entire state. Each one of these would represent a physical manifestation of McMahon's total available energy axis, with the "high-energy" end at the far southeast and the "low-energy" end at the far northwest end, this gradient driven by increasing latitude, increasing elevation, and decreasing maritime effects. One can experience this total energy gradient by driving on Route 27 from the ocean at Boothbay Harbor to Coburn Gore in the Boundary Mountains on the Quebec border or on I-95 from the New Hampshire border all the way to Houlton and then on Route 1 to Ft. Kent on the New Brunswick border.[53]

McMahon's analysis points to a secondary axis that also accounts for some of Maine's ecological diversity: differences in soils, produced by the great variation across the state in glacial history and underlying bedrock (described in chapter 2: "How Glaciation Shaped . . ."). The extreme end of this spectrum is represented by wet sites, with compact, deep organic soils, supporting species and communities tolerant of saturated conditions (e.g., softwoods such as tamarack and black spruce; shrubs such as mountain holly and sheep laurel). Moving away from this pole, soils become progressively better drained, with increasing representation by hardwoods. This was, in fact, a prominent pattern recognized by presettlement land surveyors staking out Maine from the seventeenth through nineteenth centuries (see chapter 3: "Land Surveys . . ."): waterlogged sites tend to support softwoods, whereas on better-drained sites hardwoods usually predominate,[54] a configuration seen clearly in the softwood flats and hardwood hills of the Boundary Plateau and St. John Uplands ecoregion. This contrast applies less to southern Maine, where red maple is the most common tree species in wet areas. Although not captured in McMahon's overall analysis, the influence of soils is also well illustrated by the calcareous soils of Aroostook County, which support plant species and communities that occur nowhere else in the state, including on more acidic soils in the same region.

A final diversifying factor in Maine is disturbance. As discussed previously, sites identical in climate and soils can support different communities as a result of contrasting disturbance history. This is a result of differences among species in their adaptations and responses to disturbances, such as wind, flooding, ice damage, and fire: some species (for example, hemlock, sugar maple, ovenbirds[55]) are vulnerable to such disturbances, whereas others (for example, pin cherry, paper and gray birch, alder flycatcher) thrive or are even dependent on the effects of these processes. In some sites, natural disturbances such as ice storms, floods, and fire promote such diversity. Much more common, however, are differences from place to place in anthropogenic disturbance. In southern Maine, the biological landscape varies as a function of whether an area has always been a woodlot or the time since the farm field was abandoned (see sidebar 4.4: "The Besse Farm . . ."). In northern Maine, ongoing timber harvesting is a dominant factor in the composition of local natural communities.

These examinations of environmental and biotic diversity, then, suggest the following conceptual model for why species, communities, and whole regions differ across the state. The fundamental geography of Maine engenders a tremendous diversity of regional climates, primarily in terms of temperature, that promotes strong regional ecological differentiation. Within each of these regions, variation in landform, soils, and disturbance lead to further finer-scale diversity in species and communities.[56] The suite of environmental factors intrinsic to each site acts as a *habitat filter*, allowing occupancy by only those species equipped to tolerate these conditions.

Putting Classification into Practice: Maine's Conservation Network

"Authorize an Ecological Reserves Program" — so recommended a 1993 report by the Maine State Planning Office to the 116th Legislature.[57] Seven years later, LD 477 was passed into law, establishing the legal status of such a system on Maine's Public Lots.[58] The mission of Maine's Ecological Reserves Program is "maintaining one or more natural community types or native ecosystem types in a natural condition and range of variation and contributing to the protection of Maine's biological diversity."[59] The law spells out limits: the ecological reserves cannot make up more than 15 percent of the total land under Department of Conservation (DOC) jurisdiction or 100,000 acres, whichever is less, and no more than 6 percent of the commercially viable timberland can be in the network. Land that is donated to the state or newly acquired specifically as ecological reserve land, however, does not count toward these totals (e.g., 5,186 acres of Mt. Abraham donated in 2007).

In 2001, the Maine Department of Conservation (DOC) designated about 70,000 acres of its land in thirteen areas as ecological reserves. Since then, the state-owned network has expanded to 101,000 acres in seventeen DOC reserves (including the new Moose River Reserve) and about 11,000 acres in reserves managed with identical goals by the Department of Inland Fisheries and Wildlife.[60] For a state of about 19.8 million acres, that is not a lot of land, representing less than 0.6 percent of the area, insufficient by a long shot to include the full range of the state's ecoregions and natural communities. Fortunately, the public network represents only a portion of lands managed primarily for conservation purposes, in a mostly natural condition, free of extractive human uses. In a 2005 report,[61] Andy Cutko and Rick Frisina, of

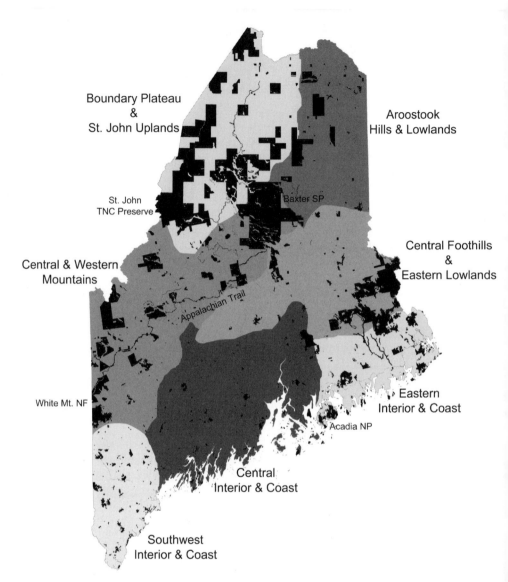

Figure 5.17 Dark blocks are lands managed primarily for conservation purposes, in a mostly natural condition. Lands include publically owned (federal, state, municipal), private (The Nature Conservancy and land trusts), and conservation easements. Lands are shown with respect to the seven ecoregions of Maine.

Table 5.1 Conservation lands of all types that do not allow logging for each of the seven major ecoregions of Maine

Ecoregion	Total Area (%)	Community Types (%)
Boundary Plateau and St. John Uplands	2.9	46
East Coast and Eastern Interior	4.7	44
Central and Western Mountains	8.1	39
Central Interior and Mid-Coast	0.8	29
South Coast and Southwestern Interior	1.0	27
Eastern Lowlands and Central Foothills	0.6	14
Aroostook Hills and Lowlands	0.9	12
Total	3	31

Note: Data are percentage of total land area in conservation and percentage of all community types protected on these lands.

Source: Andy Cutko, and Rick Frisina, *Saving All the Parts: A Conservation Vision for Maine Using Ecological Land Units* (Augusta: Maine Natural Area Program, Department of Conservation, 2005), 25–32.

the DOC's Maine Natural Areas Program, identified 679,150 acres, about 3 percent of the state, as meeting these criteria, including federal, state, The Nature Conservancy, and land trust areas (figure 5.17).[62] This compares to the goal for New England of 10 percent of the landscape by the *Wildlands and Woodlands* project and the International Union for the Conservation of Nature's benchmark of 10–15 percent.[63]

How can we objectively evaluate the extent to which this larger network protects the full range of ecological diversity in the state? And what principles should guide organizations, such as the Maine Ecological Reserves Program and The Nature Conservancy, whose raison d'être is protecting biodiversity, as they make decisions about land acquisition and management? The simple answer: use ecological classification as a template. In other words, select reserves in such a way that the network includes the complete spectrum of ecoregions and natural community types. Imagine a matrix with columns as ecoregions and rows as natural community types; each cell of this matrix represents a unique community type within an ecoregion. Some community types occur in all ecoregions; some within only a subset. Cutko and Frisina identified 412 naturally occurring combinations,[64] which in total represent the full range of landscape and natural community level diversity in the state. The goal, then, of an ecological reserves system should be to include examples — preferably multiple ones — of each of these ecological diversity units.[65]

So, how are we doing in Maine vis-à-vis this ideal, expressed by the matrix approach? According to the Cutko and Frisina report, all seven ecoregions include conservation land where harvesting is not allowed, but the percentages vary considerably, from 0.8 percent in the Central Interior and Coast to 8.1 percent in the Central and Western Mountains (table 5.1). Unfortunately, the ecoregions with the most species (southern and coastal) exhibit some of the lowest levels of protection. In the state as a whole, the Cutko and Frisina report[66] found that 70 percent of the 98 different natural community types (designated at that time) occurred two or more times, 11 percent once, and 19 percent not at all on the 679,150 acres of no-harvest conservation lands. Using the matrix approach, however, only 126 out of the 412 (31 percent) possible community type–ecoregion combinations were represented. Values varied considerably among the seven ecoregions, lowest for the Aroostook Hills and Lowlands (12 percent), highest for the Boundary Plateau and St. John Uplands (46 percent; see table 5.1). Ecoregions with a higher portion of land allocated to conservation not surprisingly tend to protect a higher percentage of naturally occurring community types (table 5.1). Most importantly, the data reveal that several ecoregions and many natural community types are strongly underrepresented, or not included at all, in the current conservation reserve network in Maine. Full coverage of these elements of ecodiversity in a natural state will require additional conservation lands.[67]

Other considerations bear on the effectiveness of the reserve network.[68] Some conservation biologists, for example, have suggested a minimum reserve size of 25,000 acres to maintain viable populations of all species and to accommodate the largest natural disturbances in New England.[69] None of the Bureau of Land's Ecological Reserves and few in the larger network are that big. Even Big Reed Forest Reserve is only 20 percent of that goal. On the other hand, most reserves occur within a landscape matrix that includes additional areas that, although subject to harvest, are managed in part with conservation goals or are legally dedicated to maintaining forestland. If these lands are added to the no-harvest areas, the total comes to about 3.7 million acres or about 17 percent of Maine. About 55 percent of this grand total is from conservation easements and 45 percent from fee ownership.[70] Effective habitat area for some species, therefore, is probably larger than that indicated by reserve size alone.[71] A key aspect of establishing an effective reserves system in Maine, then, depends on management that recognizes landscape connections among areas managed for a variety of goals by a diversity of landowners. The Maine ecoregional and natural community classification system provides a powerful tool with which to direct these conservation decisions. According to Cutko and Frisina, "The data . . . allow us to generate specific lists of natural communities not adequately protected in each region. These underlying data may be used as a 'wish list' when evaluating future conservation options to add missing natural communities to the state's portfolio of conservation lands."[72]

APPENDIX 5.1

Ecoregional Frameworks in Maine

Province	Section	Subsection (Biophysical Region)	7-Ecoregion System
Northeastern Mixed Forest	Aroostook Hills and Lowlands	Aroostook Hills	Aroostook Hills and Lowlands
Northeastern Mixed Forest	Aroostook Hills and Lowlands	Aroostook Lowlands	Aroostook Hills and Lowlands
Northeastern Mixed Forest	Maine–New Brunswick Foothills and Lowlands	Central Foothills	Central Foothills and Eastern Lowlands
Northeastern Mixed Forest	Maine–New Brunswick Foothills and Lowlands	Eastern Lowlands	Central Foothills and Eastern Lowlands
Northeastern Mixed Forest	Fundy Coastal and Interior	Eastern Coastal	East Coast and Interior
Northeastern Mixed Forest	Fundy Coastal and Interior	Eastern Interior	East Coast and Interior
Northeastern Mixed Forest	Central Maine Coastal and Embayment	Central Maine Embayment	Central Interior and Coast
Northeastern Mixed Forest	Central Maine Coastal and Embayment	Casco Bay Coast	Central Interior and Coast
Northeastern Mixed Forest	Central Maine Coastal and Embayment	Penobscot Coast	Central Interior and Coast
Adirondack–New England Mixed Forest	White Mountains	International Boundary Plateau	International Boundary Plateau and St. John Upland

Province	Section	Subsection (Biophysical Region)	7-Ecoregion System
Adirondack–New England Mixed Forest	White Mountains	St. John Upland	International Boundary Plateau and St. John Upland
Adirondack–New England Mixed Forest	White Mountains	Central Mountains	Central and Western Mountains
Adirondack–New England Mixed Forest	White Mountains	White Mountains	Central and Western Mountains
Adirondack–New England Mixed Forest	White Mountains	Mahoosuc Rangeley Lakes	Central and Western Mountains
Adirondack–New England Mixed Forest	White Mountains	Western Maine Foothills	Central and Western Mountains
Eastern Broadleaf Forest	Lower New England Section	Gulf of Maine Coastal Plain	Southwestern Interior and Coast
Eastern Broadleaf Forest	Lower New England Section	Gulf of Maine Coastal Lowland	Southwestern Interior and Coast
Eastern Broadleaf Forest	Lower New England Section	Sebago-Ossipee Hills and Plains	Southwestern Interior and Coast

Note: The first three columns show the nineteen McMahon-Bailey ecological regions (subsections) identified by province and section. The far right column shows how these units map onto the seven ecoregions described in detail in this chapter.

Sources: David T. Cleland, Jerry A. Freeouf, James E. Keys, Greg J. Nowacki, Constance A. Carpenter, W. Henry McNabb, *Ecological Subregions: Sections and Subsections of the Conterminous United States*, Gen. Tech. Report WO-76 (Washington, DC: U.S. Department of Agriculture, Forest Service, 2007); Andy Cutko and Rick Frisina, Saving all the parts: a conservation vision for Maine using ecological land units (Augusta: Maine Natural Area Program, Department of Conservation, 2005), 13.

APPENDIX 5.2
Seven Ecoregions of Maine

| | Eastern Broadleaf Forest | Northeastern Mixed Forest | | | | New England Mixed Forest | |
	Southwest Interior and Coast	Central Interior and Coast	Eastern Interior and Coast	Central Foothills and Eastern Lowlands	Aroostook Hills and Lowlands	Boundary Plateau and St. John Uplands	Central and Western Mountains
Physical Geography							
Area (million acres)	1	3.7	1.4	3.2	2.5	3.6	4.0
Prevailing elevation (feet)[a]	0–1000	0–1000	0–600	0–1000	600–1000	1000–2000	1000–4000
Landforms	coastal plain, outwash plains	rolling hills, foothills, scattered mountains	lowlands, ridges, scattered mountains	lowlands, foothills	lowlands, rolling, scattered mountains	plateau, hills, small peaks, large lakes	foothills, mountains, lakes
Climate							
Annual Freeze-free Days	155	145	142	127	104	111	111
Annual Snowfall (in.)	49	77	76	99	105	118	123
Minimum Jan Temp (°F)	14	7	8	3	-2	-2	2
Maximum July Temp (°F)	82	79	77	79	77	76	76
Annual Precipitation (in.)	46	44	47	43	39	41	48

Biota								
Küchler Types (%)[b]	Spruce-Fir	0	15	41	0	0	37	22
	Hardwood-Softwood	0	54	35	56	60	63	44
	Northern Hardwoods	69	31	24	44	40	0	34
	Appalachian Oak	31	0	0	0	0	0	0
Notable Species and Communities		*temperate:* sandplain grassland, oak-hickory, pitch pine-scrub oak, cottontail, Blanding's turtle	*major transition:* temperate, boreal, and Down East species, mudpuppy	coastal plateau peatland, coastal spruce-fir, puffins, northern water snake	greatest variety of peatlands, patterned raised bogs, *west. limit of temperate species*	eccentric bogs, rare calciphilic herbs and communities northern willows	ribbed fens, lynx, marten, *temperate species absent*	rare alpine species and communities Bicknell thrush, northern bog lemming, *temperate species absent*

[a] Scattered peaks reach elevations higher than indicated.

[b] Küchler types refers to the national classification of vegetation types.

Note: Ecoregion locations and boundaries in figure 5.15; text in sidebar 5.3.

Sources: Janet S. McMahon, *The Biophysical Regions of Maine — Patterns in the Landscape and Vegetation* (master's thesis, University of Maine, 1990); David T. Cleland, Jerry A. Freeouf, James E. Keys, Greg J. Nowacki, Constance A. Carpenter, W. Henry McNabb, *Ecological Subregions: Sections and Subsections of the Conterminous United States*, Gen. Tech. Report wo-76 (Washington, DC: US Dept. of Agriculture, Forest Service, 2007); Randall B. Boone and William B. Krohn, *Maine Gap Analysis Vertebrate Data — Parts I & II* (Orono: Maine Coop. Fish and Wild. Res. Unit, University of Maine, 1998).

The Natural Community System of Maine, Developed by the Maine Natural Areas Program, Categorized by Ecosystem Group and State Rarity Status

Community Type	State Status
Ecosystem Group: Forests	
Alder Shrub Thicket	S5
Aspen-Birch Woodland/Forest Complex	S5
Atlantic White Cedar Swamp	S2
Beech-Birch-Maple Forest	S5
Black Spruce Woodland	S2
Bluejoint Meadow	S3
Cedar-Spruce Seepage Forest	S4
Chestnut Oak Woodland	S1
Fir–Heart-Leaved Birch Subalpine Forest	S3
Hardwood Seepage Forest	S3
Hemlock-Hardwood Pocket Swamp	S2
Hemlock Forest	S4
Ironwood-Oak-Ash Woodland Forest	S3
Jack Pine Forest	S1
Jack Pine Woodland	S3
Labrador Tea Talus Dwarf-Shrubland	S2
Maple-Basswood-Ash Forest	S3
Maritime Spruce-Fir Forest	S4
Mixed Graminoid–Shrub Marsh	S5
Northern White Cedar Swamp	S4
Oak-Pine Forest	S4
Oak-Pine Woodland	S4
Pitch Pine Bog	S2
Pitch Pine Woodland	S3

Community Type	State Status
Red Maple–Sensitive Fern Swamp	S4
Red Oak–Northern Hardwoods–White Pine Forest	S4
Red Pine–White Pine Forest	S3
Red Pine Woodland	S3
Red Spruce–Mixed Conifer Woodland	S4
Spruce-Fir-Broom-moss Forest	S5
Spruce–Fir–Cinnamon Fern Forest	S4
Spruce–Fir–Wood Sorrel–Feather Moss Forest	S4
Spruce–Northern Hardwoods Forest	S5
Spruce Heath Barren	S2
Spruce Talus Woodland	S4
Tussock Sedge Meadow	S4
White Cedar Woodland	S2
White Oak–Red Oak Forest	S3
White Pine–Mixed Conifer Forest	S4
Ecosystem Group: Freshwater Shorelines	
Alder Shrub Thicket	S5
Bluebell–Balsam Ragwort Shoreline	S3
Bluejoint Meadow	S3
Bulrush Bed	S4
Cattail Marsh	S5
Cattail Marsh	S5
Circumneutral–Alkaline Water Macrophyte Suite	S2
Circumneutral Riverside Seep	S2
Dogwood-Willow Shoreline Thicket	S2
Hardwood River Terrace Forest	S2

Appendix 5.3 (*continued*)

Community Type	State Status
Lakeshore Beach	S4
Mixed Graminoid–Shrub Marsh	S5
Mixed Tall Sedge Fen	S4
Northern White Cedar Swamp	S4
Northern White Cedar Woodland Fen	S4
Pickerelweed-Macrophyte Aquatic Bed	S5
Pipewort–Water Lobelia Aquatic Bed	S5
Red Maple–Sensitive Fern Swamp	S4
Red Maple Wooded Fen	S4
Sand Cherry–Tufted Hairgrass River Beach	S2
Silver Maple Floodplain Forest	S3
Sweetgale Mixed Shrub Fen	S4
Three-way Sedge — Goldenrod Outwash Plain Pondshore	S1
Tussock Sedge Meadow	S4
Water Lily–Macrophyte Aquatic Bed	S5
Ecosystem Group: Open Uplands	
Acidic Cliff–Gorge	S4
Alpine Cliff	S1
Bilberry–Mountain Heath Alpine Snowbank	S1
Birch–Oak Talus Woodland	S3
Blueberry-Lichen Barren	S2
Boreal Circumneutral Open Outcrop	S2
Cotton Grass–Heath Alpine Bog	S2
Crowberry-Bilberry Summit Bald	S3
Diapensia Alpine Ridge	S1
Dwarf Heath–Graminoid Alpine Ridge	S2
Heath-Lichen Subalpine Slope Bog	S1

Community Type	State Status
Little Bluestem–Blueberry Sandplain Grassland	S1
Mountain Alder–Bush Honeysuckle Subalpine Meadow	S1
Pitch Pine–Heath Barren	S1
Pitch Pine–Scrub Oak Barren	S2
Rocky Summit Heath	S4
Spruce-Fir-Birch Krummholz	S3
Spruce-Heath Barren	S2
Spruce Talus Woodland	S4
Three-Toothed Cinquefoil–Blueberry Low Summit Bald	S3
Ecosystem Group: Peatlands	
Atlantic White Cedar Bog	S1
Bog Moss Lawn	S4
Deer-hair Sedge Bog Lawn	S3
Huckleberry-Crowberry Bog	S3
Leatherleaf Boggy Fen	S4
Low Sedge–Buckbean Fen Lawn	S3
Mixed Tall Sedge Fen	S4
Mountain Holly–Alder Woodland Fen	S4
Northern White Cedar Woodland Fen	S4
Pitch Pine Bog	S2
Red Maple Wooded Fen	S4
Sedge–Leatherleaf Fen Lawn	S4
Sheep Laurel Dwarf Shrub Bog	S4
Shrubby Cinquefoil–Sedge Circumneutral Fen	S2
Spruce-Larch Wooded Bog	S4
Sweetgale Mixed Shrub Fen	S4

Appendix 5.3 (*continued*)

Community Type	State Status
Ecosystem Group: Tidal	
Beach Strand	S4
Brackish Tidal Marsh	S3
Crowberry-Bayberry Headland	S3
Dune Grassland	S2
Freshwater Tidal Marsh	S2
Heath-Crowberry Maritime Slope Bog	S2
Mixed Graminoid–Forb Saltmarsh	S4
Pitch Pine Dune Woodland	S1
Rose-Bayberry Maritime Shrubland	S4
Seaside Goldenrod–Goosetongue Open Headland	S4
Spartina Saltmarsh	S3

Note: Some community types occur in multiple ecosystem groups. Definitions of state status ranks are given in sidebar 5.1. *Source:* Susan Gawler and Andrew Cutko, *Natural Landscapes of Maine: A Guide to Natural Communities and Ecosystems* (Augusta: Maine Natural Areas Program, Department of Conservation, 2010), 20–40.

6 | The Future of the Maine Woods

What Will Maine Forests Be Like in the Year 2100?

Change is an inherent characteristic of all landscapes and future change is
inevitable.— David R. Foster and Glenn Motzkin, *Northeastern Naturalist* (1998)

View from a Maine Woodlot

Janet McMahon and Chris Davis have lived on their 100 acres in Maine
since 1983. Their land stretches from the shoreline of the inner reaches of
Muscongus Bay to nearly a mile inland, encompassing a wide range of natural communities and habitats. Most of it is woods, but there are also fields,
an orchard, stonewalls, tidal flats, and, of course, their house, barn, and
lawn. The ecology of their land has changed considerably since they settled
here. The forest used to be mainly balsam fir and red spruce. Fir has been
decimated by an aphid-like insect, the balsam woolly adelgid, accidentally
introduced from Europe in the early 1900s, and spruce has experienced a
slow decline. Red oak and ash are now the major tree species. Alien honeysuckle has multiplied, its seeds and roots apparently spread inadvertently
along the dirt road by snowplowing. Moss is creeping across the lawn and
roof, probably a response to the increased rainfall over the past couple of
decades. This has also led to conversion of trickling temporary streams
to permanent ones. New animals reside in the woods. Janet and Chris
regularly hear the unmistakable repeating call of the northern cardinal,

an uncommon presence until recently. Wild turkeys, extirpated from the state in the early 1800s and reintroduced in the late 1970s, have become well established. Opossums, in their inexorable march north, have crossed the Kennebec River into the area. The absence of ticks used to be one of the great benefits of living in this region. Now, they are common.

Accompanying these biological changes, the seasons have shifted. In her work as an ecological consultant, McMahon regularly carries out aerial photography. She used to fly in the third or fourth week of May, when individual tree species could be distinguished by the timing of leaf out and the characteristic hue and texture of their spring foliage. She now must be in the air one or two weeks earlier to photograph the landscape at the right time. The same goes for the fall. It used be that McMahon would have to put her tomatoes on the porch to complete their ripening; they now ripen without such nurturing. At the nearby Watershed High School in Rockland, where she teaches science courses, McMahon had students interview residents older than seventy about how the local environment had changed. The most common observations: wetter, less snow, Japanese beetles, earlier planting, and ripe tomatoes.

The Future: Changing Environments, Changing Forests

The McMahon-Davis woodlot is a microcosm of the changing nature of the Maine Woods. The underlying physical environment and the associated biota have shifted in the last several decades. According to a wide range of models, these changes will accelerate during the rest of the twenty-first century. The causes of these transformations are complex; the driving forces do not act in isolation, but instead interact in ways that can amplify the impacts. Our understanding of these dynamics has progressed considerably, but key uncertainties remain. The goal of this final chapter is to look forward and develop probable scenarios for the future forests of Maine based on the best science available today.

We evaluate four sets of pressures that are most apt to recast the state's ecosystems: acid rain, invasive exotic pests, land use, and climate change. These forces act on species and ecosystems by altering the fundamental factors that determine how well individual organisms survive, grow, and reproduce. Such alterations will favor a different suite of species and communities than occur in the state today. Acid rain affects the levels of nutrients and deleterious chemicals in precipitation and soil. Some of the newly arriving alien plants, insects, and diseases in the state become potent competitors, predators, and pathogens, respectively, of native herbaceous plants and trees. Land use modifies the frequency, intensity, size, and type of disturbances, which alters the availability and distribution of resources required by organisms. Climate change shifts the temperature and moisture regimes governing the success of organisms and the set of species capable of persisting within a site. These four sets of forces, then, portend fundamental alteration of the ecology and character of the Maine Woods.

Forecasting is a difficult and tricky job, and we do not take it lightly. One of the most common quotations accompanying ecological prognostications these days, attributed to the physicist Niels Bohr, is that "Prediction is very difficult, especially about the future."[1] Think about it. The future trajectories of the four pressures described above will be strongly dependent on a diverse suite of not just state, not just federal, but global socioeconomic, political, and technological factors (e.g., the mix of energy sources). The impacts on Maine forests will be the result of complex causal chains, initiated by those global factors, translated as changes in the physical environment that supports organisms, and finally expressed as alterations in the species and ecosystems in the state. Fortunately, ecology and related disciplines have advanced tremendously in their ability to use what we know about the past and present to look into the future. Cautious use of these tools developed by these sciences can provide a guide to the possible future forests of Maine.

Changing Nutrient Cycles:
Acid Rain and Nitrogen Saturation

In the 1980s, *acid rain* was frequently cited as one of the most serious threats to the forests of New England. Is it a problem in Maine? Will it affect Maine's future forests? First documented as a serious issue in North America in the early 1970s at Hubbard Brook Experimental Forest in New Hampshire (less than forty miles from the Maine border),[2] acid rain is the popular term given to *acid deposition* from the atmosphere to the Earth's surface, through rain, snow, fog, and dry particles. The primary sources of acidification are sulfur dioxide (SO_2) and nitrogen oxides $(NO$ and $NO_2)$, by-products of the combustion of fossil fuels in power plants, factories, and vehicles. These gases rise into the atmosphere, where they are chemically transformed into sulfuric acid (H_2SO_4) and nitric acid (HNO_3) and transported, sometimes long distances, before deposition. The predominant source of acid rain in New England, for example, is emissions in the U.S.

Modern Ecology: Studies of Nitrogen Saturation in New England

Ecologists initiated an ambitious research project in the 1980s, pioneering two new approaches to investigating nitrogen saturation in northeastern forests. First, soil and tree attributes were investigated at 161 spruce-fir sites from Maine to upstate New York, a transect along which nitrogen deposition increased. Second, scientists set up nitrogen addition experiments in red pine, hardwood, and spruce-fir forests at Harvard Forest in Massachusetts, Mt. Ascutney in Vermont, and Bear Brook Watershed in eastern Maine. The results of these studies have reinforced the findings of studies over long periods of time at a variety of sites in the Northeast. Transect sites with higher atmospheric deposition and plots with added nitrogen had enhanced availability of soil nitrogen and aluminum and decreased supply of calcium and magnesium. Researchers expected negative impacts of experimental nitrogen saturation to take several decades, but actually found reduced tree growth and increased mortality during the first ten years. These results corroborate the role of acidic deposition in the decline of high-elevation spruce-fir forests in New England and Upstate New York. This initiative is representative of ecological science in the twenty-first century: interdisciplinary teams of researchers, large geographic scales, factoring in of land use history, and complementary observational and experimental studies of key environmental variables.[1]

NOTES

1. From J.D. Aber et al., "Nitrogen saturation in northern forest ecosystems," *BioScience* (1989) 39: 378–386; Stephen A. Norton and Ivan J. Fernandez, eds., *The Bear Brook Watershed in Maine: A Paired Watershed Experiment—the First Decade (1987–1997)* (Boston: Kluwer Academic Publishers, 1999); John Aber et al., "Nitrogen saturation in temperate forest ecosystems," *BioScience* (1998) 48: 921–934; David R. Foster and John D. Aber, eds., *Forests in Time: The Environmental Consequences of 1,000 Years of Change in New England* (New Haven, CT: Yale University Press, 2004), 259–279.

Midwest, which travel on prevailing winds and frontal systems. The result is precipitation with levels of acidity as much as 100 times normal (or two pH units).[3]

Hubbard Brook Experimental Forest researcher Charles Driscoll and colleagues summarized the wide array of impacts of acid deposition on ecosystems identified through four decades of research: acidification of soils and water bodies, altered availability of soil nutrients, increased concentrations of toxic chemicals, and damage to organisms and food webs. A more recent concern related to acid rain is the dramatic increase in deposition of nitrogen onto ecosystems as a result of human activities. Nitrogen controls the productivity of many of the forests in eastern North America. Although additional supply can initially boost growth, high levels of nitrogen over the long term can severely damage ecosystems — a process known as *nitrogen saturation* (sidebar 6.1).[4]

The effects of acid deposition and nitrogen saturation on forests are complex, involving both direct impacts on foliage and indirect effects mediated through soils. These causal pathways have been extensively investigated for sugar maple, which has suffered chronic mortality in parts of Pennsylvania, and red spruce, which has exhibited instances of *dieback* (groups of neighboring trees suffering reduced growth, loss of foliage, or mortality) in high elevations in eastern North America. In susceptible soils, acid deposition lowers pH, which can cause leaching of positively charged nutrients (*base cations*), especially calcium (Ca^{++}), to levels below the requirements of forest trees. In contrast, soil concentrations of nitrogen, aluminum, and sulfur can rise to levels well above vegetation demand. The presence of aluminum in the rooting zone inhibits the ability of fine roots to take up water and nutrients. A portion of the excess of all three of these compounds is leached out of soil into streams, lakes, and estuaries, where it can poison organisms (in the case of aluminum) or lead to *eutrophication* (in the case of nitrogen), a process that promotes excessive growth of algae, leading to oxygen depletion and mortality of fish and other species.[5]

As shown in figure 6.1, these complex soil processes are thought to act as part of a *multiple-stress syndrome*, that is, as one of several interacting stresses that together lead to decline. Acidification leaches calcium directly out of cell membranes in needles, which, in combination with soil alterations, decreases cold tolerance, leading to foliage freezing damage, but also vulnerability to other stresses, such as insect attack. A major ongoing challenge is to tease apart these complex interactions so that we can distinguish anthropogenic causes of forest damage from ones attributable to natural agents.[6]

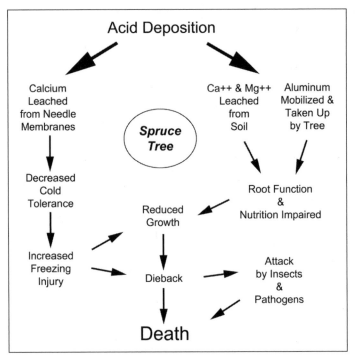

Fig 6.1 Flow diagram of the multiple stresses leading to dieback in spruce, including acid deposition. Ca⁺⁺: calcium cations; Mg⁺⁺: magnesium cations.

And there's the rub for Maine: unlike for other parts of northern New England and upstate New York, there's no conclusive evidence directly linking acid deposition and forest dieback. This makes some sense, for, as mentioned in sidebar 6.1, the levels of acid deposition in Maine are substantially lower than for the rest of the northeastern United States, even Hubbard Brook in New Hampshire.[7] On the other hand, field studies suggest that acid deposition has chemically altered forest soils (and streams and lakes) in the state. In a recent review of soil calcium status, Tom Huntington, at the U.S. Geological Survey office in Augusta, concluded that "Ca depletion is a realistic concern in Maine."[8] If deposition levels were to rise, Maine forests could well develop associated dieback problems. Even current levels over the long term could cause as yet undetected physiological stress. The nitrogen addition experiments (sidebar 6.1) carried out at Bear Brook by researchers from the University of Maine and the U.S. Forest Service clearly demonstrate the potential negative impacts of acid rain and nitrogen saturation on the soils and trees of Maine.[9]

What is the outlook for acid rain and associated forest dieback? The U.S.

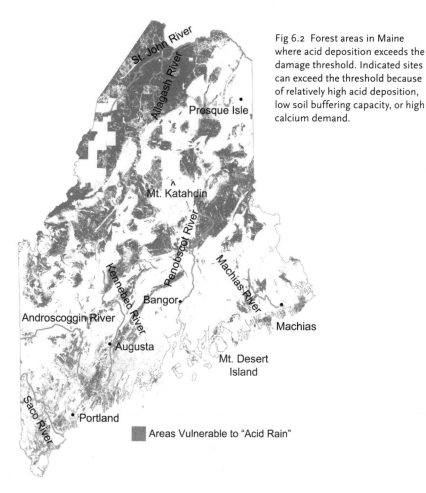

Fig 6.2 Forest areas in Maine where acid deposition exceeds the damage threshold. Indicated sites can exceed the threshold because of relatively high acid deposition, low soil buffering capacity, or high calcium demand.

Areas Vulnerable to "Acid Rain"

Clean Air Act of 1970 and its 1990 amendments have led to substantial declines in Maine and the rest of the eastern United States in deposition of sulfates but not nitrogen, a result of stringent caps on the former but not the latter.[10] Evidence for ecosystem recovery is mixed. From 1990 to 2000, sulfate levels dropped considerably and water pH rose slightly in streams at Hubbard Brook Experimental Forest and lakes across New England. On the other hand, continued declines were detected in calcium and the capacity of the soil to neutralize acidity, key indicators of chemical recovery. Ecosystem nitrate levels did not decrease during this time period. There is little evidence as yet for widespread biological recovery from damage linked to acid rain in the Northeast.[11] Although the acidity of precipitation in Maine has decreased slightly over the past twenty years, it remains more than ten times stronger than in unpolluted areas or preindustrial times. In addition, many soil effects are cumulative and the leaching of nutrients or decrease in acid neutralizing capacity will take a long time to reverse.

A recent report by Eric Miller for the Conference of New England Governors and Eastern Canadian Premiers concludes that sensitivity to acid deposition varies across Maine (figure 6.2).[12] Sites underlain by calcium-rich bedrock (e.g., limestone sites in northeastern Maine), for example, constantly produce large amount of calcium cations and thus have a high capacity to neutralize acids (much like antacids do for upset stomachs), whereas soils underlain by granite generally exhibit low buffering capacity. As a result, locations differ in the threshold of acid deposition that could cause ecosystem damage. The report concludes that 36 percent of Maine forests exceeded this damage threshold during 1999 to 2003. Compromised sites were concentrated in areas with high levels of deposition (e.g., high elevations), low levels of buffering capacity, and high levels of vegetation demand for base cations (sites recovering from intensive harvesting or with lots of sugar maple). Field studies have shown that trees growing in areas that exceed the damage threshold grow more slowly and exhibit higher levels of crown damage than those in forests below the threshold, although this work took place in Quebec, not Maine.[13]

Will acid deposition, nitrogen saturation, and the multiple-stress syndrome to which they are linked continue to alter northeastern forests or will they abate? Might they begin to affect Maine's trees? *It depends* — on the degree to which emissions, especially nitrates, change in the future. Researchers at Hubbard Brook have projected three scenarios for the New England of 2050: (1) no appreciable chemical recovery under the status quo; (2) improvement but not complete recovery with a 55 percent reduction in sulfate and 20 percent for nitrogen; and (3) chemical conditions approaching preindustrial levels with 75 percent and 30 percent decreases, respectively. These results span a wide range of possibilities, strongly dependent on regulation. The Clean Air Cross-State Air Pollution Rule, finalized by the U.S. Environmental Protection Agency in July 2011, mandates 71 percent reduction for sulfates and 52 percent for nitrogen by 2015 (from the 2005 baseline) for the eastern United States. These percentages would clear more than half of the vulnerable areas in the sensitivity map of Maine shown above and are in the ballpark of the ambitious scenario of Driscoll et al., raising the possibility of long-term recovery. These authors sum up the current and future status of acid rain in New England as follows: "Given the accumulation of acids and loss of buffering capacity in the soil, many areas are now more sensitive to acidic deposition and have developed an inertia that will delay recovery. Nevertheless, calculations from computer models show that deeper emission cuts will lead to greater and faster recovery from acidic deposition."[14] There is realistic hope, then, that the impacts of acid rain will decline in the Northeast and never become a serious forest problem in Maine.[15]

Changing Enemies:
Invasive Exotic Insects, Diseases, and Plants

The media alert came on May 13, 2010: a newfound infestation of hemlock woolly adelgid in Harpswell, a peninsula that juts into northern Casco Bay, far from the closest previous hotspot in coastal York County. "There's never been this big a jump in a natural spread," according to Maine Forest Service entomologist Allison Kanoti. "It's not surprising, but it is discouraging."[16] This tiny aphid-like insect clothes itself in a white fuzz, looking not the least bit malign. But, in fact, hemlock woolly adelgid can kill a hemlock in several years. In the southern Appalachians, entire forests of hemlocks have been wiped out, leaving hulking dead trees and open canopies.

Off the coast of Maine, not far from Harpswell, lies Monhegan Island — a place with a long history as an art colony and a reverence for nature. But the understory of parts of the island are now "too dense to permit passage" because of thorny thickets of the alien ornamental shrub, Japanese barberry. Native species grow poorly under these invaders. Barberry's seeds are now freely dispersed around the island by fruit-eating birds. Eradication requires laborious cutting and herbicide application.[17]

These are but two examples of *invasive exotics*, a group of organisms that have wrought ecological and economic damage elsewhere in the world,[18] and have the potential to alter the forests of Maine. For terrestrial ecosystems, we can neatly divide these into alien insects and diseases that use trees as hosts and exotic plants that compete with native ones. The challenge posed by invasives stems from three factors. First, because they did not evolve with these newcomers, native species are often ill-equipped for resisting them. Second, many of the aliens that become invasive are difficult to control because of prolific reproduction and rapid dispersal. Unfortunately, detection can be difficult, because most are inconspicuous, becoming obvious only after their populations have erupted. Finally, unlike single-event disturbances, such as fire or even tree harvesting, the action of invasive exotics tends to be persistent and long-term.[19]

There is truly a rogue's gallery of *exotic insects and diseases* threaten-

Fig 6.3 In clockwise order, starting with the upper-left frame: hemlock woolly adelgid, emerald ash borer, ooze from sudden oak death pathogen, and adult and larval Asian longhorned beetle.

ing the Maine Woods in the twenty-first century: brown spruce longhorn beetle, elongate hemlock scale, white pine blister rust, gypsy moth, browntail moth, hemlock woolly adelgid, Asian longhorned beetle, emerald ash borer, and sudden oak death (ramorum blight). In chapter 4 ("1920–1970 . . ."), we described the impacts on Maine forests of three virulent diseases: the chestnut blight, Dutch elm disease, and beech bark–scale insect–*Neonectria* disease. Some of the exotic insects and pathogenic fungi on Maine's horizon are every bit as lethal, and their target hosts include some of the state's major forest tree species.

The Asian longhorned beetle (ALB) is an impressive insect: large, jet-black with white splotches, and long antennae (figure 6.3). Female ALBs chew pits into tree trunks, where they deposit eggs; the larvae hatching from these eggs feed on the wood, devastating some of the major species of northern hardwood forests, including maples and birches. Since 1996, they have been found in New York City, New Jersey, Toronto, and Chicago. In 2008, a local resident in Worcester, Massachusetts, discovered a major ALB outbreak, which has led to the removal of more than 29,000 trees. In July 2010, ALBs were discovered in Boston, less than sixty miles from the Maine border. The most recent study, by Kevin Dodds and David Orwig of Harvard Forest, suggests that ALBs could spread to Maine, using red maple as a stepping stone host tree species.[20]

The emerald ash borer (EAB; figure 6.3), a beetle native to Asia, was first identified in southeastern Michigan in 2002. As of 2011, it had spread as far as east as Pennsylvania and New York. EABs kill most of the ash trees they attack—more than forty million so far. The Maine Indian Basketmakers Alliance is already preparing for the possible demise of brown ash, the chief species used for baskets. According to their executive director, Theresa Secord, the Alliance is developing plans in collaboration with tribal and other scientists for detection, prevention, and even for collection and storage of seeds for future replanting.[21]

Hemlock woolly adelgid (figure 6.3) is technically no longer just on Maine's horizon, having already infested more than 10,000 acres in southern coastal areas. Probably introduced from Japanese nursery stock into Virginia in the 1950s, it has spread across much of the distribution of its two hosts, eastern and Carolina hemlock. The tiny creatures feed en mass on the tree's phloem, stealing sugars produced by photosynthesis. Hemlock makes up about 5 percent of the state's forest inventory. A 2007 Maine Forest Service report predicted eventual infestation of the entire hemlock range in the state,[22] but recent assessments indicate that actual hemlock decline will be restricted to coastal areas.[23]

Table 6.1 shows that *invasive exotic plants* in Maine cover the gamut of growth forms, geographic origins, means of reproduction, and ecosystems. Two species are particularly notable in the context of forests. The Asiatic bittersweet vine, whose berries are a staple of holiday wreaths, is often referred to as the kudzu of the north, for its impressive ability to rapidly grow over and entangle entire stands of trees, killing them in the process (figure 6.4). Starting as an apparently innocuous ornamental in nineteenth century North America, it has since overwhelmed its native cousin.[24] Norway maple (figure 6.4), the most popular nursery tree in North America, may

The Browntail Moth: Human Scourge

According to entomologist Ephraim Porter Felt's 1906 report, "The stiff hairs of the brown tail moth caterpillar are barbed and, falling from their bodies or blowing from the empty cocoons in midsummer, may cause . . . more suffering than . . . the sting of bees or hornets. This irritation is so great . . . as to cause serious illness. . . . It has been estimated that the rental value of property in the worst infested sections about Boston has been reduced from 20% to 50% on account of the caterpillar plague."[1] Accidentally introduced from Europe to the Boston area in the 1890s, *Euproctis chrysorrhoea* spread rapidly through the eastern United States and Canada, but by the 1970s had enigmatically contracted to Cape Cod and several Maine islands. Within the last decade, the moth has spread back to mainland Maine, and has again become a serious health hazard, causing not just dermatitis but, in the worst cases, serious respiratory illness.

Why did the browntail moth range expand and contract so dramatically? What limits its range? Will it spread again? A clue to the second question comes from the scientific efforts of the Hawthorn Elementary School in Brunswick, Maine. After the cold winter of 2003, students documented high overwintering survival in caterpillars close to the ocean, with a minimum temperature of −12°F, but complete mortality five miles inland, where temperature had dipped to −18°F. Similar results were found in a recent study across the moth's range, suggesting that cold constrains the range of the browntail moth to the temperate coast. Case closed.

Or maybe not. In 1906, *Compsilura concinnata*, a European tachinid fly, was released in the United States to control outbreaks of Eurasian gypsy moth caterpillars. It did a poor job of that but went on to become a parasitoid (a parasite that kills its host) of 180 other North American butterflies and moths. Among those are

pose the most serious threat of all, although it has made only modest inroads around some towns and cities in Maine. It has been an aggressive invader in some forest stands to the south of Maine, supplanting sugar maple and reducing native plant diversity on the forest floor by combining abundant understory establishment, a high degree of shade tolerance, fast growth, and full canopy size. A recent simulation study suggests that Norway maple has the potential to become a constituent of northern hardwood forests.[25]

the giant native silk moths, several species of which have been extirpated or drastically reduced, apparently due to this tachinid fly—a prime example of why it is a bad idea to employ generalists as biological control agents. This parasitoid also likes the browntail moth. Joseph Elkinton and colleagues have recently revealed a key role for the introduced fly in the dramatic range contraction of the moth in the early twentieth century. In one experiment, the fly was found to be uncommon in moth colonies placed on the coast, whereas inland ones were decimated by the parasitoid. Although temperature may refine the current distributional boundaries of the browntail moth, the tachinid fly appears to be the primary agent controlling moth populations. How climate change might alter the range of each species and their relationship remains to be seen.[2]

NOTES

1. Ephraim Porter Felt, "The gipsy and brown tail moths," *New York State Museum Bulletin* 103, *Entomology* 25 (1906), 6.

2. Information on the browntail moth comes from Felt, "The gipsy and brown tail moths," 6; U.S. Forest Service, *Pest Alert: Browntail Moth* (Durham, NH: United States Department of Agriculture Forest Service, Northeastern Area, NA–PR–04–02); Maine Forest Service, *Browntail Moth*, Euproctis chrysorrhoea *(L.)* (Augusta: Maine Forest Service, Department of Conservation Insect and Disease Laboratory, 2008); Alex Lear, "Forest service warns of browntail moth caterpillar infestation in mid-coast," *The Forecaster* (Portland, ME), May 12, 2010; Joseph S. Elkinton et al., "Factors influencing larval survival of the invasive browntail moth (Lepidoptera: Lymantriidae) in relict North American populations," *Environmental Entomology* 37 (2008): 1429–1437; Joseph S. Elkinton, Dylan Parry, and George H. Boettner, "Implicating an introduced generalist parasitoid in the invasive browntail moth's enigmatic demise," *Ecology* 87 (2006): 2664–2672; J.S. Elkinton and G.H. Boettner, "The effects of *Compsilura concinnata*, an introduced generalist tachinid, on non-target species in North America: a cautionary tale," in R.G. Van Driesche and R. Reardon, eds., *Assessing Host Ranges for Parasitoids and Predators used for Classical Biological Control: A Guide to Best Practice* (Morgantown, WV: United States Department of Agriculture Forest Health Technology Enterprise Team, 2004), 4–14.

Unlike for invasive insects and diseases, some of the most invasive exotic plants were imported on purpose, usually as ornamentals. Our own research in Farmington suggests that the source for most woody invasives is our own yards, implying that changes in human behavior could reduce the flow of such plants.[26] Not all imported plants are bad actors, however. In fact, one-third of Maine's flora is alien, and only a small portion of those 700 species has become problematic. The so-called "tens rule" suggests that about one in ten imported species escape into the natural environment,

Table 6.1 Region of origin, growth form, reproduction, and invaded habitats for a sample of terrestrial invasive exotic plant species in Maine

Species	Region of Origin	Growth Form	Reproduction	Invaded Habitats
Purple Loosestrife	Eurasia	Root perennial[a]	Wind-dispersed seeds	Wetlands, water courses
Asian Bittersweet	Asia	Woody vine	Bird-dispersed seeds; resprouts from damaged stem	Forest edges, disturbed forest
Norway Maple	Northern Europe	Tree	Wind-dispersed seeds	Open and interior forest
Japanese Knotweed	Asia	Root perennial	Nearly any part of plant; seeds sometimes	Disturbed roadsides and water courses
Morrow & Tatarian Honeysuckle	Japan Russia	Shrub	Bird-dispersed seeds; resprouts from damaged stem	Edges, disturbed forests
Japanese Barberry	Japan	Shrub	Bird-dispersed seeds; resprouts from damaged stem	Edges & open understory
Common Buckthorn	Asia	Small tree	Bird-dispersed seeds; resprouts from damaged stem	Edges & open understory
Mile-a-Minute Vine	Asia	Root perennial vine	Seeds; resprouts from damaged stem	Open

[a] A "root perennial" is a plant that lives for more than one year, with a stem that dies back over winter and resprouts at the beginning of the growing season.

Source: Maine Natural Areas Program, "Purple loosestrife," *Maine Invasive Plants*, Bulletin No. 2508 (Augusta: Maine Department of Conservation, 2001); Jil M. Swearingen, "Purple Loosestrife," in *Least Wanted: Alien Plant Invaders of Natural Areas* (Washington, DC: Plant Conservation Alliance's Alien Plant Working Group, 2009); Maine Natural Areas Program, "Asiatic bittersweet," *Maine Invasive Plants*, Bulletin No. 2506 (Augusta: Maine Department of Conservation, 2001);

Fig 6.4 In clockwise order, starting with the upper-left frame: Asiatic bittersweet smothering a tree, purple loosestrife, Japanese barberry, and Norway maple.

Jil M. Swearingen, "Oriental Bittersweet," in *Least Wanted: Alien Plant Invaders of Natural Areas* (Wash-ington, DC: Plant Conservation Alliance's Alien Plant Work-ing Group, 2009); Maine Natural Areas Program, "Shrubby honeysuckles," *Maine Invasive Plants*, Bulletin No. 2507 (Augusta: Maine Department of Conservation, 2001); Charles E. Williams, "Exotic Bush Honeysuckles," in *Least Wanted: Alien Plant Invaders of Natural Areas* (Washington, DC: Plant Conservation Alliance's Alien Plant Working Group, 2009); L.J. Mehrhoff et al., *IPANE: Invasive Plant Atlas of New England. Department of Ecology and Evolutionary Biology* (Storrs: University of Connecticut, 2003) (retrieved on November 13, 2010, at nbii-nin.ciesin.columbia .edu/ipane).

about one in ten of those escapees become established, and about one in ten of the established species become invasive.[27] For those few species that do escape their confines and become invasive, eradication is usually difficult. Many, including Asiatic bittersweet, produce abundant seeds, widely dispersed by wind or birds, or resprout prolifically after stem destruction. Effective control usually requires some combination of pulling, cutting, burning, mowing, herbiciding — applied repeatedly. For well-established species, eradication is probably practical only at individual sites.

"Voracious, libidinous, prolific . . . eating his way across the U.S." says a 1944 article about the exotic Japanese beetle.[28] Is this an accurate vision of Maine's future regarding invasive plants, insects, and disease? Probably not. Although the ecological consequences of exotic plants have been profound in some places, such as the grasslands of western North America,[29] their impacts in Maine have been more modest, occurring mainly in wetlands, riverbanks, and highly disturbed forests, with no evidence yet of invasion of interior forests. The dense spruce forests of Monhegan Island, for example, remain largely free of invasives. When asked about nonnative plants in the old-growth forest of Big Reed Forest Reserve, Nature Conservancy Director of Stewardship, Nancy Sferra, could think of only three: European sorrel, red raspberry, and colt's foot. Harvard biologist, Peter del Tredici recently argued, in fact, that many alien invasive plants are "symptoms of degradation, not the causes of it."[30]

Invasive exotic insects and diseases, on the other hand, exhibit key traits — virulence, host specificity, and abundant host trees of major forest species — that pose serious threats to trees in Maine. Although hemlock decline from the woolly adelgid may remain confined to coastal portions of the state, a breakthrough of a species such as emerald ash borer or Asian longhorned beetle could have widespread ecological and economic consequences. The reality for many of the new wave of invasive plants and pests, according to state horticulturalist Ann Gibbs, is that Maine, with its vast, cold forests, is new and uncertain territory.[31]

Changing Disturbance:
Land Use — Forestry and Housing Development

According to the 2006 Brookings Institution report, *Charting Maine's Future*, new realities may be supplanting the standard description of Maine as a rural state with vast, relatively unfragmented forests.[32] Although Maine's population continues to grow at a modest rate (3–4 percent per decade from

1990 to 2009), housing development is expanding rapidly into previously undeveloped forest and agricultural land. In 2004, nearly 8,000 acres of forestland were harvested for conversion to nonforest use, nearly three times the 1996 level. This development is concentrated in but not confined to southern and coastal parts of the state. Even the vast, forested Unorganized Territory, with only 8,000 full-time residents, has seen housing development, especially for second homes. This new reality was punctuated by the 2009 approval (but now in the courts) of the Plum Creek Moosehead Lake project, which rezoned 17,000 acres around the lake to allow 975 housing lots and two resorts (but also stipulated conservation of 45,000 acres in fee purchase and 363,000 acres in easements).[33]

According to a recent USDA Forest Service report, 44 million acres of private forestland in the United States are likely to be developed for housing by the year 2030. The report focuses on fifteen large forested watersheds that are particularly vulnerable. Three of these are in Maine, totaling over 700,000 acres. New rural housing development in the state is partly due to emigration from towns and cities, but the bulk of it is for second homes, for which Maine ranks first in the nation. The Forest Service projects that by the year 2030 housing densities on an additional 1,000 square miles of the state will cross the threshold from rural to exurban-urban (sixty-four houses per square mile). According to Eric White, one of the authors, "If residential development continues to expand . . . , these market factors will make the retention of land in forest use increasingly difficult."[34] Similar conclusions were drawn in a 1999 projection of Maine in the year 2050: increases in urban and suburban land at the expense of forests and farmland (figure 6.5).[35]

Additional novel pressures in land use and forest practices are also building in Maine, especially revolving around the generation of new sources of energy. The lion's share of the news has been about wind energy, which has become highly controversial. As of late 2011, seven projects were operational, with 173 turbines generating a maximum of 326 megawatts (enough to power about 137,000 homes).[36] Twelve additional projects were in development. The primary ecological impacts are killing of birds and bats, direct removal of forest, the creation of roads, and disturbance of fragile habitats. A recent analysis suggests that wind farms kill far fewer birds per MW generated than do fossil fuel power plants (via pollution and climate change), but this conclusion is dependent on estimates of kill numbers, which remain uncertain. Although wind farms involve small amounts of area compared to many other land uses, their impacts are concentrated on

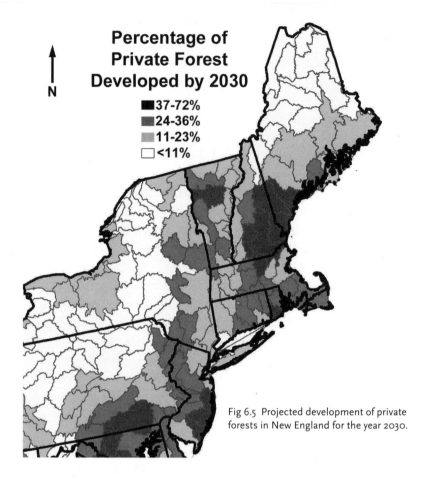

Percentage of
Private Forest
Developed by 2030

N

■ 37-72%
■ 24-36%
■ 11-23%
□ <11%

Fig 6.5 Projected development of private
forests in New England for the year 2030.

high elevation zones, some of which support rare species and communities, as described in chapter 5 (see "Katahdin: Alpine Communities"). The original proposed expansion of the Kibby Mountain project, for example, would have cleared about 200 acres, some of it "fair to good" Fir–Heart-Leaved Birch Subalpine Forest, an s3 natural community (sidebar 5.1). There are about 40,000 total acres of this community in Maine, which provide habitat for one of Maine's rarest migratory songbirds, Bicknell's Thrush. Environmental advocates were split on this proposal, with some arguing that such habitat loss is unacceptable and others accepting the loss as a reasonable tradeoff for the production of energy with a relatively low contribution to atmospheric greenhouse gases. A revised proposal with fewer turbines and reduced clearing of Fir–Heart-Leaved Birch Subalpine Forest eventually received approval. Sorting out the ecological pros and cons of wind power proposals in Maine can be exceedingly complex and plagued by insufficient research.[37]

Not so much on the radar screen, but potentially with more widespread forest impacts than wind power, is the use of biomass for energy production. Nine electric plants in Maine burn a total of more than 1.5 million tons of wood yearly, with a total generating capacity of 230 MW, and wood is also used to make pellets, which are burned for residential and commercial heating. Future volume harvested for biomass will likely hinge on the price of traditional forms of energy, as it has in the past, but also on policies that favor biomass energy as a means for reducing greenhouse gas emissions or dependence on foreign oil. A 2010 report by Manomet Center for Conservation Sciences calls into question standard assumptions about biomass energy, concluding that, compared to coal, biomass burning of harvested *live trees* releases lower levels of greenhouse gases when it is used for generating heat but not for electricity. A key point here is that the carbon footprint of biomass depends very strongly on the source of wood. Harvesting of live trees removes a sink that otherwise would have continued to take up (reduce) CO_2 from the atmosphere, whereas the use of waste wood, the primary source for the Maine biomass energy industry today, exploits a source that would have decomposed and released its carbon to the atmosphere over the long run anyway.[38] Not on the radar screen at all is the possibility of future demand for pellets (or other wood products) from the European Union. Driven by the requirement to produce 20 percent of their energy from renewable sources by 2020, European Union countries are already importing hundreds of thousands of tons of pellets from the southeastern United States. The extensive forests of Maine would be another natural source.[39]

What might this mean for Maine forests? Substantial growth of the biomass industry would likely involve the use of live wood, the likelihood of which may depend more on pure economics than a political calculus connected to reduction of carbon footprints. Such expansion would mean increased pressure on forests already subject to high levels of harvesting (see sidebar 4.5). Furthermore, many studies have demonstrated the importance of woody material left on the ground after tree trunks are removed — for nutrient return, erosion reduction, and promotion of rich communities of soil organisms.[40] All parts of trees can be burned to generate biomass energy, and, thus, widespread use of such whole-tree harvesting could compromise forest health. Projections of the balance of growth and harvest for the year 2050 for all of New England predict a growth cushion if there are no changes in demand, but unsustainable harvest levels under a high-demand scenario.[41] Given that the forests in Maine today are harvested at about the sustained yield limit (sidebar 4.5), substantial increases in biomass demand before midcentury could be met in four ways: (1) redirection of some portion of the supply currently used for pulp, lumber, or firewood, (2) increased growth with intensive silvicultural techniques (e.g., precommercial thinning, plantations, coppiced stands along power lines[42]), (3) import of wood, or (4) unsustainable harvesting, which would harm both the ecology and economics of Maine forests over the long run.[43]

In the future, the use of Maine forestland for carbon storage credits could act as an economic countercurrent to the harvesting pressures outlined above. Recent research has shown that forests in the northeastern United States sequester considerable quantities of carbon, which continues to accumulate even in old forests.[44] If federal or state governments pass legislation limiting carbon emissions but allowing emissions credit for investment in carbon storage, the vast Maine forests would be a natural source. Under such a scenario, corporations or other entities would pay for the sequestering of carbon in forestlands, which would probably mean no or light harvesting on targeted lands. Legislation favoring such land management is not presently on the horizon.

How Maine forests are managed in the future rests on who owns them — and that has changed dramatically over the past twenty years. As detailed by John Hagan, Lloyd Irland, and Andrew Whitman in a report from Manomet Center for Conservation Sciences, from 1980 to 2005 there were an astonishing 150 large land sales in Maine, totaling twenty million acres, almost equal to the entire land area of the state (figure 6.6). These transactions resulted in dramatic changes in ownership: from 1994 to 2005, forest products industry ownership of forestland declined from 59 to 16 percent, whereas the portion owned by financial investors (primarily Timber Investment Management Organizations, TIMOs, and Real Estate Investment Trusts, REITs) expanded from 3 to 40 percent (table 6.2, figure 6.7). These trends continued from 2005 to 2009, but in 2011 billionaire John Malone purchased nearly one million acres of Maine timberland from a financial group, more than tripling the amount owned by "new timber barons."[45]

These changes could have important consequences for the future forests of Maine. Surveys of owners revealed that "newer landowner types, financial investors, contractors, and new timber barons, tend to manage timberland more aggressively and with fewer biodiversity considerations" than other owners, such as the forest industry. The authors concluded that,

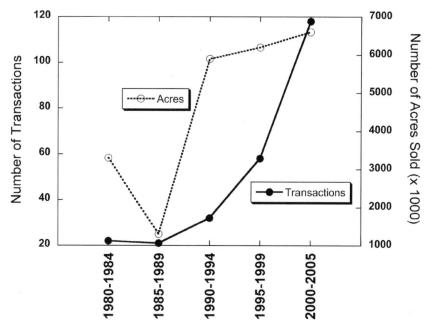

Fig 6.6 Number of land transactions and acres sold in five-year periods from 1980 to 2005 in northern New England and Upstate New York.

"Most traditional forest products companies had a long-range view of forest management and sustainability. It is not always clear that financial investors have this same view."[46] For some of these investors, this may translate into reselling of land over short time frames and consumptive uses of the forest, although there remains uncertainty about such concerns.

Table 6.2 also suggests that the last two decades have been good times for conservation land acquisitions, with nonprofit conservation group ownership rising from 0.3 to 3 percent of the total and state public lands from 7.7 to 8.7 percent. These values do not include conservation easements on private lands, which have increased by about 2 million acres (one-tenth of the state!). This period has indeed seen some historic conservation land deals, including The Nature Conservancy in Maine's purchase of 185,000 acres along the St. John River and 46,000 acres around the Debsconeag Lakes, the Appalachian Mountain Club's deal for a 37,000-acre portion of the Hundred Mile Wilderness, and the New England Forestry Foundation-Pingree family conservation easement of 762,000 acres. The ownership changes revealed by the Manomet report, in fact, raise the possibility that forest practices may follow strongly divergent future paths: enhanced biological conservation on a modestly increasing slice of the pie, but reduced

Table 6.2 Changes in land ownership of Maine timberland: 1994 to 2005

Landowner	1994		2005	
	acres	%	acres	%
Industry	6,909,725	59.2	1,818,082	15.5
Financial	371,719	3.2	3,818,596	32.6
REIT[a]	27,883	0.2	876,049	7.5
New Timber Baron	26,398	0.2	435,694	3.7
Old-line Family	2,489,683	21.3	2,447,012	20.9
Federal	201,860	1.7	206,490	1.8
State	897,947	7.7	1,023,136	8.7
Nonprofit	30,437	0.3	352,179	3.0
Tribal	253,019	2.2	243,246	2.1
Other[b]	458,194	3.9	502,699	4.3
TOTAL	11,666,865	100	11,723,183	100

[a] REIT is Real Estate Investment Trust.
[b] Other includes small ownerships by contractors, developers, individuals/families, unknown, and additional owners.

Source: Modified from John M. Hagan, Lloyd C. Irland, and Andrew A. Whitman, *Changing Timberland Ownership in the Northern Forest and Implications for Biodiversity,* Report #MCCS-FCP-2005–1 (Brunswick, ME: Manomet Center for Conservation Sciences, 2005). *Original data source:* J.W. Sewall Co., Old Town, ME.

stewardship on the vast bulk of forestland, a pattern described at the end of chapter 5.

Depending on their direction and extent, the land use changes described above could have important ecological consequences for organisms dependent on mature forests. For those species, increased housing development and harvesting would result in habitat loss and fragmentation; additional conservation lands that promote late successional forest would have the opposite effect. A large body of literature has documented the potential negative impacts of *habitat fragmentation:* increased predation from edge species, danger from roads, insufficient territory for wide-ranging animals, and the increased risk of extinction of small, isolated populations.[47] Negative effects of some forest practices and habitat fragmentation have been documented for mammals, birds, reptiles, amphibians, plants, and lichens in Maine.[48]

A key question is the extent to which these processes are broadly applicable in Maine. For less inhabited regions, the vast forest matrix is likely to

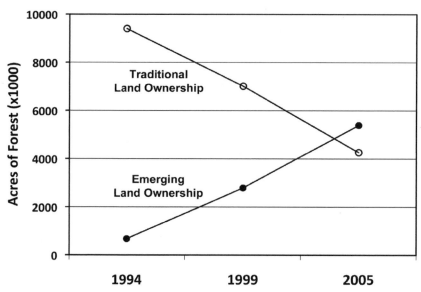

Fig 6.7 Changes in forestland ownership from 1994 to 2005. Traditional ownership: forest industry and old-line family woodlands. Emerging ownership: financial investors, REITS (real estate investment trusts), contractors, developers, and new timber barons.

mitigate such effects. In fact, the documented impacts on populations have so far been modest. According to a recent Manomet Conservation Sciences report, "Not all species — in fact the overwhelming majority of species — are not at risk as a result of modern forest practices."[49] Substantially increased levels of harvesting and housing development that transform the landscape, however, could change this, straining species closely tied to mature forests, especially those requiring large, old live and dead trees.[50] In the southern part of the state, continued residential and commercial sprawl could also further encroach on the already scant habitat available for species, such as grasshopper sparrows and the eastern cottontail, that require open grasslands and shrublands (chapter 5: "Kennebunk Plains . . .").[51] These pressures would also threaten the continuity of wood production, recreation, tourism, and the traditional rural and wildland character of Maine.

Changing Climate

The Past Century: Planet Earth

On 26 May 2010, the thermometer reached 91 in Portland, Maine, eclipsing the old record by six degrees. The first six months of 2010 were the warmest on record for Maine and the entire Earth.[52] Global warming? Maybe. As

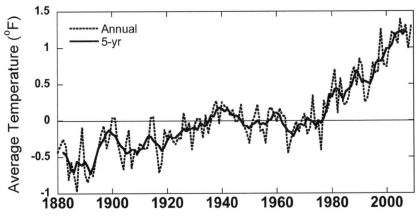

Fig 6.8 Global temperature 1880–2009, expressed as deviation from the average temperature 1951–1980.

described in chapter 2 ("Climate Change"), *weather* is a description of the meteorological conditions at some time and place, whereas *climate* is the long-term pattern of weather. The high temperature during the first half of 2010 was *weather* — extreme weather that may well have been the result of climate change, but that remains up for debate. On the other hand, more than a century of reliable thermometer records, including the hot months of 2010, has indeed led scientists to the unequivocal conclusion that *the climate of the Earth has warmed,* that is, temperatures are on average higher now than they were decades ago and a century ago. Figure 6.8, based on data from NASA, shows this rise from 1880 to 2009, an increase of nearly 1.5°F.[53] Warming has not been monotonic, but instead has exhibited yearly fluctuations, a slight midcentury cooling, and a steep increase over the past four decades.

Just how hot were those recent decades? According to the International Panel on Climate Change's (IPCC) Fourth Assessment,[54] released in 2007, eleven of the twelve years previous to the report were among the twelve warmest on record (since 1850), and "Average Northern Hemisphere temperatures during the second half of the 20th century were very likely higher than during any other 50-year period in the last 500 years and likely the highest in at least the past 1300 years." Temperatures also increased similarly during this period for the lower-troposphere, the mid-troposphere, and the ocean. Many other aspects of the Earth's physical system are also changing, in concert with global warming. Snow and ice cover have declined. Bodies of water in colder climates are ice covered for shorter periods of time. More land area is now affected by drought, driven by greater evaporation and more variable precipitation. As a result of thermal expansion and ice melt, sea level has risen. The frequency of extreme weather events has increased, a longtime prediction based on higher atmospheric energy levels. As of 2007, of the 765 observational data sets collected to test for climate change, 94 percent of them were consistent with a warming world.[55] According to Peter Thorne, previously with the British weather service and one of 300 authors of NOAA's 2010 annual State of the Climate Report, "The bottom line conclusion that the world's been warming is simply undeniable."[56]

The changing climate is driving corresponding alterations of the Earth's biological systems.[57] Two of the key bellwethers are the timing of life cycle events (*phenology*), which would be expected to occur earlier in the spring and later in the fall, and *geographic distributions,* which should move poleward in latitude and upward in altitude. Of the (remarkable) 28,671 published biological data sets addressing these hypotheses, 89 percent show changes consistent[58] with warming. Birds are migrating and laying eggs earlier in the spring; amphibians are calling and mating earlier; the timing of plant germination, leafing out, flowering, and fruiting has advanced; plankton, plants, insects, fish, amphibians, reptiles, birds, and mammals have shifted their distributions poleward, up in elevation, or deeper in the ocean. For the Northern Hemisphere, species are moving northward by an average of about four miles and upward by four meters per decade (calculated for about 1,700 species),[59] and the phenology of life-cycle activities is advancing in the spring by more than two days per decade (calculated for about 1,500 species). Surprisingly, these advancements are not always synchronized across trophic levels; a recent review of species across the UK found, for example, that secondary consumers (e.g., birds that eat herbivorous insects) lagged behind other trophic levels, including their prey, which could lead to temporal mismatch between these sets of species. Scientists have also detected higher levels of forest photosynthesis, carbon storage, and productivity, attributable at least in part to higher temperatures, longer summers, and enriched CO_2 levels, except in situations in which water is limiting.[60]

The Past Century: Maine

How about Maine? Have its climate and biological systems undergone the transformations described for the Earth as a whole? The U.S. Historical Climatology Network maintains twelve long-term temperature data sets

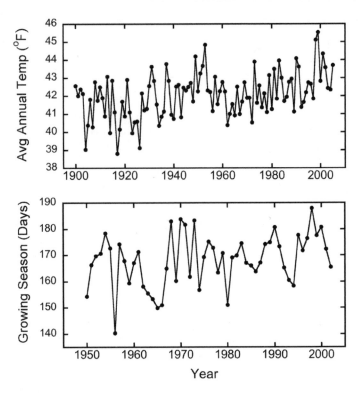

Fig 6.9 Average temperature and growing season 1900 to present for Maine.

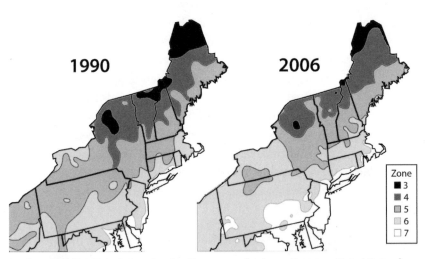

Fig 6.10 Arbor Day Foundation plant hardiness zones for the northeastern United States for 1990 and 2006. For Maine, the 2006 map shows decreases in the area attributed to the coldest zone (3) and, for the first time, coastal areas in the relatively warm zone 6.

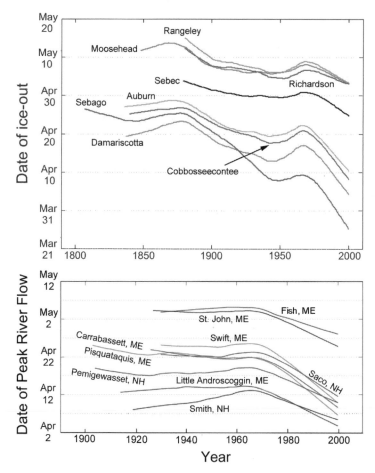

Fig 6.11 Twentieth-century advancement in the timing of (*top*) spring ice-out for eight lakes in Maine and (*bottom*) spring peak river flows in Maine and New Hampshire.

for a variety of sites around the state. Combining these series, we found that Maine's average annual temperature is indeed consistent with the global pattern, showing a similar rise over the past century (about 1.5°F), but with less increase over the last four decades. This warming has led to an increase in growing season length by about ten days from the time it was first reliably recorded in the 1950s (figure 6.9) and a rise in expected winter minimum temperatures. These changes led to a remapping of plant hardiness zones in 2006 (figure 6.10). For the entire Northeast, ice-out of lakes is occurring earlier, the timing of snowmelt-driven stream flow has advanced (figure 6.11), the ratio of winter snow to rain has declined, snowpack depth has decreased, and cloudiness, precipitation, and the frequency and intensity of extreme precipitation events have all increased.[61]

Research on the ecological impacts of this documented climate change in Maine has only recently begun. Herb Wilson and his students at Colby College did not find earlier spring migration for Maine bird species for the late versus early 1900s, in contrast to two more recent studies on birds in other parts of New England.[62] Peer-reviewed studies of other groups of organisms in the region have found advanced spring activity in response to climate change, including blooming of phytoplankton, leaf out of tree species, flowering of both wild and horticultural plants, calling of frogs, and migration of fish. A recent study revealed upward movement of the boundary between the northern hardwood and spruce-fir forests in the Green Mountains of Vermont.[63]

Professional biologists and naturalists have also noted the appearance of new temperate species in Maine and the rest of New England over the past few decades, as observed by Janet McMahon and Chris Davis on their coastal property.[64] A recent examination of Christmas Bird Count data by the Audubon Society found that out of 305 North American bird species examined, 208 (68 percent) had shifted their winter distributions northward over the past forty years, a pattern correlated with warmer January temperatures.[65] This movement has brought bird species, such as the Carolina wren, previously seen very sporadically, into Maine every year. "Twenty years ago, I remember people driving hours to see the one Carolina Wren in the state," said Jeff Wells, an ornithologist based in southern Maine. "Now, every year I get two or three just in my area. Obviously, things have changed."[66] For Maine, the association of warming and migration of species into the state has not been subject to explicit scientific tests, but the many species following this pattern (as opposed to two moving south[67]), the documented level of warming, and the vast background of research linking species biology and ambient temperature provides circumstantial evidence of a causal connection. Interestingly, there is no documentation as yet of species displacement northward out of the state as a result of warming conditions.

The Next Century

What should we expect for the next century? The general circulation models that have identified the causes of recent warming can also be employed to project climate for the next decades and beyond. Each model starts with an assumption about future greenhouse gas emissions, ranging from

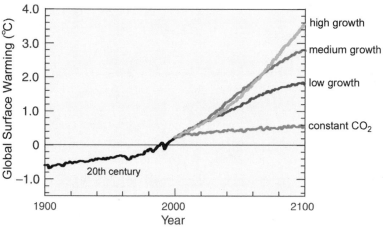

Fig 6.12 Projections for global warming to the year 2100. Lines are averages from multiple models for surface warming (relative to the average for the years 1980–1999), drawn as continuations of the twentieth century record from thermometers. Four different future scenarios for atmospheric greenhouse gas concentrations are used: no change from year 2000 (an obviously unrealistic but still instructive scenario) and low, medium, and large increases, corresponding to SRES scenarios A2, A1B, and B1. From 2000 to 2010, global emissions most closely followed the A2 scenario.

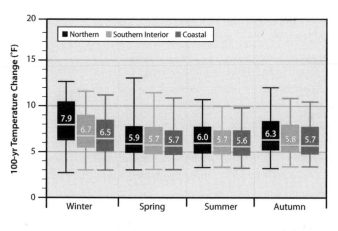

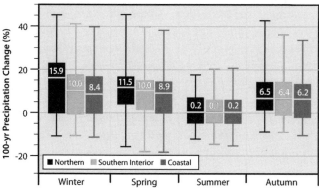

Fig 6.13 Projections for twenty-first-century changes in temperature and precipitation for four seasons for three climate divisions of Maine. Median values are solid horizontal lines in boxes. Top and bottom of boxes are 25th and 75th percentiles (from forty-two model simulations). Vertical lines depict minimum and maximum values. Models run using the moderate IPCC scenario A1B.

Why Has the Earth Warmed?

Why has the Earth recently warmed? A climate scientist would ask that question in the following way: Which physical factors (*forcings*) are responsible for increasing the net heat in the climate system? Net heat increases when incoming energy is higher than outgoing energy in the Earth-atmosphere system. As described in chapter 2 (figure 2.3), before human influences, net heat on the planet was largely controlled by solar irradiance, which increases incoming energy (positive forcing); greenhouse gases, which reduce outgoing energy (positive forcing); volcanic activity, which blocks incoming energy (negative forcing); and albedo, which increases outgoing energy (negative forcing). Accordingly, the planet has tended to warm under conditions of higher solar irradiance and greenhouse gases and lower volcanic activity and albedo.[1]

Over the past century, humans have introduced additional forcings. *Black carbon* particles, for example, formed through incomplete fossil fuel combustion, increase heat in the Earth system by absorbing it in the atmosphere and decreasing albedo where it has fallen on ice and snow (positive forcing). Atmospheric *sulfate aerosols* from pollution block incoming energy (negative forcing). Finally, humans have increased levels of atmospheric greenhouse gases (positive forcing); carbon dioxide, for example, has risen from about 280 parts per million (ppm) in preindustrial times to 385 ppm in 2008.[2]

To assess which natural and anthropogenic forcings are driving recent global warming, climate scientists examine how the magnitude of each has changed as temperatures have risen. Forcings that have increased the net energy in the system have likely played a role in temperature increases, whereas stable ones have probably been unimportant. This matching of forcings with warming (called *fingerprinting*) is implemented using models that simulate the climate system (*general circulation models*). These mathematical, computer-based models are built on the fundamental laws of energy and thermodynamics, developed over the past two centuries by physicists. Although challenges remain in fully replicating all climate processes, these models have become sophisticated, realistic, and robust, capable of accurately replicating past and current climate conditions, including the last century of warming. Forcings are added to the climate models individually and in combination, and, for each run, a set of predicted temperatures is produced. If the predicted temperature series matches the actual temperature pattern for the past century, then scientists know that the forcing used in that model can explain global warming.

Repeated tests of these scenarios at the major climate research centers around the world have unanimously identified a major role for anthropogenic greenhouse gas emissions in current global warming. In simulations of land surface temperatures at the National Center for Atmospheric Research, for example, natural forcings were mainly responsible for the early twentieth-century rise in temperature, volcanic and sulfate aerosols for the mid-century cooling, and anthropogenic greenhouse gases for warming in the second half of the twentieth century. Anthropogenic global warming has also been detected in fingerprint studies of increased ocean heat content, higher downward longwave radiation, greater warming at night than during the day, upper atmosphere cooling, the rise in the altitude of the troposphere, and other global climate attributes.[3]

A scientific consensus has emerged on anthropogenic climate change, expressed by the hundreds of scientists contributing to the 2007 IPCC report. A recent survey found that 97 percent of actively publishing climatologists agreed with these conclusions, as did nearly all the professional scientific societies of the world and the national academies of sciences in the G8+5 nations.[4] Typical is the statement adopted by the American Association for the Advancement of Science, the world's largest general scientific society: "The scientific evidence is clear: global climate change caused by human activities is occurring now."[5]

NOTES

1. The explanations in this section derive largely from J. Hansen et al., "Earth's energy imbalance: confirmation and implications," *Science* 308 (2005): 1431–1435; S. Solomon et al., eds., *Contribution of Working Group I to the Fourth Assessment Report of the Intergovernmental Panel on Climate Change, 2007* (Cambridge, UK: Cambridge University Press, 2007); *The Causes of Global Climate Change*, Science Brief 1 (Arlington, VA: Pew Center on Global Climate Change, 2008).

2. T.J. Blasing, *Recent Greenhouse Gas Concentrations, July 2010*, Carbon Dioxide Information Analysis Center, Oak Ridge National Laboratory, U.S. Department of Energy (retrieved on August 8, 2010, at cdiac.ornl.gov/pns/current_ghg.html).

3. Hansen et al., "Earth's energy imbalance," 1431–1435; Solomon et al., *Contribution of Working Group I*; G.A. Meehl et al., "Combinations of natural and anthropogenic forcings in twentieth-century climate," *Journal of Climate* 17 (2004): 3721–3727; B.D. Santer et al., "Identification of anthropogenic climate change using a second-generation reanalysis," *Journal of Geophysical Research-Atmospheres* 109 (2004): D21104; Rolf Philipona et al., "Radiative forcing —measured at Earth's surface—corroborate the increasing greenhouse effect," *Geophysical Research Letters* 31 (2004): LO3202; S. Levitus, J. Antonov, and T. Boyer, "Warming of the world ocean, 1955–2003," *Geophysical Research Letters* 32 (2005): Lo2604; T. Takemura et al., "Time evolutions of various radiative forcings for the past 150 years estimated by a general circulation model," *Geophysical Research Letters* 33 (2006): L19705; J. Lastovicka et al., "Global change in the upper atmosphere," *Science* 314 (2006): 1253–1254.

4. William R. L. Anderegga et al., "Expert credibility in climate change," *Proceedings of the National Academy of Sciences of the United States of America*, published online before print on June 21, 2010 (retrieved on August 5, 2010, at www.pnas.org/content/early/2010/06/04/1003187107.full.pdf+html). Statements of scientific societies and academy of sciences can be accessed at Wikipedia, "Scientific opinion on climate change" (en.wikipedia.org/wiki/Scientific_opinion_on_climate_change_).

5. AAAS Board Statement on Climate Change, Approved by the Board of Directors American Association for the Advancement of Science, December 9, 2006.

substantial reductions to large increases, produced through differing scenarios about political, socioeconomic, and technological change. All of these models project substantial further global warming over the next century, ranging from 3 to 7°F (see figure 6.12). They also forecast more heat waves, fewer cold spells, higher but more variable precipitation, a higher frequency of heavy precipitation events, less snow and ice cover, ocean acidification, more intense hurricanes and typhoons, and higher sea level. Projections envisage a similar future for Maine, as shown in figure 6.13: higher temperature, especially in the winter and in northern Maine; more precipitation, especially in the winter; and more drought in the summer (because of higher temperatures). In other words, we can expect hotter summers with more humidity and drought and wetter winters with more rain, less snow, and earlier ice out.[68]

Given the response of species to recent temperature increases of 1.5°F, it should not be surprising that large ecological impacts are projected for the more extreme warming scenarios described above. These models develop forecasts using anticipated climate change and biological relationships of species or ecosystems to climate variables (sidebar 6.4). The results envisage substantial migrations of species, ecosystem types, and even biomes across the world by the end of the twenty-first century — more with higher-emissions scenarios, less with more optimistic ones. Ecosystems are forecast to transform in additional ways: longer growing seasons; increased forest growth, production, and carbon storage (except where increased drought is expected); increased frequency, severity, and acreage of wildfire; higher levels of insect and disease outbreaks; and species extinctions.[69]

Some of the projections for Maine are jarring. Temperate oak-hickory forests in Kennebec County and Virginia pine in southern parts of the state by the end of the twenty-first century, for example. Figure 6.14 shows forecasts for four representative Maine species: red spruce and magnolia warbler are currently common in central and northern parts of the state; white oak and tufted titmouse occur mainly in southern and central Maine. Suitable habitat for the two northern species is projected to shrink in Maine (and increase in Canada), whereas that for the two temperate species is predicted to spread across the state. As shown in figure 6.14, the extent of these expected migrations depends strongly on assumptions about the degree of warming. Geographic shifts are anticipated for many other species, a sample of which is shown in table 6.3.[70] White pine, the eponymous tree of the Pine Tree State, should find the warmer, drier conditions in Maine — similar to those during its heyday 8–12,000 years ago (chapter 2)

Table 6.3   A sample of species expected to decrease and increase in abundance and range in Maine by the end of the twenty-first century as a result of climate change

Species Expected to Decrease	Species Expected to Increase
Balsam fir	Pignut hickory
Northern white cedar	Flowering dogwood
Paper birch	Sweet birch
Yellow-bellied sapsucker	Red-bellied woodpecker
Black-capped chickadee	Carolina wren
Common loon	Mallard duck
Canada lynx	Bobcat
American marten	Fisher
Moose	White-tailed deer

Source: Louis Iverson, Anantha Prasad, and Stephen Matthews, "Modeling potential climate change impacts on the trees of the northeastern United States," *Mitigating and Adapting Strategies for Global Change* 13 (2008): 487–516; N.L. Rodenhouse et al., "Potential effects of climate change on birds of the Northeast," *Mitigating and Adapting Strategies for Global Change* 13 (2008): 517–540; Catherine Burns et al., "Biodiversity," in G.L. Jacobson et al., eds., *Maine's Climate Future: An Initial Assessment* (Orono: University of Maine, April 2009), 30–34.

— much to its liking.[71] Cumulative species-specific changes such as these are expected to lead to major realignments of forest ecosystems in Maine. Northern coniferous and northern hardwood forests are projected to shrink throughout New England as suitable habitat moves northward and upslope in mountains, replaced by temperate oak-hickory hardwood forests, forecast to move north as much as 100 miles (figure 6.15).[72]

Interacting Stresses and Forest Dieback

In the late 1990s, white pine in southern Maine suffered severe dieback. Crowns thinned and yellowed, and up to 50 percent of trees died in some stands. What caused this sudden episode of mortality? Bill Livingston, a professor of forest science at the University of Maine, is a kind of Sherlock Holmes when it comes to trees and the mysterious circumstances surrounding their demise. Through analyses of tree rings, soils, and weather records, Livingston and his colleagues uncovered the culprit — actually, the *culprits*, for it turns out that a complex series of interacting factors led to this episode of dieback. White pine established en masse on abandoned

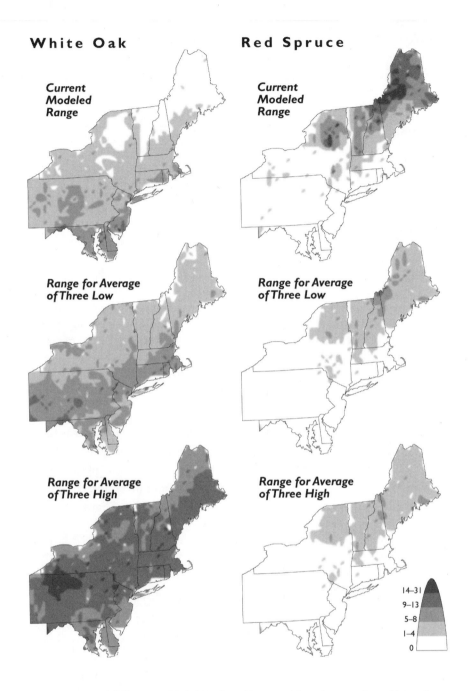

White Oak

Current Modeled Range

Range for Average of Three Low

Range for Average of Three High

Red Spruce

Current Modeled Range

Range for Average of Three Low

Range for Average of Three High

14–31
9–13
5–8
1–4
0

Fig 6.14 Projected shifts in suitable habitat for red spruce, white oak, magnolia warbler, and tufted titmouse in response to predicted climate change. For each species, maps show modeled ranges for today (*left*) and projected suitable habitat for the year 2100 for the average of three models using low future emissions (IPCC B1 scenario; *center*) and the average of three models using high future emissions (IPCC A1fi scenario; *right*).

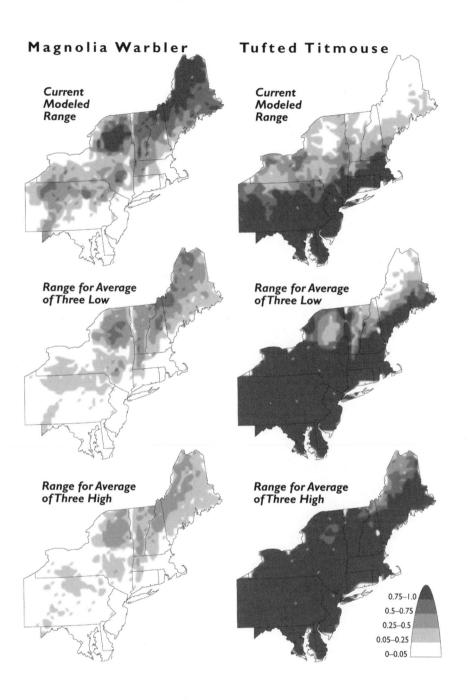

Magnolia Warbler

Current Modeled Range

Range for Average of Three Low

Range for Average of Three High

Tufted Titmouse

Current Modeled Range

Range for Average of Three Low

Range for Average of Three High

0.75–1.0
0.5–0.75
0.25–0.5
0.05–0.25
0–0.05

Projecting Future Species Distributions: Bioclimatic Envelope Versus Mechanistic Models

Two main approaches have been used to project how climate change might alter species ranges. *Bioclimatic envelope models* estimate geographical shifts in suitable habitat ("bioclimatic envelope") for a species based on forecasted changes in climate. Take sugar maple, for example. Louis Iverson, Anantha Prasad, and Stephen Mathews at the U.S. Forest Service developed a detailed quantitative description (a model) of the relationship between its current distribution and climate-related variables. They then used this relationship to project sugar maple's future range for several climate change scenarios (moderately to much warmer). As conditions warm in the model, the suitable climate boundaries for this species move north—and so presumably will sugar maple itself, decreasing its abundance in the southern half of Maine.[1]

Bioclimatic envelope models are correlative, that is, they use the correlation of a species' abundance to current climate patterns to predict shifts in suitable habitat caused by future climate change. In contrast, *mechanistic models* are based on how climate factors affect the functioning of an individual organism: its photosynthetic rate, thermoregulation, frost survival, dispersal, etc. Relationships of these processes with climate variables (e.g., maximal monthly temperature, monthly precipitation, etc.) are employed to project how simulated future climate patterns will alter geographic ranges, acting through individual organisms. A mechanistic model for sugar maple by Xavier Morin and colleagues, for example, also projects a northward shift in its range, resulting from drought-induced mortality at the southern boundary and enhanced fruit ripening and flower frost survival beyond the northern boundary.[2] Envelope models have been criticized for assuming that species distributions are controlled entirely by climate (rather than other processes such as competition) and for not considering whether species will be capable of migrating at rates sufficient to keep pace with geographical shifts in suitable habitat. On the other hand, the success of mechanistic models is highly dependent on a detailed understanding of the physiology and ecology of a species, knowledge available for only a fraction of species—which explains why envelope models are more commonly used than mechanistic ones. There have been important recent strides in developing guidelines for the appropriate use of each type of model and in combining these two approaches.[3]

NOTES

1. Louis Iverson, Anantha Prasad, and Stephen Matthews, "Modeling potential climate change impacts on the trees of the northeastern United States," *Mitigation and Adaptation Strategies for Global Change* 13 (2008): 487–516; Louis R. Iverson et al., "Estimating potential habitat for 134 eastern US tree species under six climate scenarios," *Forest Ecology and Management* 254 (2008): 390–406; USDA Forest Service, Northern Research Station, *Climate Change Atlas* (retrieved on August 15, 2010, at www.nrs.fs.fed.us/atlas).

2. Xavier Morin, Carol Augsburger, and Isabelle Chuine, "Process-based modeling of species' distributions: what limits temperate tree species' range boundaries?" *Ecology* 88 (2007): 2280–2291; Xavier Morin, David Viner, and Isabelle Chuine, "Tree species range shifts at a continental scale: new predictive insights from a process-based model," *Journal of Ecology* 96 (2008): 784–794.

3. M. Luoto et al., "Uncertainty of bioclimate envelope models based on the geographical distribution of species," *Global Ecology and Biogeography* 14 (2005): 575–584; Juha Pöyryl et al., "Species traits are associated with the quality of bioclimatic models," *Global Ecology and Biogeography* 17 (2008): 403–414; Colin M. Beale, Jack J. Lennon, and Alessandro Gimona, "Opening the climate envelope reveals no macroscale associations with climate in European birds," *Proceedings of the National Academy of Sciences* 105 (2008): 14908–14912; Richard P. Duncan, Phillip Cassey, and Tim M. Blackburn, "Do climate envelope models transfer? A manipulative test using dung beetle introductions," *Proc. R. Soc. B* 276 (2009): 1449–1457; Stefan Dullinger, Thomas Dirnböck, and Georg Grabherr, "Modelling climate change-driven treeline shifts: relative effects of temperature increase, dispersal and invisibility," *Journal of Ecology* 92 (2004): 241–252; Risto K. Heikkinen et al., "Methods and uncertainties in bioclimatic envelope modelling under climate change," *Progress in Physical Geography* 30 (2006): 1–27; M.B. Arajo et al., "Validation of species-climate impact models under climate change." *Global Change Biology* 11 (2005): 1504–1513; Michael Kearney and Warren Porter, "Mechanistic niche modelling: combining physiological and spatial data to predict species ranges," *Ecology Letters* 12 (2009): 334–350; Richard G. Pearson and Terence P. Dawson, "Predicting the impacts of climate change on the distribution of species: are bioclimate envelope models useful?" *Global Ecology & Biogeography* 12 (2003): 361–371.

farmland in southern Maine from the late-1800s to the mid-1900s, on sites not typically occupied by this species previous to land clearing. These open sites were ideal for juvenile white pines, but not for older trees decades later, which by then grew in crowded conditions, with roots restricted to shallow depths by plow pans, soil compaction, and natural barriers. As a result, the 1995 drought hit pole-sized white pines hard, killing many. With their defenses compromised, weakened trees were then easy prey for insect herbivores (*Ips* bark beetles) and fungal pathogens (*Armillaria* root rot).[73]

Around the same time, a similar chain of causal events appears to have led to dieback of balsam fir in eastern Maine. Allison Kanoti, formerly a graduate student in Livingston's lab and now a Maine state forest entomologist, discovered greatly reduced fir growth stemming from an outbreak of

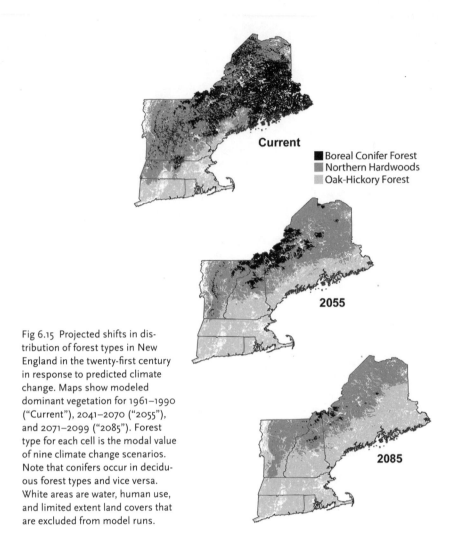

Current

■ Boreal Conifer Forest
■ Northern Hardwoods
▨ Oak-Hickory Forest

2055

2085

Fig 6.15 Projected shifts in distribution of forest types in New England in the twenty-first century in response to predicted climate change. Maps show modeled dominant vegetation for 1961–1990 ("Current"), 2041–2070 ("2055"), and 2071–2099 ("2085"). Forest type for each cell is the modal value of nine climate change scenarios. Note that conifers occur in deciduous forest types and vice versa. White areas are water, human use, and limited extent land covers that are excluded from model runs.

the alien balsam woolly adelgid, which was apparently caused by several consecutive warm winters.[74] Mortality in the already-stressed balsam fir was then exacerbated by the severe drought of 2001. Similar episodes of multiple-stress forest decline have been documented in Maine for beech (beech bark disease, high temperatures, and drought), white birch (low snow cover, freezing injury, and drought), and ash (low snow cover, freezing injury, and drought).[75]

These case studies demonstrate that the four sets of stresses described in this chapter — acid rain, exotic invasive organisms, land use, and climate change — will act synergistically rather than independently. These studies show further that species and communities will not always respond to environmental change in a gradual, incremental way, but instead may experience episodes of abrupt collapse when thresholds of stress are crossed. Similar forest declines are playing out worldwide. In a recent review of drought-induced forest dieback, Craig Allen, of the U.S. Geological Survey, and a group of international coauthors, found that most of their eighty-six case studies involved the cumulative effects of drought and other agents of mortality, especially insects and diseases. In the southwestern United States, for example, a combination of multiyear drought and pine bark beetles led to mortality as high as 90 percent in Rocky Mountain pinyon pine over more than 3,000,000 acres. The authors raise the possibility that the incidence of catastrophic forest decline is increasing across the globe, linked to climate change, which is pushing temperature and moisture variability beyond the boundaries of historic conditions.[76]

The four sets of stresses identified in this chapter are apt to interact in two ways. First, they could amplify each other in the sense that one stress will raise the intensity of others. Warming, for example, has direct effects on forest trees, but also can spur outbreaks of insect and fungal pests or allow their spread to areas that were previously too cold, which could occur in the future for hemlock woolly adelgid. Similarly, manipulations of forests, such as harvesting, are likely to accelerate the rate of climate change impacts. Climate change alone is unlikely to kill many mature trees; in most cases, trees can persist for centuries even when conditions are not ideal. Once cut, however, they are easily replaced by a next generation of different tree species better adapted to the new climate regime. Increased levels of harvesting or housing development may, likewise, provide opportunities in forests for colonization by invasive plants.

Second, stresses will likely act cumulatively in the sense that the action of one stress on a tree will lower the threshold for deleterious effects of others. Such interactive impacts are illuminated by *Manion's Spiral* model, which proposes that trees die and forests decline generally as a result of a series of three types of stresses: *predisposing factors* increase tree vulnerability, *inciting stresses* trigger mortality and decline, and *contributing factors* wreak further mortality on seriously compromised trees.[77] Manion visualized a spiral, with predisposing factors propelling trees into a whorl of stresses, circling inward, leading to the center point — death. Most stresses can play any of these three roles, which might differ from one dieback episode or tree species to another. We saw this clearly in the cases of dieback in Maine forests, where mortality was the result not of just one environmental agent

Species We Might Be Missing:
Northern Maine and High Elevations

Figure 6.14 presents a simplified picture of species gradually and smoothly retreating northward as the climate warms. This may be the case for some species like Canada lynx (figure 6.16), which occurs continuously from relatively undeveloped northern Maine through the boreal forest to the tundra. With long legs and broad paws, it is exquisitely adapted to moving through deep snow and preying on snowshoe hare. But in shallow snow, these advantages are diminished, as is the abundance of its main prey. Thus, the lynx's range is projected to gradually move northward out of Maine as winters warm and snowfall declines. The same fate probably awaits the American (pine) marten (figure 6.16), which is adapted to hunting beneath the snow, and is thought to require annual snowfall of at least seventy-six inches. These vacated areas in northern Maine will likely be filled by relatives of the lynx and marten, the bobcat and the fisher (figure 6.16), respectively, whose current ranges encompass more southerly areas with relatively little snowfall.[1]

Such coordinated latitudinal procession is unlikely to play out for species restricted to the cool environments of the highest elevations in the state. Alpine species (see chapter 5: "Mount Katahdin: Alpine Communities"), such as Lapland rosebay, *Diapensia*, the Katahdin arctic butterfly, the American pipit, Bicknell's thrush, and the northern bog lemming, for example, may well be extirpated from the state as their suitable habitat disappears off the top of the mountains they inhabit. Unlike for lynx and marten, the nearest conspecifics will be not just north in Canada but far away in the northern boreal forest or tundra. Northern boreal populations of these alpine species should have the geographical flexibility to move with climate change. Species around the world entirely restricted to narrow habitats, however, are among the most vulnerable to extinction as a result of warming and habitat dislocation.[2]

Fig 6.16 Photographs and range maps, in clockwise order, from upper left: Canada lynx, American (pine) marten, fisher, and bobcat.

NOTES

1. Summary of this possible species replacement: Catherine Burns et al., "Biodiversity," in G.L. Jacobson et al., eds., *Maine's Climate Future: An Initial Assessment* (Orono: University of Maine, April 2009), 31. The snowfall threshold for Canada lynx has been estimated at 106 inches of snow per year, L.L. Hoving et al., "Broad-scale predictors of Canada lynx occurrence in eastern North America," *Journal of Wildlife Management* 69 (2005): 739–751; see also P. Gonzalez et al., *Potential Impacts of Climate Change on Habitat and Conservation Priority Areas for Lynx canadensis (Canada Lynx)* (Washington, DC: Watershed, Fish, Wildlife, Air, and Rare Plants Staff, National Forest System, Forest Service, U.S. Department of Agriculture, and Arlington, VA: NatureServe, 2007); K. Nowell, *"Lynx Canadensis,"* in *IUCN Red List of Threatened Species*, version 2010.3 (Washington, DC: International Union for Conservation of Nature and Natural Resources, 2010) (retrieved on September 22, 2010, at www.iucnredlist. org). Fishers and martens: W.B. Krohn, K.D. Elowe, and R.B. Boone, "Relations among fishers, snow, and martens: development and evaluation of two hypotheses," *The Forestry Chronicle* 71 (1995): 97–105.

2. C.D. Thomas et al., "Extinction risk from climate change," *Nature* 427 (2004): 145–148.

but instead of a series of pressures that progressively weakened trees, with the role of each stress depending on tree species or circumstances. Dieback, then, may play a role in the alteration of Maine forests, as a response that integrates the manifold stresses impinging on forests, instigating the decline of species poorly suited to the new environmental regime and the rise of those better adapted to these novel conditions.

The Future of the Maine Woods: Forecasts and Uncertainty

This book has sketched the history of Maine forests over 15,000 years up to the present. The enduring theme has been change — ecological change at multiple time scales driven by glacial retreat, shifting climate, and, more recently, human culture. Most periods of ecological stability were short-lived, disrupted by resumed climate change, disturbance, and influx of novel species. It may not be an exaggeration to describe the forests of Maine, and many others as well, as in a constant state of recovery from the most recent perturbation, whether natural or anthropogenic. The studies and models presented in this chapter suggest that Maine forests will continue to change in the future.

Projecting the details of that ecological future is daunting. Not only are the environmental drivers exceedingly complex, but they are themselves dependent on a set of socioeconomic and technological trajectories that are if anything even more challenging to predict. One might rightfully ask, "how successful would we have been fifty years ago in predicting the condition of forests in Maine today?" Although a good forecaster might have envisaged a future of exotic forest pests and land use pressures, given decades of such experience already, few could have imagined transformative phenomena such as climate change. After all, even Nobel Prize–winning scientists of those times thought that "there was no risk that humanity could do real harm to anything so gigantic as the earth."[78]

On the other hand, the content of this book is a testament to how far the ecological and earth sciences have progressed since 1960, developing an impressive capacity to (1) capture fine-scale environmental data from space, water, land, and organisms, (2) reveal complex patterns across the Earth's landscape and into the deep past, (3) carry out multifactor, large-scale experiments, manipulating key variables such as CO_2, temperature, and acid deposition; and (4) construct models that can reasonably replicate observed patterns over space and time — projecting into the future. The first general circulation model (GCM) of the Earth was not built until

Scientific Uncertainty and Human Society

All scientific ideas are subject to revision; we should never be absolutely sure that the truth has been reached. Old ideas should be tested continually, in an effort to tear them down and replace them with better ones. Ideas that survive this constant attack will be especially robust. Experience shows that if we then behave as if these surviving ideas are true, we will succeed—in curing diseases, finding clean water, building things that stand up . . . and so on.
—Richard B. Alley, *The Two-Mile Time Machine:*
   Ice Cores, Abrupt Climate Change, and Our Future

the late 1960s, when certain conceptual and computational thresholds were reached. The first GCM global warming projection models were published in the 1980s; they have done a good job of predicting temperature change. But those models are primitive versions of the descendent GCMs employed today, which are capable of replicating subtle aspects of energy exchange and climate across the Earth.[79]

Three sets of ecological pressures appear to pose the greatest potential for changing Maine's forests in the future: climate change, invasive exotic insects, and land use. For climate change, the prospective tool of choice has become projection models. Natural and social scientists have urged caution in the use of these models to predict the future state of forests because of their uncertainties,[80] but they can play an important role in planning, as long as they are seen for what they are: "first approximations indicating the potential magnitude and broad pattern of future impacts,"[81] that is, a sense of probable direction, boundaries, and scale of change. Projections are unanimous that aspects of the fundamental character of Maine forests will change. First and foremost, the landscape seems destined to become less northern and more temperate. This means a decline in the abundance of particular species associated with northern climates and an increase in more southern species. The highest emissions scenarios suggest that the nature of Maine could become similar to that in places as far south as Maryland. That is possible. On the other hand, we might instead expect totally novel climates and ecosystems, unlike those found anywhere today. In chapter 2 ("Lessons Learned"), we presented the compelling argument by

Jack Williams and his colleagues that climates and communities prevailed during the Holocene that have no analogs today. These authors have demonstrated that such no-analog climates and species associations are likely also to arise in a future marked by climate change.[82]

The pace of climate-mediated forest change is especially difficult to peg, and may differ from one species to another. Mobile, shorter-lived species, such as small birds, might track changing climates closely, whereas species such as trees, with restricted migration capacity and long lives, might lag behind changing environments, perhaps more so than indicated in some of the projections. A chief observation of this book, for example, is that regional distributions of major forest types in Maine have shifted very little from presettlement to today (chapter 5: "Dividing up the State . . ."), despite considerable environmental change, suggesting a degree of ecological inertia that is difficult to build into projection models. The most recent studies of range changes in response to warming reveal the complexity of these issues: a study of hundreds of species across the world detected rapid latitudinal and elevational migration, but a detailed study of 92 tree species in the eastern United States found little evidence for climate-mediated migration.[83]

Predicting the future role of forest dieback in Maine is challenging. Two already well-established pests in the state — beech bark–*Neonectria* disease and spruce-budworm — should continue to cause chronic mortality in beech and periodic acute dieback in spruce-fir forests, respectively. These could be exacerbated by the new stresses discussed in this chapter. The arrival and spread of Asian longhorned beetle, emerald ash borer, or exotic pests not yet on the radar screen, could further alter the relative abundances of tree species, given the species-specific host preferences of these vectors. An even more complex issue is the extent to which interactions among multiple stresses, each with their own independent uncertainties, will lead to forest decline for particular species, forest types, and regions of the state. Bill Livingston, the forest pathologist whose work we described earlier in this chapter, does not envision catastrophe: "The studies I have made actually have emphasized to me the ability of trees to survive; once established in the dominant strata, trees will survive a large amount of stress. I don't expect to see a large amount of die-offs, just on the margins or on the poor sites for trees. I expect that species change will occur more through shifts in the composition of natural regeneration than in massive die-offs of current trees." Whether multiple-stress dieback becomes prevalent may hinge on climate change. If the climate warms and increased drought and fire ensues according to more dire scenarios, levels of stress may surpass historic conditions, leading to widespread forest decline. With more modest climate change, diebacks may be little more than what Livingston and other researchers have chronicled in the recent past.

Current trends in land use — specifically, the dichotomy between areas dedicated to mature forests versus those subject to frequent harvesting — point to diverging paths for the composition and structure of Maine forests. In preserves, we should see continued succession toward long-lived climax tree species such as sugar maple, with fewer early successional species such as aspens — similar to conditions during presettlement times. In most of the landscape, however, a mix of tree species will likely prevail, as frequent harvesting and housing development maintain an abundance of species adapted to disturbance — similar to conditions prevailing over the past century. Although it is tempting to peg the maintenance of biological diversity on the preserved portion of the forest, land use on the much larger private forest may be just as crucial, despite high levels of disturbance, because of its role in maintaining a continuous forest matrix. In fact, of all the potential drivers of change, development of forestland probably ranks first as the biggest threat to radically transforming the Maine Woods — by diminishing the forested landscape itself. Once land cover shifts from forest to nonforest, it seems very unlikely that it will revert. For these reasons, development is the most certain enemy of forests in Maine.

The ecological changes envisioned here will affect more than just the well-studied, charismatic species upon which we have focused. Alterations of the tree layer, which forms the structural foundation of terrestrial systems in places like Maine, will ripple out to entire ecosystems, influencing the flow of energy, the cycling of nutrients, and the lives and populations of most species, including humans. The natural infrastructure for wood production and recreation, for example, will need to reconfigure in proportion to changes in land use, climate, and ecosystems. According to two of the most well-respected ecologists in the United States, Gene Likens and Jerry Franklin, "At stake in the northern forest are ecosystem services and social values that citizens hold dear and rely on: clean, dependable water supplies; clean air; healthy soils to nourish trees for forest products; habitats to support wildlife; and accessible landscapes used by millions of people for hiking, birding, fishing, hunting, and a myriad of other activities for recreation and inspiration."[84]

We don't know what the future will bring.[85] There are bound to be surprises. Even the forecasts for Maine forests do not converge on a single bright line into the future. Instead, they identify a range of possible scenarios, clear at first, and then fading into the haziness of the decades ahead. This reflects not only their intrinsic uncertainty, but also the reality that many of the underlying storylines about how human civilization will change are as yet unwritten. These scenarios are scary because some of the forecasts envision outcomes that organisms, including humans, would find challenging. On the other hand, the multiplicity of possible paths also suggests that there is no ineluctable future, and that humans can still choose a storyline that will protect the attributes we cherish most about the Maine Woods.

Notes

1. The Maine Woods

Epigraph from Henry David Thoreau, *The Maine Woods* (New York: Penguin Books, 1988; first published by Ticknor and Fields, 1864), 88.

1. Woodrow B. Thompson, Carol B. Griggs, Norton G. Miller, Robert E. Nelson, Thomas K. Weddle, and Taylor M. Kilian, "Associated terrestrial and marine fossils in the late-glacial Presumpscot Formation, southern Maine, USA, and the marine reservoir effect on radiocarbon ages," *Quaternary Research* 75 (2011): 552–565. A similar trove of fossils was also found in 1976 about 500 feet northwest of the hospital site, research from which was published in F. Hyland, W.B. Thompson, and R. Stuckenrath Jr., "Late Wisconsinan wood and other tree remains in the Presumpscot Formation, Portland, Maine," *Maritime Sediments* 14 (1978): 103–120 and R.S. Anderson, N.G. Miller, R.B. Davis, and R.E. Nelson, "Terrestrial fossils in the marine Presumpscot Formation: implications for late Wisconsinan paleoenvironments and isostatic rebound along the coast of Maine," *Canadian Journal of Earth Sciences* 27 (1990): 1241–1246.

2. Information on postglacial Maine environment and forests from Thompson et al., "Associated terrestrial and marine fossils"; H.W. Borns Jr. et al., "The deglaciation of Maine, U.S.A.," in J. Ehlers and P.L. Gibbard, eds., *Quaternary Glaciations — Extent and Chronology, Part II: North America* (Amsterdam, Netherlands: Elsevier, 2004), 89–109; Thomas K. Weddle and Michael J. Retelle, "Deglacial history and relative sea-level changes, Northern New England and adjacent Canada," *Geological Society of America*, Special Paper 351 (2001); Ronald B. Davis and George L. Jacobson Jr., "Late glacial and early Holocene landscapes in northern New England and adjacent areas of Canada," *Quaternary Research* 23 (1985): 341–368; George Jacobson Jr. and Ronald B. Davis, "The real forest primeval: the evolution of Maine's forests over 14,000 years," *Habitat: Journal of the Maine Audubon Society* 5 (1988): 26–29; D.H. DeHayes et al., "Forest responses to changing climate: lessons from the past and uncertainty for the future," in R.A. Mickler, R.A. Birdsey, and J. Hom, eds., *Responses of Northern U.S. Forests to Environmental Change*, Ecological Studies 139 (New York: Springer, 2000), 495–540. For additional information, see chapter 2.

3. Generally, we will use the word "Indians" to describe the peoples who settled and, before European contact, occupied the lands that became Maine. This follows the guidance of James Axtell: "In the Americas, 'Indians' is short, carries no

pejorative baggage despite Columbus, and is reasonably clear, unless we do comparative, world, or immigration history." James Axtell, *Native and Newcomers: the Cultural Origins of North America* (New York: Oxford University Press, 2001). We will use the word "Paleoindians" for those native peoples who lived in Maine during postglacial times but have unclear continuity with present-day Indian tribes.

4. The most recent evidence suggests minimum first presence of Paleoindians 12–13,000 years ago. Jonathan C. Lothrop et al., "Paleoindians and the Younger Dryas in the New England-Maritimes Region," *Quaternary International* 242 (2011): 546–569. See also David Sanger, "The original Native Mainers," *Habitat: Journal of the Maine Audubon Society* 5 (1988): 37–41; American Friends Service Committee, *The Wabanakis of Maine and the Maritimes* (Philadelphia, PA: American Friends Service Committee, 1989), A3–A10 and Section D: Fact Sheets; Bruce J. Bourque, "Prehistoric Indians of Maine," in Richard W. Judd, Edwin A. Churchill, and Joel W. Eastman, eds., *Maine: The Pine Tree State from Prehistory to the Present* (Orono: University of Maine Press, 1995), 12–30; Bruce J. Bourque, *Twelve Thousand Years: American Indians in Maine* (Lincoln: University of Nebraska Press, 2001), 73–109.

5. Edwin A. Churchill, "English beachheads in seventeenth-century Maine," in Richard W. Judd, Edwin A. Churchill, and Joel W. Eastman, eds., *Maine: The Pine Tree State from Prehistory to the Present* (Orono: University of Maine Press, 1995), 51–75. Around this time, the French established beachheads in eastern coastal Maine, especially at present day Castine; Alaric Faulkner and Gretchen Fearon Faulkner, "Acadian settlement, 1607–1700," in Richard W. Judd, Edwin A. Churchill, and Joel W. Eastman, eds., *Maine: The Pine Tree State from Prehistory to the Present* (Orono: University of Maine Press, 1995), 76–96.

6. Gorham Historical Society (retrieved on July 20, 2011, at www.gorham historical.com/chronology).

7. Bill McKibben, "An Explosion of Green," *Atlantic Monthly* 275 (April 1995): 61–83.

8. The Nature Conservancy in Maine (retrieved on July 20, 2011, at www.nature. org/ourinitiatives/regions/northamerica/unitedstates/maine/placesweprotect/ mount-agamenticus.xml); Mount Agamenticus Conservation Region (retrieved on July 20, 2011, at www.agamenticus.org).

9. Plant Hardiness Zones provide gardeners with a way to select plants that are sufficiently hardy for their local climate. Lois Berg Stack, *Plant Hardiness Zone Map of Maine*, University of Maine Cooperative Extension Bulletin No. 2242 (Orono: University of Maine, 2006), based on the national *Hardiness Zone Map*, Arbor Day Foundation (retrieved on May 5, 2011, at www.arborday.org/media/ zones.cfm) and *USDA Plant Hardiness Zone Map*, Miscellaneous Publication 1475 (Washington, DC: United States Department of Agriculture, 1990).

10. History of the Moosehead region is from Everett L. Parker, *Kineo: Moosehead Sentinel from Native Americans to Hotel Grandeur* (Greenville, ME: Moosehead Communications, 2004); George J. Varney, "History of Greenville, Maine,"

in *A Gazetteer of the State of Maine* (Boston: B. B. Russell, 1886) (retrieved on November 20, 2009, at history.rays-place.com/me/greenville-me.htm). Aspects related to logging are mainly from Philip T. Coolidge, *History of the Maine Woods* (Bangor, ME: Furbush-Roberts, 1963), 53–94; Lawrence C. Allin and Richard W. Judd, "Creating Maine's Resource Economy, 1783–1861," in Richard W. Judd, Edwin A. Churchill, and Joel W. Eastman, *Maine: The Pine Tree State from Prehistory to the Present*, 262–288; D.C. Smith, *A History of Lumbering in Maine, 1861–1960* (Orono: University of Maine Press, 1972).

11. Henry David Thoreau, *The Maine Woods* (New York: Penguin Books, 1988; original edition: Ticknor & Fields, 1864), 238.

12. *Katahdin* means "mountain" in Abenaki. Accordingly, we will sometimes refer to the mountain as simply "Katahdin" and elsewhere as "Mt. Katahdin."

13. Total land area, including lakes and waterways, is 21.3 million acres.

14. DeLorme, *Maine Atlas and Gazetteer*, 13th edition (Yarmouth, ME: DeLorme, 2007); Christopher S. Cronan et al., "An assessment of land conservation patterns in Maine based on spatial analysis of ecological and socioeconomic indicators," *Environmental Management* 45 (2010): 1076–1095; William H. McWilliams et al., *The Forests of Maine: 2003*, Resources Bulletin NE-164 (New Town, PA: U.S. Department of Agriculture, Forest Service, Northeastern Research Station, 2005); Susan C. Gawler et al., *Biological Diversity in Maine: an Assessment of Status and Trends in the Terrestrial and Freshwater Landscape* (report prepared for the Maine Forest Biodiversity Project) (Augusta: Maine Natural Areas Program, Department of Conservation, 1996); Susan C. Gawler, and Andrew R. Cutko, *Natural Landscapes of Maine: A Classification of Vegetated Natural Communities and Ecosystems* (Augusta: Maine Natural Areas Program, Department of Conservation, 2010); Natural Communities and Ecosystems webpage, Maine Natural Areas Program, Department of Conservation, State of Maine (retrieved on March 8, 2010, at www.maine.gov/doc/nrimc/mnap/features/community.htm); Stack, *Plant Hardiness Zone Map of Maine.*

15. H.W. Borns Jr. et al., "The deglaciation of Maine, U.S.A.," in J. Ehlers and P.L. Gibbard, eds., *Quaternary Glaciations — Extent and Chronology, Part II: North America* (Amsterdam, Netherlands: Elsevier, 2004).

2. From Rocks to Ice to Forests

Epigraph from Craig Childs, "Springtime quakes are part of life on Earth," *Los Angeles Times*, April 14, 2010.

1. Information on Alfred Wegner and the eventual acceptance of his theory is from Patrick Hughes, "The meteorologist who started a revolution," *Weatherwise* 47 (1994): 1–29 (retrieved on October 20, 2009, at www.pangaea.org/wegener .htm).

2. Alfred Wegener, "Die Entstehung der Kontinente," *Geologische Rundschau* 3

(1912): 276–292; Alfred Wegener, *Die Entstehung der Kontinente und Ozeane (The Origin of Continents and Oceans)* (Braunschweig: Friedrich Vieweg & Sohn Akt. Ges., 1922); Alfred Wegener (translated from the 4th German edition by John Biram), *The Origin of Continents and Oceans* (New York: Dover, 1966). Firsthand accounts by the many scientists who developed plate tectonics theory are provided in Naomi Oreskes, ed., *Plate Tectonics: an Insider's History of the Modern Theory of the Earth* (Boulder, CO: Westview Press, 2003). For a good account of the theory as of 2005, see Stephen Marshak, *Earth: Portrait of a Planet*, 2nd edition (New York: W.W. Norton, 2005), chapter 3.

3. In Greek mythology, Iapetus was a Titan, the son of Uranus and Gaea, ancestor of humans through his son, Prometheus. Source: *Encyclopaedia Britannica*, Volume 14, 11th edition (New York: Encyclopaedia Britannica, 1910), 215 (retrieved on November 18, 2009, at books.google.com).

4. Description of the origins of the land of Maine are mainly from Cees R. van Staal and Robert D. Hatcher Jr., "Global setting of Ordovician orogenesis," *The Geological Society of America* Special Paper 466 (2010); C.R. Van Staal, "Pre-Carboniferous metallogeny of the Canadian Appalachians," in W.D. Goodfellow, ed., *Mineral Resources of Canada: a Synthesis of Major Deposit-Types, District Metallogeny, the Evolution of Geological Provinces, and Exploration Method* (Ottawa, ON: Mineral Deposit Division, Geological Association of Canada, and the Geological Survey of Canada, 2007); Robert G. Marvinney, *Simplified Bedrock Geologic Map of Maine* (Augusta: Maine Geological Survey, Department of Conservation, State of Maine, 2002). Additional information is from more "popular" accounts in C.R. Van Staal, "The Northern Appalachians," in R.C. Selley et al., eds., *Encyclopedia of Geology*, Volume 4 (Oxford, England: Elsevier, 2005), 81–91; Atlantic Geoscience Society, *The Last Billion Years: a Geological History of the Maritime Provinces of Canada* (Halifax, NS: Nimbus Publishers, 2001); David L. Kendall, *Glaciers and Granite: a Guide to Maine's Landscape and Geology* (Camden, ME: Down East Books, 1987), 157–176; Dean Bennett, *Maine's Natural Heritage: Rare Species and Unique Natural Features* (Camden, ME: Down East Books, 1988), 6–17; Stephen Pollock and Jan E. Pierson, "Foundation of the past: an introduction to Maine's bedrock geology," *Habitat: Journal of the Maine Audubon Society* 5 (1988): 18–21.

5. Dr. Doug Reusch, University of Maine at Farmington, personal communication.

6. James Zachos et al., "Trends, rhythms, and aberrations in global climate 65 Ma to present," *Science* 292 (2001): 686–693 (quotation: 686).

7. Good examples are provided by Richard B. Alley, *The Two-Mile Time Machine: Ice Cores, Abrupt Climate Change, and Our Future* (Princeton, NJ: Princeton University Press, 2000), 83–128.

8. Browse through a few issues of the journals *Science* or *Nature* and you're bound to come across articles and arguments on these issues. Despite the heat of this debate, there is robust consensus on the fundamentals of climate change, recent global warming, and the large role of human activities in that warming. See,

for example, R.K. Pachauri and A. Reisinger, eds., *Climate Change 2007: Synthesis Report*, Fourth Assessment Report (Geneva, Switzerland: International Panel on Climate Change, 2007).

9. When you brag that winter tends to be colder in Maine than in Florida, you're talking about climate; when you complain about a particularly cold winter day in Maine, you're talking about weather. As Robert Heinlein said, "Climate is what you expect, weather is what you get" (*Time Enough for Love* [New York: Putnam, 1973], 371. In our discussions in this section, we are always dealing with climate — average or prevailing weather over some relatively long period in a particular place, usually Maine.

10. Discussion of this theory of climate control relies on J.D. Hays, J. Imbrie, and N.J. Shackleton, "Variations in the Earth's orbit: pacemakers of the ice ages," *Science* 194 (1976): 1121–1132; John Imbrie and Katherine P. Imbrie, *Ice Ages: Solving the Mystery* (Cambridge, MA: Harvard University Press, 1986); R.A. Muller and G.J. MacDonald, "Glacial cycles and astronomical forcing," *Science* 277 (1997): 215–218; Zachos et al., "Trends, rhythms, and aberrations"; Alley, *The Two-Mile Time Machine*, 83–128.

11. In a recent publication, Dr. Doug Reusch of the University of Maine at Farmington argues that one of the most prominent climate shifts in Earth's history, from hot to cold, 34 million years ago, was the result of uplift and subduction of a large land area near Australia and New Zealand. This is a good example of the extent to which geologists think the shuffling around of continents on the Earth's surface can modify the climate. Douglas N. Reusch, "New Caledonian carbon sinks at the onset of Antarctic glaciation," *Geology* 39 (2011): 807–810.

12. International Stratigraphic Chart, International Commission on Stratigraphy (retrieved on March 8, 2010, at www.stratigraphy.org/upload/ISChart2009 .pdf).

13. When eccentricity is such that the Earth is closer to the Sun during the summer than the winter, seasonality is enhanced; when the opposite occurs, seasonality is reduced.

14. Imbrie and Imbrie, *Ice Ages*.

15. Milankovitch's mathematical theory of climate change builds on the earlier work of James Croll, *Climate and Time in their Geological Relations: A Theory of Secular Changes of the Earth's Climate* (New York: D. Appleton and Company, 1893; 1st edition, 1875) (retrieved on May 8, 2010, at Google Books).

16. Carl Wunsch, "Quantitative estimate of the Milankovitch-forced contribution to observed Quaternary climate change," *Quaternary Science Reviews* 23 (2004): 1001–1012.

17. H.W. Borns Jr. et al., "The deglaciation of Maine, U.S.A.," in J. Ehlers and P.L. Gibbard, eds., *Quaternary Glaciations — Extent and Chronology, Part II: North America* (Amsterdam, Netherlands: Elsevier, 2004), 89–109. Additional information from Harold W. Borns, "The making of Maine: the advance and retreat of the

last great glacier," *Habitat: Journal of the Maine Audubon Society* 5 (1988): 22–25; Thomas K. Weddle and Michael J. Retelle, "Deglacial history and relative sea-level changes: northern New England and adjacent Canada," *Geological Society of America*, Special Paper 351 (2001); Woodrow B. Thompson, Carol B. Griggs, Norton G. Miller, Robert E. Nelson, Thomas K. Weddle, and Taylor M. Kilian, "Associated terrestrial and marine fossils in the late-glacial Presumpscot Formation, southern Maine, USA, and the marine reservoir effect on radiocarbon ages," *Quaternary Research* 75 (2011): 552–565.

18. United States Geological Survey, Water Science for Schools Webpage (retrieved on November 20, 2009, at ga.water.usgs.gov/edu/waterproperties.html).

19. Information on the Younger Dryas from Borns et al., *Quaternary Glaciations — Extent and Chronology*; W.S. Broecker, D. Peteet, D. Rind, "Does the ocean-atmosphere system have more than one stable mode of operation?" *Nature* 315 (1985): 21–25; Alley, *The Two-Mile Time Machine*, 110–118; W.S. Broecker, G.H. Denton, R.L. Edwards, H. Cheng, R.B. Alley, and A.E. Putnam, "Putting the Younger Dryas cold event into context," *Quaternary Science Reviews* 29 (2010): 1078–1081.

20. Information on Maine during the Younger Dryas: W.A. Newman, A.N. Genes, and T. Brewer, "Pleistocene geology of north-eastern Maine," in H.W. Borns Jr., P. LaSalle, and W.B. Thompson, eds., *Late Pleistocene History of Northern New England and Adjacent Quebec*, Special Paper 197 (Boulder, CO: Geological Society of America, 1985), 59–70; Borns et al., "The deglaciation of Maine, U.S.A." Estimate of the abrupt rise in temperature at the end of the Younger Dryas: personal communication, Ann Dieffenbacher-Krall, whose estimates are based on fossil tree pollen and chironomid head capsules, indicating changes in the assemblages of these species at the end of the Younger Dryas.

21. Most of the plucking occurred due to freeze-thaw at the bottom of the ice sheet. As ice flows across obstructions, the increased pressure causes it to thaw, move easily over the obstruction because of lubrication from the water, and then refreeze when the pressure is released. This then adheres the rock to the ice allowing it to be plucked. Thanks to David Foster for this explanation.

22. The extreme pressure exerted by glaciers melts their undersides, which allows them to move readily downhill (north to south for the Laurentide Ice Sheet).

23. Descriptions of the signatures of glaciers on Maine's current landscape are mainly from W.B. Thompson and H.W. Borns Jr., *Surficial Geologic Map of Maine* (Augusta: Maine Geological Survey, 1985) updated by Robert G. Marvinney, 2003; Dabney W. Caldwell, *Roadside Geology of Maine* (Missoula, MT: Mountain Press, 1998); Maine's Ice Age Trail webpage, Geological Survey of Maine and Climate Change Institute, University of Maine (retrieved March 10, 2010, at www.maine.gov/doc/nrimc/mgs/explore/surficial/facts/jul07.htm). Additional information from Kendall, *Glaciers and Granite*.

24. W.B. Thompson and H.W. Borns Jr., *Surficial Geologic Map of Maine* (Augusta: Maine Geological Survey, 1985) updated by Robert G. Marvinney, 2003.

25. Robert M. Thorson, *Stone by Stone: The Magnificent History in New England's Stone Walls* (New York: Walker and Company, 2002), 45.

26. Readers will notice two types of past dates used in this chapter. Based on the natural decay of ^{14}C to ^{12}C, paleoecologists often obtain the ages of fossil organic material to calculate *radiocarbon years* before present. However, because the global atmospheric concentration of these two forms of carbon varies naturally over time, radiocarbon years usually do not equal *calendar years*. To calculate actual calendar age, radiocarbon years must be calibrated taking into account this changing ratio of carbon isotopes. Some scientific publications report raw radiocarbon years, but here we use real time, that is, actual years before present. For reference, we also provide radiocarbon years in figure 2.7.

27. Description of late Pleistocene and Holocene changes in environment and vegetation come mainly from Margaret B. Davis, "Pleistocene biogeography of temperate deciduous forests," *Geoscience and Man* 13 (1976): 13–26; Ronald B. Davis and George L. Jacobson Jr., "Late glacial and early Holocene landscapes in northern New England and adjacent areas of Canada," *Quaternary Research* 23 (1985): 341–368; G.L. Jacobson Jr., T. Webb III, and E.C. Grimm, "Patterns and rates of vegetation change during the deglaciation of eastern North America," in W.F. Ruddiman and H.W. Wright Jr., eds., *North America and Adjacent Oceans during the Last Deglaciation*, DNAG v. K-3 (Boulder, CO: The Geology Society of North America, 1987), 277–288; George Jacobson Jr. and Ronald B. Davis, "The real forest primeval: the evolution of Maine's forests over 14,000 years," *Habitat: Journal of the Maine Audubon Society* 5 (1988): 26–29; George Jacobson Jr. and Ann Dieffenbacher-Krall, "White pine and climate change: insights from the past," *Journal of Forestry* 93 (1995): 39–42; R. Scott Anderson et al., "History of late- and post-glacial vegetation and disturbance around Upper South Branch Pond, northern Maine," *Canadian Journal of Botany* 64 (1986): 1977–1986; R. Scott Anderson et al., "Gould Pond, Maine: late-glacial transitions from marine to upland environments," *Boreas* 21 (1992): 359–371; Stephen T. Jackson et al., "Mapped plant-macrofossil and pollen records of late Quaternary vegetation change in eastern North America," *Quaternary Science Reviews* 16 (1997): 1–70; D.H. DeHayes et al., "Forest responses to changing climate: lessons from the past and uncertainty for the future," in R.A. Mickler, R.A. Birdsey, and J. Hom, eds., *Responses of Northern U.S. Forests to Environmental Change*, Ecological Studies 139 (New York: Springer, 2000), 495–540. Additions from Heather Almquist et al., "The Holocene record of lake levels of Mansell Pond, central Maine, USA," *The Holocene* 11 (2001): 189–201; Molly Schauffler and George Jacobson, "Persistence of coastal spruce refugia during the Holocene in northern New England, USA, detected by stand-scale pollen stratigraphies," *Journal of Ecology* 90 (2002): 235–250; Bryan Shuman et al., "Evidence for the close climatic control of New England vegetation history," *Ecology* 85 (2004): 1297–1310; Ann M. Dieffenbacher-Krall and Andrea M. Nurse, "Late-glacial and Holocene record of lake levels of Mathews Pond and Whitehead

Lake, northern Maine, USA," *Journal of Paleolimnology* 34 (2005): 283–310; P.E. Newby, J.B. Bradley, A. Spiess, B. Shuman, and P. Leduc, "A Paleoindian response to Younger Dryas climate change," *Quaternary Science Reviews* 24 (2005): 141–154; Jonathan C. Lothrop, Paige E. Newby, Arthur E. Spiess, and James W. Bradley, "Paleoindians and the Younger Dryas in the New England-Maritimes Region," *Quaternary International* 242 (2011): 546–569.

28. The pollen of these two pine species is very difficult to distinguish, although one of the authors (AMB) and colleagues have developed a simple statistical approach using several measurements of pollen grains: Andrew M. Barton, Andrea M. Nurse, Katelyn Michaud, and Sarah W. Hardy, "Use of CART Analysis to Differentiate Pollen of Red Pine (*Pinus resinosa*) and Jack Pine (*P. banksiana*) in New England," *Quaternary Research* 75 (2011): 18–23.

29. This is in contrast to sites from which glaciers have retreated in place such as Alaska, where the development of forests is largely a successional process because temperatures have not changed as dramatically (yet) as they did during early postglacial times. The patterns of primary succession described here are mainly from research on those deglaciated sites: L. Crocker and J. Major, "Soil development in relation to vegetation and surface age at Glacier Bay, Alaska," *Journal of Ecology* 43 (1955): 427–448; D.B. Lawrence, "Glaciers and vegetation in south-eastern Alaska," *American Scientist* 46 (1958): 89–122; F.S. Chapin, L.R. Walker, and L.C. Sharman, "Mechanisms of primary succession following deglaciation at Glacier Bay, Alaska," *Ecological Monographs* 64 (1994): 149–175; Christopher L. Fastie, "Causes and ecosystem consequences of multiple pathways of primary succession at Glacier Bay, Alaska," *Ecology* 76 (1995): 1899–1916.

30. See P.E. Newby, J.B. Bradley, A. Spiess, B. Shuman, and P. Leduc, "A Paleoindian response to Younger Dryas climate change," *Quaternary Science Reviews* 24 (2005): 141–154.

31. Newby et al. "A Paleoindian response."

32. M.B. Davis, R.W. Spear, and L.C.K. Shane, "Holocene climate of New England," *Quaternary Research* 14 (1980): 240–250; Ray W. Spear, Margaret B. Davis, and Linda C. K. Shane, "Late Quaternary history of low- and mid-elevation vegetation in the White Mountains of New Hampshire," *Ecological Monographs* 64 (1994): 85–109; T. Webb, III, et al., 1993. "Vegetation, lake levels, and climate in eastern North America for the past 18,000 years," in H.E. Wright Jr., et al., eds., *Global Climates Since the Last Glacial Maximum* (Minneapolis: University of Minnesota Press, 1993), 415–467.

33. For a summary of the evidence, see Bryan Shuman et al., "Evidence for the close climatic control of New England vegetation history," *Ecology* 85 (2004): 1297–1310; D. Sanger, H. Almquist, and A.C. Dieffenbacher-Krall, "Mid-Holocene cultural adaptations to Central Maine," in D.H. Sandweiss and K.A. Maasch, eds., *Climate Change and Cultural Dynamics: A Global Perspective on Holocene Transitions* (London: Academic Press, 2006), 435–456. Ann Dieffenbacher-Krall (per-

sonal communication), of the University of Maine Climate Change Institute, isn't so sure about the role of fire, at least as far north as Maine, where her lake sediment studies have uncovered only modest evidence of charcoal during those early Holocene times. She thinks that the evidence is stronger for increased fire in southern New England.

34. Shuman et al., "Evidence for the close climatic control."

35. Information on the hemlock decline from Davis, "Pleistocene biogeography of temperate deciduous forests"; Najat Bhiry and Louise Filion, "Mid-Holocene hemlock decline in eastern North America linked with phytophagous insect activity," *Quaternary Research* 45 (1996): 312–320; Janice L. Fuller, "Ecological impact of the mid-Holocene hemlock decline in southern Ontario, Canada," *Ecology* 79 (1998): 2337–2351; Jean Nicolas Haas and John H. McAndrews, "The summer drought related hemlock *(Tsuga canadensis)* decline in eastern North America 5,700 to 5,100 years ago," in Katherine A. McManus, Kathleen S. Shields, Dennis R. Souto, eds., *Proceedings: Symposium on Sustainable Management of Hemlock Ecosystems in Eastern North America, GTR-NE-267* (Newtown Square, PA: U.S. Department of Agriculture, Forest Service, Northeastern Research Station, 2000), 81–88; Sanger et al., *Climate Change and Cultural Dynamics*; Shuman et al., "Evidence for the close climatic control"; David R. Foster et al., "A climatic driver for abrupt mid-Holocene vegetation dynamics and the hemlock decline in New England," *Ecology* 87 (2006): 2959–2966.

36. K. Gajewski, "Late Holocene pollen stratigraphy in four northeastern United States lakes," *Geographie Physique et Quaternaire* 41 (1987): 377–386; K. Gajewski, "Late Holocene climate changes in eastern North America estimated from pollen data," *Quaternary Research* 29 (1987): 255–262; Almquist et al, "The Holocene record of lake levels of Mansell Pond"; Dieffenbacher-Krall and Nurse, "Late-glacial and Holocene record of lake levels."

37. M. Schauffler and G. L. Jacobson Jr., "Persistence of coastal spruce refugia during the Holocene in northern New England, USA, detected by stand-scale pollen stratigraphies," *Journal of Ecology* 90 (2002): 235–250.

38. Matts Lindbladh et al., "A late-glacial transition from *Picea glauca* to *Picea mariana* in southern New England," *Quaternary Research* 67 (2007): 502–508.

39. E. Lucy Braun, *The Deciduous Forests of Eastern North America* (Philadelphia: Blakiston, 1950).

40. Davis, "Pleistocene biogeography of temperate deciduous forests"; Davis and Jacobson, "Late glacial and early Holocene landscapes"; Jacobson et al., "Patterns and rates of vegetation change during the deglaciation." The most recent treatment of this is D.S. Barrington and C.A. Paris, "Refugia and migration in the Quaternary history of the New England flora," *Rhodora* 109 (2007): 369–386.

41. P.A. Delcourt and H. R. Delcourt, *Long-Term Forest Dynamics of the Temperate Zone: A Case Study of Late-Quaternary Forests in Eastern North America* (New York: Springer-Verlag, 1987).

42. Richard Pearson, "Climate change and the migration capacity of species," *Trends in Ecology and Evolution* 21 (2006): 111–113.

43. Jason S. McLachlan, James S. Clark, and Paul S. Manos, "Molecular indicators of tree migration capacity under rapid climate change," *Ecology* 86 (2005): 2088–2098.

44. Pollen provides a sample of entire plant communities occurring in the past, whereas fossil remains of animal communities is very spotty. Some studies, however, have used fossils to identify large numbers of species from specific taxa, such as chironomids (Dorte Koster, John Lichter, Peter D. Lea, and Andrea Nurse, "Historical eutrophication in a river-estuary complex in mid-coast Maine," *Ecological Applications* 17 [2007]: 765–778) and beetles (Woodrow B. Thompson et al., "Associated terrestrial and marine fossils in the late-glacial Presumpscot Formation, southern Maine, USA, and the marine reservoir effect on radiocarbon ages," *Quaternary Research* 75 [2011]: 552–565). Discovering remnants of ancient animals requires great effort and luck, as well as tolerance of red herrings, such as the modern goat bone found by one of Maine's foremost archeologists, Arthur Spiess, in the early 1980s; *Boston Globe*, "Historical find — or just a goat bone?" September 15, 1983.

45. B. Gary Hoyle et al., "Late Pleistocene mammoth remains from Coastal Maine, USA," *Quaternary Research* 61 (2004): 277–288.

46. Arthur Spiess, "Comings and goings: Maine's prehistoric wildlife," *Habitat: Journal of the Maine Audubon Society* 5 (1988): 30–33.

47. John E. Guilday, "Differential extinction during late-Pleistocene and recent times," in P.S. Martin and H.E. Wright Jr., eds., *Pleistocene Extinctions: The Search for a Cause* (New Haven, CT: Yale University Press, 1967), 121–140; Paul S. Martin, "Prehistoric overkill," in P.S. Martin and H.E. Wright Jr., eds. *Pleistocene Extinctions: The Search for a Cause* (New Haven, CT: Yale University Press, 1967), 75–120.

48. Guilday, "Differential extinction during late-Pleistocene"; Richard A. Kiltie, "Seasonality, gestation time, and large mammal extinctions," in Paul S. Martin and Richard G. Klein, eds., *Quaternary Extinctions: A Prehistoric Revolution* (Tucson: University of Arizona Press, 1984), 299–314; R. Dale Guthrie, "New carbon dates link climatic change with human colonization and Pleistocene extinctions," *Nature* 441 (2006): 207–209.

49. Martin, "Prehistoric overkill"; John Alroy. "A multispecies overkill simulation of the end-Pleistocene megafaunal mass extinction," *Science* 292 (2001): 1893–1896.

50. Paul L. Koch and Anthony D. Barnosky, "Late Quaternary extinctions: state of the debate," *Annual Reviews of Ecology, Evolution, and Systematics* 37 (2006): 215–250.

51. Richard Firestone et al., "Evidence for an extraterrestrial impact 12,900 years

ago that contributed to the megafaunal extinctions and Younger Dryas cooling," *Proceedings of the National Academy of Sciences* 104 (2007): 16016–16021.

52. Jacquelyn L. Gill et al., "Pleistocene megafaunal collapse, novel plant communities, and enhanced fire regimes in North America," *Science* 326 (2009): 1100–1103.

53. Josh Donlan et al., "Re-wilding North America," *Nature* 436 (2005): 913–914.

54. William Stolzenburg, "Where the wild things were," *Conservation* 7 (2006): 28–34.

55. For information on early peoples and nonhuman animals of Maine, see Robson Bonnichsen, "The coming of the fluted-point people," *Habitat: Journal of the Maine Audubon Society* 5 (1988): 34–36; Spiess, "Comings and goings"; Arthur E. Spiess and Robert A. Lewis, *The Turner Farm Fauna: 5000 Years of Hunting and Fishing in Penobscot Bay, Maine,* Occasional Publication in Maine Archaeology No. 11 (Augusta: Maine Archaeological Society, 2001); Bruce J. Bourque, *Twelve Thousand Years: American Indians in Maine* (Lincoln: University of Nebraska Press, 2001), 1–101; Newby et al., "A Paleoindian response; Jonathan C. Lothrop, Paige E. Newby, Arthur E. Spiess, and James W. Bradley, "Paleoindians and the Younger Dryas in the New England-Maritimes Region," *Quaternary International* 242 (2011): 546–569.

56. D.W. Prentiss, "Description of an extinct mink from the shellheaps of the Maine coast," *Proceedings of the United States National Museum* 26 (1903): 887–888. Recent evidence points to the sea mink as a species distinct from other native weasels; Rebecca A. Sealfon, "Dental divergence supports species status of the extinct sea mink (Carnivora: Mustelidae: *Neovison macrodon*)," *Journal of Mammalogy* 88 (2007): 371–383.

57. Spiess and Lewis, *The Turner Farm Fauna,* 161.

58. Heather Almquist-Jacobson and David Sanger, "Holocene climate and vegetation in the Milford drainage basin, Maine, U.S.A., and their implications for human history," *Vegetation History and Archeobotany* 4 (1995): 211–222.

59. Alley, *The Two-Mile Time Machine,* 131.

60. Shuman et al., "Evidence for the close climatic control."

61. Emily W.B. Russell, Ronald B. Davis, R. Scott Anderson, Thomas E. Rhodes, and Dennis S. Anderson, "Recent centuries of vegetational change in the glaciated north-eastern United States," *Journal of Ecology* 81 (1993): 647–664.

62. For example, see Robert J. Whelan, *The Ecology of Fire* (Cambridge: Cambridge University Press, 1995); R.J. Naiman and H. Décamps, "The ecology of interfaces: riparian zones," *Annual Review of Ecology and Systematics* 28 (1997): 621–658.

63. John W. Williams, Bryan N. Shuman, and Thompson Webb III, "Dissimilarity analyses of late-Quaternary vegetation and climate in eastern North America," *Ecology* 82 (2001): 3346–3362.

3. The Presettlement Forest of Maine

Epigraph from E. Lucy Braun, *The Deciduous Forests of Eastern North America* (Philadelphia: Blakiston, 1950), 38, 338

1. Information on the Popham Colony from William D. Williamson, *The History of the State of Maine*, Volume 1 (Hallowell, ME: Glazier, Masters and Company, 1832), 196–203; [James Davies?], "Relation of a voyage to Sagadahoc, 1607–1608," in Henry S. Burrage, ed., *Early English and French Voyages, Chiefly from Hakluyt, 1534–1608* (New York: Charles Scribner's Sons, 1906), 397–419; Henry S. Burrage, *The Beginning of Colonial Maine, 1602–1658* (Portland, ME: Marks Printing House, 1914), 63–99; Edwin Churchill, "The European discovery of Maine," in Richard W. Judd, Edwin A. Churchill, and Joel W. Eastman, eds., *Maine: The Pine Tree State from Prehistory to the Present* (Orono: University of Maine Press, 1995), 31–50; Andrew J. Wahll, ed., *Sabino: Popham Colony Reader, 1602–2000* (Bowie, MD: Heritage Books, 2000), 72–102; Jeffrey P. Brain, "The Popham Colony: an historical and archeological brief," *Maine Archeological Society Bulletin* 43 (2003): 1–28. For Champlain, see Samuel de Champlain, *The Voyages and Explorations of Samuel de Champlain*, Volume I (New York: Allerton Book Co., 1904), chapters 3 and 4, 83–103. For Waymouth, see James Rosier, "True relation of Waymouth's voyage, 1605," in Henry S. Burrage, ed., *Early English and French Voyages, Chiefly from Hakluyt, 1534–1608* (New York: Charles Scribner's Sons 1906), 353–394.

2. Davies, "Relation of a voyage to Sagadahoc 1607–1608," 419.

3. Williamson, *The History of the State of Maine*, 196–203; Davies, "Relation of a voyage to Sagadahoc, 1607–1608"; Burrage, *The Beginning of Colonial Maine, 1602–1658*; Churchill, "The European discovery of Maine"; Wahll, *Sabino: Popham Colony Reader, 1602–2000*; Brain, "The Popham Colony: an historical and archeological brief."

4. John Josselyn, *New-England's Rarities Discovered* (London: Printed for G. Widdowes at the Green Dragon in St. Pauls Church-yard, 1672), 4 (retrieved on August 8, 2011 at books.google.com); Henry Wadsworth Longfellow, *Evangeline, a Tale of Acadie* (London: David Bogue, 6th edition; originally 1847), 1.

5. Estimated population size was 20,000; agriculture was apparently largely confined to the Saco River area and south; and evidence is lacking so far for widespread use of fire. See W.A. Patterson, III, and K.E. Sassaman, "Indian fires in the prehistory of New England," in G.P. Nicholas, ed., *Holocene Human Ecology in Northeastern North America* (New York: Plenum Publishing Co, 1988), 107–135; Harald E.L. Prins, "Turmoil on the Wabanaki frontier," in Richard W. Judd, Edwin A. Churchill, and Joel W. Eastman, eds., *Maine: The Pine Tree State from Prehistory to the Present* (Orono: University of Maine Press, 1995), 97–119; American Friends Service Committee, *The Wabanakis of Maine and the Maritimes* (Philadelphia, PA: American Friends Service Committee, 1989), A3–A10 and Section D: Fact Sheets; Bruce J. Bourque, "Prehistoric Indians of Maine," in Richard W. Judd, Edwin A.

Churchill, and Joel W. Eastman, eds., *Maine: The Pine Tree State from Prehistory to the Present* (Orono: University of Maine Press, 1995), 12–30; Bruce J. Bourque, *Twelve Thousand Years: American Indians in Maine.* (Lincoln: University of Nebraska Press, 2001); Phillip W. Conkling, *Islands in Time: A Natural and Cultural History of the Islands of the Gulf of Maine* (Camden, ME: Down East Books, 1999), 59; William Cronon, *Changes in the Land: Indians, Colonist, and the Ecology of New England* (New York: Hill and Wang, 1983). Examples of documented ecological impacts elsewhere in eastern North America include the following: G.E. Stinchcomb, T.C. Messner, S.G. Driese, L.C. Nordt, and R.M. Stewart, "Pre-colonial (A.D. 1100–1600) sedimentation related to prehistoric maize agriculture and climate change in eastern North America," *Geology* 39 (2011): 363–366; Kurt A. Fesenmyer and Norman L. Christensen Jr., "Reconstructing Holocene fire history in a southern Appalachian forest using soil charcoal," *Ecology* 91(2010): 662–670; Michael K. Foster and William Cowan, eds., *In Search of New England's Native Past: Selected Essays by Gordon M. Day* (Amherst: University of Massachusetts Press, 1998); W.A. Patterson, III, and K.E. Sassaman, "Indian fires in the prehistory of New England," in G.P. Nicholas, ed., *Holocene Human Ecology in Northeastern North America* (New York: Plenum Publishing Co, 1988), 107–135.

6. *Presettlement* forest is sometimes confusingly termed *pre-European settlement* forest, the latter usage being grammatically contradictory.

7. Silviculture is the branch of forestry concerned with the growing of trees and forest stands. Laura S. Kenefic et al., "Reference stands for silvicultural research: a Maine perspective," *Journal of Forestry* 103(2005): 363–367; Robert S. Seymour, Alan S. White, and Philip deMaynadier, "Natural disturbance regimes in northeastern North America evaluating silvicultural systems using natural scales and frequencies," *Forest Ecology and Management* 155 (2002): 357–367.

8. Although less now than in the past, some scientists equate presettlement forests with climax forests, the ultimate stable endpoint of succession, a theory whose acceptance has waxed and waned over the past decades but which has played an important role in the development of the science of ecology.

9. Even our most intensive national process (U.S. Forest Service's Forests Inventory Analysis) only measures forest parameters at one location per 6,000-plus acres and every five years.

10. See for example, William L. Balee, *Advances in Historical Ecology* (New York: Columbia University Press, 1998); Dave Egan and Evelyn A. Howell, *The Historical Ecology Handbook: A Restorationist's Guide to Reference Ecosystems* (Washington, DC: Island Press, 2001).

11. L.C. Wroth, ed., *The Voyages of Giovanni da Verrazzano, 1524–1528* (New Haven, CT: Yale University Press, 1970); B.F. DeCosta, *Ancient Norumbega or the Voyages of Simon Ferdinando and John Walker to the Penobscot River 1579–1580* (Albany, NY: John Munsell's Sons, 1890), New England Historical Genealogical Society, April 1890, 7.

12. For early usage see John Evelyn, *Sylva: Or a Discourse of Forest-trees and Propagation of Timber in his Majesties Dominion* (London: The Royal Society, 1664); Phillip Miller, *The Gardeners Dictionary*, 4th edition (London: John Rivington, 1754); Humphrey Marshall, *Arbustrum Americanum: The American Grove* (Philadelphia: J. Cruickshank, 1785); Charles Cogbill witness tree database. For modern usage see Fay Hyland and Ferdinand H. Steinmetz, *The Woody Plants of Maine, Their Occurrence and Distribution*, University of Maine Studies, 2nd. Ser. No. 59 (Orono: University of Maine, 1944).

13. The expanding English Royal Navy needed the ultra-functional white pine for masts and spars as the traditional supply of Scots pine from the Baltic was controlled by others and supposedly dwindling. R.G. Albion, *Forests and Sea Power* (Cambridge, MA: Harvard University Press, 1926); Charles P. Carroll, *The Timber Economy of Puritan New England* (Providence, RI: Brown University Press, 1973). Interestingly, George Waymouth's voyage (1605) is sometimes attributed the distinction of having first introduced white pine to England. Confusingly a century later white pine was given the name "Lord Weymouth pine" or "Weymouth pine" in Britain, apparently because Lord Weymouth (spelled with an *e*) imported it in about 1705 and planted it widely on his estate. Neither of these names was used in early American records. Wikipedia entry for Thomas Thynne, 1st Viscount Weymouth, (retrieved on August 18, 2011); Albion, *Forests and Sea Power*; Carroll, *The Timber Economy of Puritan New England*.

14. H.P. Biggar, ed., W.F. Ganong, translator, *The Works of Samuel de Champlain, Volume 1: Voyages du Sievr de Champlain* (Toronto: The Champlain Society, 1922), 270–342, gives both the French original and an English translation. The Verrazzano and Champlain narratives were not written in English, but the French vernacular names (*sapin* for "fir," generally meaning spruce, or *noyer* for "nut-tree" or "walnut," which must be hickory in modern terms) were no less ambiguous in their literal translations.

15. There is much difference of opinion over the exact spot of this first description from inland Maine. Recently David Morey published an analysis of the evidence from Rosier's account of the voyage, persuasively arguing that the inland "river" ascended by Waymouth was actually Penobscot Bay extending into the Penobscot River itself. Then they started their trek at the Rockport/Camden line. Others traditionally propose that it was from the St. George River in Warren. In either case, the trip was into the southern Camden Hills near the Goose River. David C. Morey, editor and annotator, *The Voyage of* Archangell, *James Rosier's Account of the Waymouth Voyage of 1605, A True Relation* (Gardiner, ME: Tilbury House, 2005). For the St. George River interpretation see Rosiers Relation of Waymouth's Voyage of 1605 in Burrage, ed., *Early English and French Voyages*, 355–398, 369–370, 381–389, quotation for Camden hills: 384–385.

16. Translations of Champlain vary. Champlain's original French (in Biggar) reads, *"I'y vis peu de sapins, mais bien quelques pins à vn costé de la riuiere: Tous*

chesnes à l'autre" (page 291), meaning "I saw a few firs (old usage), but also some pines on one side and the river; all oaks on the other." This has been added to by Otis as "lofty pines" and "massive oaks," while Eckstrom translates *"sapin"* as "spruce," and others use the literal "fir" (page 56). See Biggar and Ganong, *The Works of Samuel de Champlain, Vol. 1, Voyages dv Sievr de Champlain*, 291; Charles P. Otis, translator, *Voyages of Samuel de Champlain* (Boston: The Prince Society, 1880), 34–68, quotation on page 42.

17. Information on location of Champlain's observations and the historic condition of the forest in Fannie H. Eckstrom, "Champlain's visit to the Penobscot," *Sprague's Journal of Maine History* 1 (1913): 56–65; "Rosier's relation of Waymouth's voyage of 1605," in Burrage, *Early English and French Voyages*, 355–394, quotation: 386–387.

18. The maps of both E. Lucy Braun's "original forest pattern" and of A.W. Kuchler's "potential natural vegetation" indicate a mixed hardwood covering most of Maine. In contrast, nowhere in the early descriptions were forests in these coastal and riverine settings dominated by northern hardwood forest. See Braun, *Deciduous Forests of Eastern North America*, attached map; A.W. Kuchler, *Potential Natural Vegetation of the Conterminous United States*, Special Publication No. 36 (Washington, DC: American Geographical Society, 1964), loose map.

19. Maine State Archives, Land Office Records, Field Notes, Box 3 3–2.

20. Adams and Perham field notes, Bingham Papers, Historical Society of Pennsylvania, Philadelphia, PA.

21. Maine Historical Society, *Documentary History of the State of Maine*, vol. 8, "Northeastern Boundary arbitration Statement on the part of the United States of the case referred to the King of the Netherlands" (Portland, ME: The Lafavor-Tower Co., 1902), 64; Alec McEwen, *In Search of the Highlands: Mapping the Canada-Maine Boundary* (Fredericton, NB: Acadiensis Press, 1839, 1988), 43–47; R. Judd, "The Aroostook War, 1828–42," in Judd et al., *Maine: The Pine Tree State*, 345–353.

22. W. F. Odell in Maine Historical Society, "Northeastern Boundary arbitration," appx. 56, 405–406, 416–417.

23. Map in Maine Historical Society. A plan of the road from Belfast to the Western Line of the Waldo Patent taken at the Request of Henry Knox, Esq., and Charles Barret, Esq., by James Malcom, Sept. 24th, 1974; published in D. Albion, *Searsmont: The Old Towns of Quantabacook 1764–1796* (Camden, ME: Camden Herald Publ. Co., 1978).

24. For details of the grants, town founding, settlement, and relationship between the settlers, including squatters, and the owners, see Alan Taylor, *Liberty Men and the Great Proprietors: The Revolutionary Settlement on the Maine Frontier, 1760–1820* (Chapel Hill: University of North Carolina Press, 1990); Charles E. Clark, *The Eastern Frontier: The Settlement of Northern New England, 1610–1763* (Hanover, NH: University of New England Press, 1970); Edwin A. Churchill,

"English beachheads in seventeenth-century Maine," in Judd et al., *Maine: The Pine Tree State*, chapter 3.

25. Ephraim Ballard was the husband of Martha Ballard, author of the well-known diary, published as a book (Laurel Thatcher Ulrich, *A Midwife's Tale: The Life of Martha Ballard, Based on Her Diary, 1785–1812* [New York: Vintage Books, 1990]), and produced as a film (*The Midwife's Tale*, written and produced by Laurie Kahn-Leavitt, directed by Richard P. Rogers, aired on Public Broadcasting Corporation's *American Experience*, 1998). Martha's diary entries for this period indicate Ephraim going off to survey. Ephraim received help with this survey from Lemuel Perham (see chapter 4, "The Perham Farm . . .").

26. C.V. Cogbill, John Burk, and G. Motzkin, "The forests of presettlement New England, USA: spatial and compositional patterns based on town proprietor surveys," *Journal of Biogeography* 29 (2002): 1279–1304.

27. Interestingly, many vernacular Maine names for trees in old records have changed (appendix 3.3). For example, Norway pine (today's red pine) was named after its similarity to Scots Pine from the country of Norway. Contrary to local lore, it was not named after Norway, Maine, as the name was first used in 1732 in Scarborough, Maine, while the town was not named until 1797.

28. The "species" of presettlement surveys are a simplification of today's taxonomic species. The surveys do not include some rare species or difficult distinctions such as the spruces. The name "red spruce" did not actually exist until 1880. Before that time, it was lumped with black spruce both by surveyors and taxonomists. See Cogbill et al., "The forests of presettlement New England."

29. White oak is found in the Kennebec valley as far north as Waterville. Bur oak (a close relative of white oak) grows in the Sebasticook and Penobscot valleys. In 1848, on the Penobscot this white oak had only "a few trees . . . yet standing, and . . . an old settler [said] that they were formerly abundant." Aaron Young, "The forests of York County," *The Maine Farmer and Mechanic*, May 5, 1848.

30. These regions are easily separated, except along the mid-coast where three zones meet. Here the secondary hardwood-conifer transition is marked by a mixing of species and does not represent any range limits. There is also a magnification of the transition as seen today in the Camden Hills, where the most southern oak forest is found on south-facing slopes and the most northern spruce forest is found on adjacent north-facing slopes.

31. Young, "The forests of York County."

32. C.G. Lorimer, "The presettlement forest and natural disturbance cycle of northeastern Maine," *Ecology* 58 (1977): 139–148.

33. Lori Mitchener, unpublished data.

34. Maine State Archives, Land Office Records, Field Notes Box 3 3–6; Maine State Archives.

35. Determination of fire regimes based on one point in time can be subject to large errors, depending on when the last disturbance occurred. The dry sandy

substrate, composition of fire-adapted species (e.g., black spruce, red pine, aspen), and multiple dates of the fires all indicate that this was a long-standing pattern in this extreme northwestern part of Maine.

36. There are important challenges associated with these data: Cores are available at relatively few sites, species differ in pollen production and dispersal distance, and core sites differ in the distance from which they collect pollen. With these caveats in mind, we report studies appropriate for the questions addressed here. See S.T. Jackson, "Pollen source area and representation in small lakes in the northeastern United States," *Review of Palaeobotany and Palynology* 63 (1990): 53–76; M.B. Davis, "Palynology after Y2K — understanding the source area of pollen in sediments," *Annual Review of Earth and Planetary Sciences* 28 (2000): 1–18.

37. We tallied only "larger" microscopic particles of charcoal (> 250 μm), which are indicative of the local fire regime, as opposed to smaller particles, which float long distances, potentially from quite distant fire sources.

38. R. Scott Anderson, George L. Jacobson Jr., Ronald B. Davis, and Robert Stuckenrath, "Gould Pond, Maine: Late-glacial transitions from marine to upland environments," *Boreas* 21 (1992): 359–371.

39. R. Scott Anderson, Ronald B. Davis, Norton G. Miller, and Robert Stuckenrath, "History of late- and post-glacial vegetation and disturbance around Upper South Branch Pond, northern Maine," *Canadian Journal of Botany* 64 (1986): 1977–1986.

40. Molly Schauffler and George Jacobson, "Persistence of coastal spruce refugia during the Holocene in northern New England, USA, detected by stand-scale pollen stratigraphies," *Journal of Ecology* 90 (2002): 235–250.

41. From Longfellow, *Evangeline, a Tale of Acadie.* Echoing the first stanza, "This is the forest primeval," Longfellow thus concludes his poem with this beginning of his last stanza.

42. A sign at Big Reed Pond relates the important role also of Stephen Wheatland (1897–1987), the Pingree family agent for sixty years, who resisted pressure to cut old growth around the pond.

43. The results and conclusions from years of research at Big Reed Forest Reserve are mainly from M.J. Moesswilde, "Age structure, disturbance, and development of old growth red spruce stands in northern Maine" (master's thesis, University of Maine, 1995); Unna Chokkalingam, "Spatial and temporal patterns and dynamics in old-growth northern hardwood and mixed forests of northern Maine" (PhD dissertation, University of Maine, 1998); Unna Chokkalingam and Alan S. White, "Structure and spatial patterns of trees in old-growth northern hardwood and mixed forests of northern Maine," *Plant Ecology* 156 (2001): 139–160; Schauffler and Jacobson, "Persistence of coastal spruce refugia during the Holocene"; Shawn Fraver, "Spatial and temporal patterns of natural disturbance in old-growth forests of northern Maine, USA" (PhD dissertation, University of Maine, 2004); Shawn Fraver and Alan S. White, "Disturbance dynamics of old-

growth *Picea rubens* forests of northern Maine," *Journal of Vegetation Science* 16 (2005): 597–610; Shawn Fraver and Alan S. White, "Identifying growth releases in dendrochronological studies of forest disturbance," *Canadian Journal of Forest Research* 35 (2005): 1648–1656; Erika Rowland, "Disturbance pattern of multiple temporal scales in old-growth conifer–northern hardwood stands in northern Maine, USA" (PhD dissertation, University of Maine, 2006); Shawn Fraver, Robert S. Seymour, James H. Speer, and Alan S. White, "Dendrochronological reconstruction of spruce budworm outbreaks in northern Maine, USA," *Canadian Journal of Forest Research* 37 (2007): 523–529; Shawn Fraver, Alan S. White, and Robert S. Seymour, "Natural disturbance in an old-growth landscape of northern Maine, USA," *Journal of Ecology* 97 (2009): 289–298; Erica L. Rowland and Alan S. White, "Topographic and compositional influences on disturbance patterns in a northern Maine old-growth landscape," *Forest Ecology and Management* 259 (2010): 2399–2409. See also Robert S. Seymour, Alan S. White, and Philip de-Maynadier, "Natural disturbance regimes in northeastern North America: evaluating silvicultural systems using natural scales and frequencies," *Forest Ecology and Management* 155 (2002): 357–367; Laura S. Kenefic et al., "Reference stands for silvicultural research: a Maine perspective," *Journal of Forestry* 103(2005): 363–367; Charles V. Cogbill, *Evaluation of the Forest History and Old-Growth Nature of the Big Reed Pond Preserve, T8 R10 and T8 R11 WELS, Maine* (report prepared for the Maine Nature Conservancy, 1985). Thanks to The Nature Conservancy in Maine for constant support and funding for studies in their Big Reed Forest Reserve.

44. Dbh, diameter at breast height, is the standard way to measure the size of a tree. "Breast height" is defined as 4.5 feet or 1.4 meters.

45. Several species (quaking aspen, big-tooth aspen, black cherry, red pine, balsam poplar, and basswood) are unknown from the Reserve, even though they are common elsewhere in Maine, even in the harvested areas around the Reserve.

46. Lee E. Frelich, *Forest Dynamics and Disturbance Regimes* (New York: Cambridge University Press, 2002).

47. Zebulon Bradley and Edwin Rose in 1833 (T8 R10 WELS) and Bradley and William Parrott in 1840 (T8 R11 WELS) conducted the presettlement surveys of the vicinity of Big Reed Forest Reserve. Maine State Archives, Land Office Records, Field notes 12:15, 16, 20, 27; 63: 93–106.

48. For witness trees see Maine State Archives, Land Office Records, Field notes 12:15, 16, 20, 27; 63: 93–106. for narrative description of west line of T8 R10 see Maine State Archives, Land Office Records, Field notes 12:15.

49. Maine State Archives, Augusta, ME. Land Office Records, Field Notes 12:15, 16 20, 27; 63: 93–106. The 1976 proportions of the 2 mile line segment are 39 percent mixed wood, 30 percent hardwoods, 28 percent softwoods, and 3 percent cedar swamp. Survey cruise notes and aerial photographic type mapping (see figure 3.11) done in 1976 for the Seven Islands Company, by the James Sewell Company, Old Town, Maine.

50. Descriptions of Maine's most important remnant old-growth stands can be found in Critical Areas Program, *Natural Old-Growth Forest Stands in Maine and Its Relevance to the Critical Areas Program*, Planning Report 77 (Augusta, ME: State Planning Office, 1983).

51. This work was carried out chiefly by the Maine Critical Areas Program, now the Maine Natural Areas Program, which culminated in the publication of Critical Areas Program, *Natural Old-Growth Forest Stands in Maine and Its Relevance to the Critical Areas Program*.

52. John M. Hagan and Andrew A. Whitman, "Late-successional forest: a disappearing age class and implications for biodiversity," Report FMSN 2004–2 (Brunswick, ME: Manomet Center for Conservation Sciences, 2004).

53. The Cobbosseecontee stand was selectively logged in the 1995 and severely damaged by the ice storm of 1998. The stand measurements were done before these disturbances, and even today many of the characteristics are only mildly changed from before the disturbances.

54. This age structure implies that the typical tree is half the age of the maximum biological longevity and that the average age in a stand is roughly half the maximum age observed. For example, the average age at North Turner Brook is 240 years, 57 percent of the maximum 423 years (figure 3.17), while the four old-growth stands' median age varies from 51 percent to 71 percent of the maximum (table 3.3).

55. All four remnant stands had scattered small "bits" (cubical fragments) or BBs (rounded eroded chunks) of charcoal in the upper levels of the mineral soil. The charcoal's lack of abundance, location below the decayed forest floor, and small size and eroded condition all indicate a long-past fire history on the site. Certainly fires did not cause the initiation of the current stands. Interestingly, despite detailed searches at Big Reed, soil charcoal, all deeply buried, has been found at only a few isolated sites. The sample of remnant stands were selected for their lack of a history of catastrophic disturbance, and fire can be further eliminated as a recent factor at these sites.

56. Francis Parkman, *Francis Parkman's Works: A Half-Century of Conflict* (Boston: Little, Brown, 1907).

57. The density of trees in the presettlement forest of northwestern Maine can be calculated from the measured distance between some 291 corner points and witness trees in original surveys (1845–1850). The average point to tree distance of 12.8 feet infers a forest density of 172.5 witness trees per acre. This compares to a modern density of 206 trees > 4" dbh per acre over the whole Big Reed forest or an average of 258 trees per acre for the four old-growth stands (see figure 3.3). This discrepancy could be due to smaller minimum tree sizes used by early surveyors. Data from Charles Cogbill witness tree database.

58. Consistent with this conclusion, surveyors frequently mentioned small saplings (or "staddles"), but only rarely referred to "great" trees, their term for large trees.

59. The oldest hemlock from Big Reed Forest Reserve by Chokkalingam in 1997; red spruce from North Turner Brook cored in Baxter State Park by Cogbill in 1990; cedar from Duboullie Lake (T15 R9 WELS) by Kern in 1983.

60. Hagan and Whitman, "Late-successional forest." The sample reported here certainly underestimates the number of very old trees, given the high frequency of old but partially hollow trees. The oldest northern white cedar (over 300 years), for example, had a hollow area about one-half its radius!

61. Presettlement land surveys of the area near the Big Reed Forest Reserve, in 1833 and 1840 reveal the occurrence of some large disturbances, including blow-down and fire. In T8 R11, "The land upon the north west is much burnt over [in about 1816] (19.5 percent of the town lines) leaving some pine and hardwood ridges which have escaped." Two sizeable areas, totaling 1.8 percent of the lines, were noted as "old growth mostly fallen down" including "much down" on the western line of T8 R10. Information from the following: Maine State Archives, Land Office Records, Field notes 12:15, 16 20, 27; 63: 93–106; State Land Records, Field Notes 16: 92; State Land Records, Field notes 89: 512–513; Philip T. Coolidge, *History of the Maine Woods* (Bangor, ME: Furbish-Roberts, 1963); A.H. Wilkins, *Ten Million Acres of Timber* (Woolwich, ME: TBW Books, 1978); Maine State Archives, Land Office Records, Field Notes 12:15, 16 20, 27; 63: 93–106; Maine State Archives, Land Office Records, Field Notes 16: 88.

62. For example, Charles V. Cogbill, "Dynamics of the boreal forests of the Laurentian Highlands, Canada," *Canadian Journal of Forest Research* 15 (1985): 252–261; Edward A. Johnson, *Fire and Vegetation Dynamics: Studies from the North American Boreal Forest* (Cambridge: Cambridge University Press, 1992); Deborah Elliot-Fisk, "The Taiga and Boreal Forest," in Michael G. Barbour and William Dwight Billings, eds., *North American Terrestrial Vegetation*, 2nd edition (Cambridge: Cambridge University Press, 2001).

63. Unlike old, fire-scarred trees, with readily dated past fires, pollen cores generally cannot provide the resolution necessary to evaluate the true prevalence and importance of past fire, with some exceptions, such as Erika Rowland's cores from hollows in Big Reed Forest Reserve.

64. C.A. Copenheaver, A.S. White, and W.A. Patterson III, "Vegetation development in a southern Maine pitch pine–scrub oak barren," *Journal of the Torrey Botanical Society* 127 (2000): 19–32.

65. For a discussion of the much lower impact of Indian burning in Maine compared to southern New England, see W.A. Patterson, III, and K.E. Sassaman, "Indian fires in the prehistory of New England," in G.P. Nicholas, ed., *Holocene Human Ecology in Northeastern North America* (New York: Plenum Publishing Co, 1988), 107–135.

66. For example, Patterson and Sassaman, "Indian fires in the prehistory of New England"; William Cronon, *Changes in the Land: Indians, Colonist, and the Ecology of New England* (New York: Hill and Wang, 1983).

67. Emily W.B. Russell, Ronald B. Davis, R. Scott Anderson, Thomas E. Rhodes, and Dennis S. Anderson, "Recent centuries of vegetational change in the glaciated north-eastern United States," *Journal of Ecology* 81 (1993): 647–664; Schauffler and Jacobson, "Persistence of coastal spruce refugia during the Holocene."

4. From European Settlement to Modern Times

Epigraph from Rene Dubos, *The Wooing of Earth* (New York: Charles Scribers' Sons, 1980).

1. The history of Farmington and the Perham family: Francis G. Butler, *A History of Farmington, Franklin County, 1776–1885* (Farmington, ME: Press of Knowlton, McLeary, and Co., 1885), 551 and 39; Thomas Parker, *History of Farmington, Maine, From Its Settlement to the Year 1846* (Farmington, ME: J.S. Swift, 1846), 18.

2. Personal communication, Dr. Doug Reusch, associate professor of geology, University of Maine at Farmington.

3. David Sanger, "The original Native Mainers," *Habitat: Journal of the Maine Audubon Society* 5 (1988): 37–41; American Friends Service Committee, *The Wabanakis of Maine and the Maritimes* (Philadelphia, PA: American Friends Service Committee, 1989), A3–A10 & Section D: Fact Sheets; Bruce J. Bourque, "Prehistoric Indians of Maine," in Richard W. Judd, Edwin A. Churchill, and Joel W. Eastman, eds., *Maine: The Pine Tree State from Prehistory to the Present* (Orono: University of Maine Press, 1995), 12–30; Bruce J. Bourque, *Twelve Thousand Years: American Indians in Maine* (Lincoln: University of Nebraska Press, 2001), 73–109; William Cronon, *Changes in the Land: Indians, Colonist, and the Ecology of New England* (New York: Hill and Wang, 1983). One of the largest Penobscot villages was located where the Sandy River empties into the Kennebec, near present-day Norridgewock, until it was decimated by the mid-1700s by disease and an attack: David L. Ghere, "Diplomacy and war on the Maine frontier, 1678–1759," in Judd et al., eds., *Maine: The Pine Tree State from Prehistory to the Present*, 130, 138.

4. Butler, *A History of Farmington*, 45.

5. This line, surveyed in 1789, was the northwest corner of the Kennebec Purchase, now located on the Farmington/New Sharon town line in Franklin County, as described in chapter 3. Today, these species are supplemented with more light-demanding, disturbance-associated species, such as aspens, pin cherry, white birch, and red maple.

6. Butler, *History of Farmington*, 550.

7. Silas (Sr.), Hannah, and their granddaughter, Georgiana (daughter of Silas D. and Mary H.), who died in 1847 at the age of two, are the only family members buried in the tiny Perham cemetery on the farm.

8. According to the 1950 U.S. Agricultural Census, the Perhams grew Indian corn, potatoes, beans, peas, wheat, and oats and owned two horses, two working oxen, three milk cows, eight other cattle, and fifty sheep.

9. Evidence of the life of the Perham family comes from genealogical descriptions in Butler, *History of Farmington*, 549–552; 1850 U.S. Agricultural Census; maps of Farmington, including 1910 U.S. Geological Survey topographic map; descriptions from real estate deeds; personal communication with neighbors, especially Burdina Hardy, Henry Hardy, and Oliver Osborne; observations of the land by the current owners, Andrew Barton and Sarah Sloane.

10. U.S. Productions of Industry Census 1850–1880.

11. The size of the ownership changed over time, and was about 120 acres by the mid-1900s.

12. History of Farmington is largely from Parker, *History of Farmington*, and Butler, *A History of Farmington*.

13. Philip T. Coolidge, *History of the Maine Woods* (Bangor, ME: Furbish-Roberts, 1963), 19.

14. Richard G. Wood, *A History of Lumbering in Maine, 1820–1861*, Maine Studies No. 33 (Orono: University of Maine Press, 1935), preface written by David C. Smith.

15. Coolidge, *History of the Maine Woods*, 22; Lawrence C. Allin and Richard W. Judd, "Creating Maine's resource economy, 1783–1861," in Judd et al., eds., *Maine: The Pine Tree State from Prehistory to the Present*, 268.

16. Gordon M. Day, "The Indian as an ecological factor in the northeastern forest," *Ecology* 34 (1953): 329–346.

17. Harald E.L. Prins, "Turmoil on the Wabanaki frontier," in Judd et al., eds., *Maine: The Pine Tree State from Prehistory to the Present*, 97–119; Phillip W. Conkling, *Islands in Time: A Natural and Cultural History of the Islands of the Gulf of Maine* (Camden, ME: Down East Books, 1999), 59; Cronon, *Changes in the Land*, 85–89.

18. Cronon, *Changes in the Land*, 85–91.

19. English settlements: Edwin A. Churchill, "English beachheads in seventeenth-century Maine," in Judd et al., eds., *Maine: The Pine Tree State from Prehistory to the Present*, 51–75. French settlements: Alaric Faulkner and Gretchen Fearon Faulkner, "Acadian settlement, 1607–1700," in Judd et al., eds., *Maine: The Pine Tree State from Prehistory to the Present*, 76–96.

20. Coolidge estimates 1–2 percent: Coolidge, *History of the Maine Woods*, 3, 31–32. See also Allin and Judd, "Creating Maine's resource economy, 1783–1861," 268–270; John S. Springer, *Forest Life and Forest Trees: Comprising Winter Camp-Life Among the Loggers* (New York: Harper and Brothers, Publishers, 1851), 39–40.

21. Warfare in Maine, then part of Massachusetts: Prins, "Turmoil on the Wabanaki frontier," 114–118; Ghere, "Diplomacy and war on the Maine frontier, 1678–1759," 120–142; Richard D. Brown and Jack Tager, *Massachusetts: A Concise History* (Amherst: University of Massachusetts Press, 2000), 45–47; Conkling, *Islands in Time*, 63.

22. Alan Taylor, *Liberty Men and the Great Proprietors: The Revolutionary Settle-*

ment on the Maine Frontier, 1760–1820 (Chapel Hill: University of North Carolina Press, 1990). For the classic discussion of this phenomenon in southern New England, see Cronon, *Changes in the Land*, 77, 160–166.

23. U.S. Population Census Records, 1790–1840.

24. Wood, *A History of Lumbering in Maine, 1820–1861*, 48; David C. Smith, *A History of Lumbering in Maine, 1861–1960*, Maine Studies No. 93 (Orono: University of Maine Press, 1972), 5; Conkling, *Islands in Time*, 25; Coolidge, *History of the Maine Woods*, 31–32; Taylor, *Liberty Men and the Great Proprietors*. Even after statehood in 1820, half of the yet ungranted townships in Maine were "owned" by Massachusetts and together with the half "owned" by Maine were sold by a jointly run Land Office. Massachusetts continued to hold land until 1853; Wood, *A History of Lumbering in Maine, 1820–1861*, 49.

25. Estimate of forest area in 1600: Lloyd C. Irland, *Maine's Forest Area, 1600–1995: Review of Available Estimates*, Miscellaneous Publication 736 (Augusta: Maine Agricultural and Forest Experiment Station, University of Maine, 1998). Estimates of land clearing: Coolidge, *History of the Maine Woods*; Wood, *A History of Lumbering in Maine, 1820–1861*; Irland, *Maine's Forest Area, 1600–1995*.

26. Robert M. Thorson, *Stone by Stone: The Magnificent History in New England's Stone Walls* (New York: Walker and Company, 2002).

27. Wood, *A History of Lumbering in Maine, 1820–1861*, 24, 28, 182; Richard W. Judd, Edwin A. Churchill, and Joel W. Eastman, *Maine: The Pine Tree State from Prehistory to the Present* (Orono: University of Maine Press, 1995), 266–270; Coolidge, *History of the Maine Woods*, 33–50; Gordon G. Whitney, *From Coastal Wilderness to Fruited Plain: A History of Environmental Change in Temperate North America from 1500 to the Present* (Cambridge, Cambridge University Press, 1994), 214; Richard W. Judd, *Aroostook: A Century of Logging in Northern Maine* (Orono: University of Maine Press, 1989), 39.

28. History of uses of wood for purposes other than export lumber: Wood, *A History of Lumbering in Maine, 1820–1861*, 24, 28, 182; Judd, *Aroostook*, specifically 84–88; Judd et al., *Maine: The Pine Tree State from Prehistory to the Present*, 266–270; Coolidge, *History of the Maine Woods*, 33–50. Fuelwood use: Whitney, *From Coastal Wilderness to Fruited Plain*, 209–216; Lloyd C. Irland, *Wildlands and Woodlots: The Story of New England's Forests* (Hanover, NH: University Press of New England, 1982), 5–6.

29. Coolidge, *The History of the Maine Woods*, 31–32, 48, 50. For details on the watersheds and tributaries of the St. John River, see Judd, *Aroostook*, 1–19.

30. Henry David Thoreau, *The Maine Woods* (New York: Penguin Books, 1988; first published by Ticknor and Fields, 1864), 111

31. Coolidge, *The History of the Maine Woods*, 50.

32. The percentages of cleared land in this figure are lower than those reported elsewhere, because we include only land that was actually cleared and open, used as croplands, hayfields, pasture, orchards, etc. Most articles quote total farmland,

which also includes farm woodlands, which make up a substantial portion of total farm area. For example, in 1880, the peak year of clearance, woodland made up over 40 percent of the total 6.55 million acres in Maine farms. Although the general trends revealed here are robust, the specifics must be viewed with caution because of changing definitions of farmland categories over time, a point made forcefully in recent articles: Navin Ramankutty, Elizabeth Heller, and Jeanine Rhemtulla, "Prevailing myths about agricultural abandonment and forest regrowth in the United States," *Annals of the Association of American Geographers* 100 (2010): 502–512; SoEun Ahn et al., *Agricultural Land Changes in Maine: A Compilation and Brief Analysis of Census of Agriculture Data, 1850–1997*, Technical Bulletin 182 (Orono: Maine Agricultural and Forest Experiment Station, 2002).

33. Population data: U.S. Population Censuses, 1840–1880. Land clearance data: Coolidge, *History of the Maine Woods*, 307; U.S. Agricultural Censuses, 1850–1880; Ahn et al, *Agricultural Land Changes in Maine*. Other sources: Judd et al., *Maine: The Pine Tree State from Prehistory to the Present*; Judd, *Aroostook*, 7.

34. A good example being the Telos cut, which redirected water flow from the St. John to the Penobscot watershed.

35. Development of the lumber industry: Coolidge, *History of the Maine Woods*, 53–91; Wood, *A History of Lumbering in Maine, 1820–1861*, 23–48; Smith, *A History of Lumbering in Maine, 1861–1960*, 13; Allin and Judd, "Creating Maine's resource economy, 1783–1861," 268–273; Whitney, *From Coastal Wilderness to Fruited Plain*, 173–176. For the later development of the logging industry and advent of railroads in Aroostook County, see Judd, *Aroostook*, 129–172.

36. U.S. Bureau of the Census (Charles S. Sargent), *Tenth Census: Report on the Forests of North America*, (Washington, DC: Government Printing Office, 1884), map between pages 496 and 497. For an excellent description of the early development of the logging industry in this region, see Judd, *Aroostook*, 2–3, 47–79.

37. Report of the Commissioner appointed to superintend the Survey of the North American Railway (Augusta, ME, 1851), 6, quoted in Woods, *A History of Lumbering in Maine, 1820–1861*, 29.

38. Coolidge provides estimates of harvest for 1839, 1849, 1859, 1869, and 1879. We averaged these five for an average annual harvest level (about 500 million board feet) for the entire period and multiplied it by 40 years.

39. Emily W. B. Russell et al., "Recent centuries of vegetational change in the glaciated north-eastern United States," *Journal of Ecology* 81 (1993): 647–664; Heather Almquist-Jacobson and David Sanger, "Holocene vegetation and climate in the Milford drainage basin, Maine, U.S.A. and their implications for human history," *Vegetation History and Archeobotany* 4 (1995): 211–222; Molly Schauffler and George Jacobson, "Persistence of coastal spruce refugia during the Holocene in northern New England, USA, detected by stand-scale pollen stratigraphies," *Journal of Ecology* 90 (2002): 235–250; Ray W. Spear, Margaret B. Davis, and Linda C.K. Shane, "Late Quaternary history of low- and mid-elevation vegetation in the

White Mountains of New Hampshire, *Ecological Monographs* 64 (1994): 85–109; D. Köster, J. Lichter, P.D. Lea, and A. Nurse, "Historical eutrophication in a river-estuary complex in mid-coast Maine," *Ecological Applications* 17 (2007): 765–778.

40. Wood, *A History of Lumbering in Maine, 1820–1861*, 27–47, 135; Coolidge, *History of the Maine Woods*, 65; Smith, *A History of Lumbering in Maine, 1861–1960*, 12, 26; Allin and Judd, "Creating Maine's resource economy, 1783–1861," 268–270; Wayne M. O'Leary, Lawrence C. Allin, James B. Vickery, and Richard W. Judd, "Traditional industries in the age of monopoly, 1865–1930," in Judd et al., *Maine: The Pine Tree State from Prehistory to the Present*, 410–411; Irland, *Woodlands and Woodlots*, 28; Whitney, *From Coastal Wilderness to Fruited Plain*, 173–176.

41. Maine Census Data, Fogler Library, University of Maine (retrieved on March 31, 2011 at http://www.library.umaine.edu/census/default.htm).

42. Frederick H. Dyer, *A Compendium of the War of the Rebellion* (Des Moines, IA: The Dyer Publishing Company, 1908), 11.

43. O'Leary et al., "Traditional industries in the age of monopoly, 1865–1930," 404–410; Richard W. Judd, *Common Lands, Common People: The Origins of Conservation in Northern New England* (Cambridge, MA: Harvard University Press, 1997), 15–39; Coolidge, *The History of the Maine Woods*, 67, 81–84; Irland, *Wildlands and Woodlots*, 66–68, 116. A paper by Paul Frederic traces the paths of most outmigrants from Industry, Maine, to three primary areas: the Great Plains, the Upper Midwest, and California; Paul B. Frederic, "Going west! Leaving 19th century rural northern New England: diaspora from the town of Industry, Maine," in Dimitrios Konstadakopulos and Soterios Zoulas, *100 Years in America: Historical Determinants and Images of the Identity and Culture of Diasporas from Southeastern Europe*, Proceedings of the Conference held at the Hellenic College, Brookline, MA, October 11, 2008 (Bristol, England: The University of the West of England, 2010), 85–96.

44. Not all farmers struggled during this period; especially those on more productive lands thrived and expanded their holdings by purchasing the land surrendered by outmigrants, who in many cases tended more marginal lands. An excellent example of this process is provided by the Frederic farm, established in 1795 in Starks, a town whose population shrunk from 1,559 to 636 between 1840 and 1900. The Frederics stayed put, their farm thrived, and their land holdings burgeoned during that time and well into the next century. As Glen, father of current owner Paul Frederic and retired geography professor, said, "this was the best time to do business because you didn't need much money." The classic book on this phenomenon is for Vermont: Hal S. Barron, *Those Who Stayed Behind: Rural Society in Nineteenth-Century New England* (Cambridge: Cambridge University Press, 1984).

45. Judd, *Common Lands, Common People*, 15–39; O'Leary et al., "Traditional industries in the age of monopoly, 1865–1930," 404–410; Smith, *A History of Lumbering in Maine, 1861–1960*, 7.

46. R.M. DeGraaf and M. Yamasaki, *New England Wildlife, Habitats, Natural History and Distribution* (Hanover, NH: University Press of New England, 2001), 11.

47. O'Leary et al., "Traditional industries in the age of monopoly, 1865–1930," 411. See also Judd, *Aroostook*, 104.

48. Smith, *A History of Lumbering in Maine, 1861–1960*, 233; Judd, *Aroostook*, 173–199.

49. A main contributor to this success was the transition from mechanical to chemical pulping, a much more efficient process.

50. Smith, *A History of Lumbering in Maine, 1861–1960*, 233–261, 381–387; Irland, *Wildlands and Woodlots*, 29, 152–153.

51. Smith, *A History of Lumbering in Maine, 1861–1960*, 233–261; Coolidge, *The History of the Maine Woods*, 345–356; Nathan R. Lipfert, Richard W. Judd, and Richard R. Wescott, "New industries in an age of adjustment, 1865–1930" in Judd et al., *Maine: The Pine Tree State from Prehistory to the Present*, 427–429; R.S. Seymour, "The red spruce–balsam fir forest of Maine: evolution of silvicultural practice in response to stand development patterns and disturbances," in M.J. Kelty, B.C. Larson, and C.D. Oliver, C.D., eds., *The Ecology and Silviculture of Mixed-Species Forests* (Norwell, MA: Kluwer, 1992), 217–244; A. Cary, *Report of Austin Cary, Third Annual Report, Maine Forest Commissioner* (Augusta: Maine Forest Commission, 1896), 15–203; M. Westveld, "Observations on cutover pulpwood lands in the Northeast," *Journal of Forestry* 26 (1928): 649–664.

52. A. Cary, *Report of Austin Cary*, 100. In the U.S. Bureau of the Census 1884 *Report on the Forests of North America*, Charles Sargent, Professor of Arboriculture at Harvard, wrote, "The original pine and spruce forests of the state [Maine] have been practically destroyed"; U.S. Bureau of the Census, *Tenth Census*.

53. George Perkins Marsh, address delivered before the Agricultural Society of Rutland County, Sept. 30, 1847, in *The Evolution of the Conservation Movement, 1850–1920* (Washington, DC: Library of Congress, American Memory) (retrieved on March 30, 2011, at http://lcweb2.loc.gov/ammem/amrvhtml/conshome.html, 17–18.

54. Henry David Thoreau, *Walden; or Life in the Woods* (Boston: Ticknor and Fields, 1854); George Perkins Marsh, *Man and Nature* (New York: C. Scribner, 1864).

55. Smith, *A History of Lumbering in Maine, 1861–1960*, 333–353; Char Miller, *Gifford Pinchot and the Making of Modern Environmentalism* (Washington, DC: Island Press, 2001); Steven Fox, *The American Conservation Movement: John Muir and His Legacy* (Madison: University of Wisconsin Press, 1985).

56. Judd, *Common Lands, Common People*, 10.

57. Judd, *Common Lands, Common People*, 93; Judd, *Aroostook*, 204.

58. Lipfert et al., "New industries in an age of adjustment, 1865–1930," 434.

59. Concerns about heavy cutting from this paragraph: Judd, *Common Lands, Common People*, 93, 197–228; Lipfert et al., "New industries in an age of adjustment, 1865–1930," 430–438; Cary, *Report of Austin Cary*, 15–203.

60. Frances Wiggin, for example, argued for reserves in special places, such as the Rangeley Lakes, Moosehead Lake area, West Branch of Penobscot, Mt. Katahdin, and the Allagash River; Smith, *A History of Lumbering in Maine, 1861–1960*, 366.

61. Judd, *Common Lands, Common People*, 45, 90–120; Richard W. Judd, "The Maine Woods: a legacy of controversy," The Margaret Chase Smith Essay, *Maine Policy Review* 16 (2007): 8–10; Smith, *A History of Lumbering in Maine, 1861–1960*, 333–382; Coolidge, *The History of the Maine Woods*, 127–140, 278–280, 607–609; Irland, *Wildlands and Woodlots*, 130–132; Lipfert et al., "New industries in an age of adjustment, 1865–1930," 429–431, 437; Neil Rolde, *The Interrupted Forest: A History of Maine's Wildlands* (Gardiner, ME: Tilbury House, 2001), 270–272, 295–296; Judd, *Aroostook*, 207–221.

62. Judd, *Common Lands, Common People*, 90–120; Smith, *A History of Lumbering in Maine, 1861–1960*, 333–382; Coolidge, *The History of the Maine Woods*, 127–140, 278–280, 607–609; Irland, *Wildlands and Woodlots*, 130–132; Lipfert et al., "New industries in an age of adjustment, 1865–1930," 429–431.

63. Smith, *A History of Lumbering in Maine, 1861–1960*, 392–399, 422–428; Irland, *Woodlands and Woodlots*, 31–33; Coolidge, *The History of the Maine Woods*, 141–155.

64. Smith, *A History of Lumbering in Maine, 1861–1960*, 427.

65. Irland, *Woodlands and Woodlots*, 32; see also Seymour, "The red spruce–balsam fir forest of Maine," 217–244; Judd, *Aroostook*, 201–202, 250–255.

66. Lipfert et al., "New industries in an age of adjustment, 1865–1930," 438; Richard H. Condon, "Maine out of the mainstream, 1945–1965," in Judd et al., *Maine: The Pine Tree State from Prehistory to the Present*, 535–538; Irland, *Wildlands and Woodlots*, 28–29.

67. Farmland abandonment: Lipfert et al., "New industries in an age of adjustment, 1865–1930," 438; Condon, "Maine out of the mainstream, 1945–1965," 535–538. Pollen evidence: Schauffler and Jacobson, "Persistence of coastal spruce refugia during the Holocene"; Köster et al., "Historical eutrophication in a river-estuary complex in mid-coast Maine."

68. Even earlier than the exotic species reported here were diseases such as smallpox, brought in by early European explorers and settlers, which decimated Indian tribes in Maine and throughout the eastern United States (see this chapter: "Before 1781 . . .").

69. Russell et al., "Recent centuries of vegetational change"; Almquist-Jacobson and Sanger, "Holocene vegetation and climate in the Milford drainage basin"; Schauffler and Jacobson, "Persistence of coastal spruce refugia during the Holocene"; Köster et al., "Historical eutrophication in a river-estuary complex in mid-coast Maine."

70. Susan Freinkel, *American Chestnut: The Life, Death, and Rebirth of a Perfect Tree* (Berkeley, CA: University of California Press, 2007), 82.

71. Lorin L. Dame and Henry Brooks, *Handbook of the Trees of New England* (Boston: Athenaeum Press, 1901), 95–97; Coolidge, *History of the Maine Woods,* 260; U.S. Forest Service, Northeastern Area, *Forest Health Protection — Dutch Elm Disease* (retrieved on April 10, 2011at www.na.fs.fed.us/fhp/ded/).

72. William Livingston, personal communication; Matthew T. Kasson and William H. Livingston, "Spatial distribution of *Neonectria* species associated with beech bark disease in northern Maine," *Mycologia,* 101 (2009): 190–195; Coolidge, *History of the Maine Woods,* 258.

73. Irland, *Wildlands and Woodlots,* 170–171; Rolde, *The Interrupted Forest,* 313–329; DeGraaf and M. Yamasaki, *New England Wildlife,* 10, 341, 353.

74. Irland, *Wildlands and Woodlots,* 1.

75. Rolde, *The Interrupted Forest,* 25, 343–344, 349. By the end of the spraying program, chemical insecticides had been replaced by a bacterial one, Bt (*Bacillus thuringiensis*), which is relatively benign environmentally.

76. These machines, 18 feet between their tracks, were so expensive to purchase and operate that they were often run 24 hours a day; Judd, *Aroostook,* 251.

77. Mitch Lansky, *Beyond the Beauty Strip: Saving What Is Left of Our Forests* (Gardiner, ME: Tilbury House Publishers, 1992), 139.

78. Sources for changes during this period: Irland, *Wildlands and Woodlots,* 1, 9, 11, 30, 43, 46, 51, 81–83, 117; Lloyd C. Irland, "Maine forests: a century of change, 1900–2000," *Maine Policy Review* 9 (2000): 66–77; Lloyd C. Irland, *The Northeast's Changing Forest* (Petersham, MA: Harvard University Press and Harvard Forest, 1999); Lansky, *Beyond the Beauty Strip;* Judd, "The Maine Woods: a legacy of controversy"; Richard H. Condon and William D. Berry, "The tides of change, 1967–1988" in Judd et al., *Maine: The Pine Tree State from Prehistory to the Present,* 563–568; Northern Forest Lands Council, *Finding Common Ground: Conserving the Northern Forest* (Concord, NH: Northern Forest Lands Council, 1994); Christopher M. Klyza and Stephen C. Trombulak, eds., *The Future of the Northern Forest* (Hanover, NH: University Press of New England, 1994); Rolde, *The Interrupted Forest;* Mike LeVert, Charles S. Colgan, and Charles Lawton, "Are the economics of a sustainable Maine forest sustainable?" *Maine Policy Review* 16 (2007): 26–37; David Dobbs and Richard Ober, *The Northern Forest* (White River Junction, VT: Chelsea Green, 1995), xiii–xxvi, 59–159, 319–342; R.J. Lilieholm, L.C. Irland, and J.M. Hagan, "Changing socio-economic conditions for private woodland protection," in S.C. Trombulak, and R. Baldwin, eds., *Multi-Scale Conservation Planning* (New York: Springer-Verlag, 2010).

79. Nancy Allen, "Cutting out clearcutting," *Synthesis/Regeneration* 10, Spring 1996 (retrieved on April 30, 2011at http://www.greens.org/s-r/10/10–17.html).

80. Lansky, *Beyond the Beauty Strip,* 9. See also Dobbs and Ober, *The Northern Forest,* 117–136; Judd, *Aroostook,* 255–261; Rolde, *The Interrupted Forest;* Erika L. Rowland, Alan S. White, and William H. Livingston, *A Literature Review of the Effects of Intensive Forestry on Forest Structure and Plant Community Composition*

at the Stand and Landscape Levels, Miscellaneous Publication 754 (Orono: Maine Agricultural and Forest Experiment Station, 2005). For a perspective on changing pressures and policies in Maine's Unorganized Territories during these times, see Paul B. Frederic, "Public policy and land development: the Maine Land Use Regulation Commission," *Land Use Policy* 8 (1991): 50–62.

81. Forest-based jobs declined the most in the 1970s and 1980s, but starting in the 1990s and continuing to today, the mill workforce has been hit hard due to global competition. Lloyd Irland, "Maine's forest industry: from one era to another," in Richard Barringer, ed., *Changing Maine: 1960–2010* (Gardiner, ME: Tilbury House, 2004), 363–387.

82. Paul K. McCann, *Timber! The Fall of Maine's Paper Giant* (Ellsworth, ME: Ellsworth American, 1994), 65–66.

83. Additional purchases amounted to 96,486 in New York, 67,088 in New Hampshire, and 22,426 acres in Vermont.

84. Northern Forest Lands Council, *Finding Common Ground*; Carl Reidel, "The political process of the Northern Forest Lands Study," in Klyza and Trombulak, eds., *The Future of the Northern Forest*, 93–111.

85. Sources for quantitative data on changing land ownership: Suming Jin and Steven A. Sader, "Effects of forest ownership and change on forest harvest rates, types and trends in northern Maine," *Forest Ecology and Management* 228 (2006): 177–186; John M. Hagan, Lloyd C. Irland, and Andrew A. Whitman, *Changing Timberland Ownership in the Northern Forest and Implications for Biodiversity*, Report No. MCCS-FCP-2005–1 (Brunswick, ME: Manomet Center for Conservation Sciences, 2005). Sources for the aftermath of the Diamond International sale and the changing incentives for land ownership: Dobbs and Ober, *The Northern Forest*, xiii–xxvi, 59–159, 319–342; Sara A. Clark and Peter Howell, "From Diamond International to Plum Creek: the era of large landscape conservation in the northern forest," *Maine Policy Review* 16 (2007): 56–65; Irland, "Maine forests: a century of change, 1900–2000"; Irland, *Wildland and Woodlots*, 11; Irland, "Maine's forest industry"; LeVert et al., "Are the economics of a sustainable Maine forest sustainable?"; Dobbs and Ober, *The Northern Forest*, 119.

86. There are now many textbooks and edited volumes in the well-developed field of conservation biology. Three good ones are Malcolm Hunter Jr. and James Gibbs, *Fundamentals of Conservation Biology*, 3rd edition (Oxford, England: Blackwell Publishing, 2007), Martha J. Groom, Gary K. Meffe, and C. Ronald Carroll, *Principles of Conservation Biology*, 3rd edition (Sunderland, MA: Sinauer, 2006), Malcolm L. Hunter Jr., ed., *Maintaining Biodiversity in Forested Ecosystems* (Cambridge: Cambridge University Press, 1999).

87. Lansky, *Beyond the Beauty Strip*, 101–106, 372; Jamie Sayen, "Northern Appalachian wilderness: the key to sustainable natural and human communities in the northern forest," in Klyza and Trombulak, eds., *The Future of the Northern Forest*, 177–197; Stephen C. Trombulak, "Ecological health and the northern for-

est," *Vermont Law Review* 19 (1995): 283–333; Malcolm L. Hunter Jr., "The diversity of New England forest ecosystems," in J.A. Bissonette, ed., *Is Good Forestry Good Wildlife Management?* Miscellaneous Publication No. 689 (Orono: Maine Agricultural Experimental Station, 1986), 35–47. Data on old-growth sites is compiled in Mary Byrd Davis, *Old Growth in the East: A Survey*, an online edition, updated to 2008 (retrieved on May 2, 2011, at www.primalnature.org/ogeast/survey.html).

88. Gro Flatebo, Carol R. Foss, and Steven K. Pelletier, *Biodiversity in the Forests of Maine: Guidelines for Land Management*, UMCE Bulletin No. 7147 (Orono: University of Maine Cooperative Extension, 1999), 9. See also Susan C. Gawler et al., *Biological Diversity in Maine: An Assessment of Status and Trends in the Terrestrial and Freshwater Landscape* (report prepared for the Maine Forest Biodiversity Project) (Augusta: Maine Natural Areas Program, Department of Conservation, 1996); Hunter and James, *Fundamentals of Conservation Biology*; Rowland, et al., *A Literature Review of the Effects of Intensive Forestry*.

89. Complete lack of regeneration is rare in Maine, a feature of our forests that contrasts sharply with some other regions, like the West, where regenerating tree species without planting or seeding can be impossible.

90. Seymour, "The red spruce–balsam fir forest of Maine."

91. Irland, *Wildlands and Woodlots*; Irland, *The Northeast's Changing Forest*; Judd, "The Maine Woods: a legacy of controversy."

92. Maine Forest Service, Maine Forest Practices Act (retrieved on April 20, 2011, at www.maine.gov/doc/mfs/pubs/htm/fpa_04.html); Maine Forest Service, MFS Rule, Chapter 23, "Timber harvesting standards to substantially eliminate liquidation harvesting," 2005; Maine Forest Service, MFS Rule, Chapter 20, "Forest regeneration and clearcutting standards," 1999.

93. Maine Silvicultural Activities Reports, Department of Conservation, Maine Forest Service, 1988–2009; Steven A. Sader, Matthew Bertrand, and Emily H. Wilson, "Satellite change detection of forest harvest patterns on an industrial forest landscape," *Forest Science* 49 (2003): 341–353; Maine Forest Service, *An Evaluation of the Effects of the Forest Practices Act* (Augusta: Maine Forest Service, Department of Conservation, 1995); Maine Forest Service, *The State of the Forest and Recommendations for Forest Sustainability Standards* (Augusta: Maine Forest Service, Department of Conservation, 1999).

94. *The Final Report of the Maine Forest Certification Advisory Committee*, The Maine Forest Certification Initiative, January 28, 2005, 7.

95. Ibid, 16.

96. These values differ somewhat from those given in Clark and Howell, "From Diamond International to Plum Creek." The explanation for this appears to be differences in the years for which the acquisitions were coded. For this table, acquisitions were always coded for the year in which it was first made and not for subsequent conveyances (to the State of Maine, for example).

97. Clark and Howell, "From Diamond International to Plum Creek"; Chris-

topher S. Cronan et al., "An assessment of land conservation patterns in Maine based on spatial analysis of ecological and socioeconomic indicators," *Environmental Management* 45 (2010): 1076–1095.

98. Lansky, *Beyond the Beauty Strip*, 365.

99. This variety of expressions can be found in Dobbs and Ober, *The Northern Forest*; Robert J. Lilieholm, "Forging a common vision for Maine's North Woods," *Maine Policy Review* 16 (2007): 12–25; Elizabeth Dennis Baldwin, Laura S. Kenefic, and Will F. LaPage, "Alternative large-scale conservation visions for northern Maine: Interviews with decision leaders in Maine," *Maine Policy Review* 16 (2007): 78–91; Robert F. Baldwin, Stephen C. Trombulak, Karen Beazley, Conrad Reining, Gillian Woolmer, John R. Nordgren, and Mark Anderson, "Importance of Maine for ecoregional conservation planning," *Maine Policy Review* 16 (2007): 66–77.

100. Associated Press, "Burt's Bees founder Quimby wants to donate her land for national park," *Bangor Daily News*, March 28, 2011. In the early 1990s, Restore: The North Woods proposed a Maine Woods National Park and Preserve of 3.2 million acres, which has been debated ever since.

101. Judd, "The Maine Woods: a legacy of controversy."

102. Irland, *Maine's Forest Area, 1600–1995.*

103. Bill McKibben, "An explosion of green," *Atlantic Monthly*, April 1995 (online version retrieved on April 20, 2011 at www.theatlantic.com/politics/environ/green.htm). See also Lilieholm et al., "Changing socio-economic conditions for private woodland protection." For a lucid description of changes leading up to the modern forests of all of New England, see David R. Foster et al., *Wildlands and Woodlands* (Petersham, MA: Harvard Forest, 2010) and www.wildlandsand woodlands.org.

104. If unmerchantable, albeit never-cut, forests on upper slopes of mountains are included, the total old growth probably would approach 20,000 acres.

105. Current forest acres: G.L. McCaskill et al., *Maine's Forest Resources, 2008,* Res. Note NRS-53 (Newtown Square, PA: U.S. Department of Agriculture, Forest Service, Northern Research Station. 2010). Farmland clearing, forest recovery, forest area: U.S. Agricultural Censuses and Ahn et al., *Agricultural Land Changes in Maine*; Irland, *Maine's Forest Area, 1600–1995.* Old growth: Mary Byrd Davis, "Maine," in *Old Growth in the East: A Survey* (Online Ed.) (retrieved on May 31, 2011 at www.primalnature.org); C.V. Cogbill, "Black growth and fiddlebutts: the nature of old-growth red spruce," in M.B. David, ed., *Eastern Old-Growth Forests: Prospects for Rediscovery and Recovery* (Washington, DC: Island Press, 1996). Old-growth acreage does not include peatlands or alpine areas.

106. Gawler et al., *Biological Diversity in Maine*, 15, 20, 71.

107. Russell et al., "Recent centuries of vegetational change."

108. A set of plots around the state is inventoried every five years, one-fifth of them each year. This Forest Inventory and Analysis program collects a wide range

of data, including details about the sizes, numbers, and acreage of tree species and forest types, other plant species, and land ownership. Data used here are Tables A7 and A9 from the 2003 results: G.L. McCaskill et al., *Maine's Forest Resources*, 2008. Additional details can be found at Northeastern Forest and Inventory Analysis, USDA Forest Service (www.fs.fed.us/ne/fia/states/me/ME5yr.html).

109. Rowland et al., *A Literature Review of the Effects of Intensive Forestry* (Orono: Maine Agricultural and Forest Experiment Station, 2005).

5. The Length and Breadth and Height of Maine

Epigraph from Dean Bennett, *Maine's Natural Heritage: Rare Species and Unique Natural Features* (Camden, ME: Down East Books, 1988), inside cover page.

1. Captain John Smith, *A Description of New England* (1616), 22. Electronic version (Paul Royster, editor, 2006) (retrieved on March 8, 2010, at DigitalCommons@University of Nebraska — Lincoln, digitalcommons.unl.edu/).

2. The title for this chapter is taken from Stanley B. Atwood, *The Length and Breadth of Maine* (Orono: University of Maine Press, 2004).

3. Nicholas Jardine, James A. Secord, and Emma C. Spary, *Cultures of Natural History* (Cambridge: Cambridge University Press, 1996), 298.

4. Description from S.C. Gawler et al., *Biological Diversity in Maine: An Assessment of Status and Trends in the Terrestrial and Freshwater Landscape* (report prepared for the Maine Forest Biodiversity Project) (Augusta: Maine Natural Areas Program, Department of Conservation, 1996), appendices; Malcolm L. Hunter Jr., Aram J.K. Calhoun, and Mark McCollough, *Maine Amphibians and Reptiles* (Orono: University of Maine Press, 1999), 161–164; Susan C. Gawler, and Andrew R. Cutko, *Natural Landscapes of Maine: A Classification of Vegetated Natural Communities and Ecosystems* (Augusta: Maine Natural Areas Program, Department of Conservation, 2010); Personal communication with staff of The Nature Conservancy in Maine; The Nature Conservancy Kennebunk Plains webpage (retrieved March 8, 2010, at www.nature.org/wherewework/northamerica/states/maine/preserves/art20991.html).

5. David L. Kendall, *Glaciers and Granite: A Guide to Maine's Landscape and Geology* (Camden, ME: Down East Books, 1987), 81–82.

6. Diane Ebert May and Ronald B. Davis, *Alpine Tundra Vegetation of Maine Mountains and Its Relevance to the Critical Areas Program*, Planning Report 36 (Augusta: Maine Critical Areas Program, State Planning Office, 1978); Dean Bennett, *Maine's Natural Heritage: Rare Species and Unique Natural Features* (Camden, ME: Down East Books, 1988), 71–73; Gawler and Cutko, *Natural Landscapes of Maine*.

7. Details on these two species and this community are from Gawler and Cutko, *Natural Landscapes of Maine*.

8. The Birds of North America Online, American Pipit webpage (retrieved on March 8, 2010, at bna.birds.cornell.edu/bna/species/095/articles/introduction);

Maine Department of Inland Fisheries and Wildlife, Katahdin Arctic web-page (retrieved on March 8, 2010, at www.maine.gov/IFW/wildlife/species/endangered_species/katahdin_arctic/index.htm).

9. Information on the Orono Bog Boardwalk comes from material at the site and Orono Bog Boardwalk webpage (retrieved on March 8, 2010, at www.orono bogwalk.org/index.htm).

10. Originally, two lots of 320 acres were allotted per township and, two years later, four lots of the same size, one for the settled minister, one for the ministry, one for schools, and one reserved for later disposition by the legislature. After statehood, the State of Maine increased the total provision to 1,000 acres, a law that came into force in 1832 (Stanley B. Atwood, *The Length and Breadth of Maine* [Augusta, ME: Kennebec Journal Print Shop, 1946], 31). After 1850, rights to har-vest timber and grass were sold, mainly to private forest products companies, whose land surrounded the lots (Neil Rolde, *The Interrupted Forest: A History of the Maine's Wildlands* [Gardiner, ME: Tilbury House Publishers, 2001], 332–335).

11. Information from personal observations; personal communication with Bill Haslam; *Flagstaff Region Management Plan* (Augusta: Bureau of Parks and Lands, Department of Conservation, State of Maine, June 12, 2007), part C, 103.

12. S.G. Nilsson et al., Tree-dependent lichens and beetles as indicators in con-servation forests. *Conservation Biology* 9 (1995): 1208–1215; Andrew A. Whitman and John M. Hagan, "An index to identify late-successional forest in temperate and boreal zones," *Forest Ecology and Management* 246 (2006): 144–154; William Purvis, *Lichens* (Washington, DC: Smithsonian Institution Press, 2000).

13. Details on the preserve goals and natural history are from *Upper St. John River Forest Management Plan* (Brunswick, ME: The Nature Conservancy in Maine, April 2003, updated September 2009); Gawler and Cutko, *Natural Land-scapes of Maine*; Focus Areas of Statewide Ecological Significance webpage, Maine Natural Areas Program (retrieved on March 8, 2010, at www.maine.gov/doc/nrimc/mnap/focusarea/st_john.pdf); Maine Department of Inland Fisher-ies and Wildlife, *Maine's Comprehensive Wildlife Conservation Strategy* (Augusta: Maine Department of Inland Fisheries and Wildlife, 2005), appendix 7. Addi-tional information on Furbish's lousewort from S.C. Gawler, D.C. Waller, and E.S. Menges, "Environmental factors affecting establishment and growth of *Pe-dicularis furbishae*, a rare endemic of the St. John River Valley, Maine," *Bull. Torrey Bot. Club* 114 (1987): 280–292; Eric S. Menges, "Population viability analysis for an endangered plant," *Conservation Biology* 4 (1990): 52–62.

14. Ada Graham and Frank Graham Jr., *Kate Furbish and the Flora of Maine* (Gardiner, ME: Tilbury House, 1995).

15. Captain John Smith, *Map of New England* (1614) (retrieved from Digi-talCommons@University of Nebraska — Lincoln at digitalcommons.unl.edu/, March 8, 2010). Referred to as Snadoun hill in his map and Snodon hill in his manuscript, Smith, *A Description of New England*, iii.

16. Kendall, *Glaciers and Granite*, 219.

17. Species composition data from Gawler and Cutko, *Natural Landscapes of Maine*; whip-poor-will distribution from The Birds of North America Online, Whip-poor-will webpage (retrieved on March 8, 2010, at bna.birds.cornell.edu/bna/species/620/articles/introduction). Atlantic white-cedar distribution: USDA Natural Resources Conservation Service Plants Database (retrieved on March 8, 2010, at plants.usda.gov/java/profile?symbol=CHTH2).

18. Michael Begon, Colin R. Townsend, and John L. Harper, *Ecology: From Individuals to Ecosystems*, 4th edition (Oxford: Blackwell, 2006), 20–24.

19. The earliest formal biome classification in North America was by C. Hart Merriam, chief of the Division of Ornithology and Mammalogy at the U.S. Department of Agriculture, who, in the 1890s, described the dramatic changes in environment and biota from the depths of the Grand Canyon to the peaks of the San Francisco Mountains just south of the canyon. He further drew a parallel between this altitudinal series and the changes in "life zone" from the southwestern deserts to the arctic. From C.H. Merriam and L. Steineger, *Results of a Biological Survey of the San Francisco Mountain Region and the Desert of the Little Colorado, Arizona*, North American Fauna Report 3 (Washington, DC: Division of Ornithology and Mammalia, U.S. Department of Agriculture, 1890); C. Hart Merriam, "The geographic distribution of animals and plants in North America," in *Yearbook of the United States Department of Agriculture 1894* (Washington, DC: U.S. Department of Agriculture, 1895): 203–214.

20. Robert H. Whittaker, *Communities and Ecosystems*, 2nd edition (New York: Macmillan, 1975).

21. R.G. Davies et al., "Topography, energy and the global distribution of bird species richness," *Proc. R. Soc. Lond., B, Biol.Sci.* 274 (2007): 189–1197; L.B. Buckley and W. Jetz, "Environmental and historical constraints on global patterns of amphibian richness," *Proc. R. Soc. Lond., B, Biol. Sci.* 274 (2007): 1167–1173; H. Kreft and W. Jetz, "Global patterns and determinants of vascular plant diversity," *Proc. Natl Acad. Sci. U.S.A.* 104 (2007): 5925–5930. For an exception for precipitation, see Robert R. Dunn et al., "Climatic drivers of hemispheric asymmetry in global patterns of ant species richness," *Ecology Letters* 12 (2009): 324–333.

22. For a comparison of several ecoregional classifications, see Robert S. Thompson et al., "Topographic, bioclimatic, and vegetation characteristics of three ecoregion classification systems in North America: comparisons along continent-wide transects," *Environmental Management* 34 (2005), Suppl. 1: S125–S148.

23. Roger G. Bailey, *Ecoregions of the United States* (Map) (Washington, DC: United States Department of Agriculture, Forest Service, 1994); Roger G. Bailey, *Ecosystem Geography* (New York: Springer-Verlag, 1996); David T. Cleland, Jerry A. Freeouf, James E. Keys, Greg J. Nowacki, Constance A. Carpenter, and W. Henry McNabb, *Ecological Subregions: Sections and Subsections of the Conter-*

minous United States, Gen. Tech. Report WO-76 (Washington, DC: United States Department of Agriculture, Forest Service, 2007).

24. A.W. Küchler, *Potential Natural Vegetation of the Conterminous United States*, American Geographical Society, Special Publication No. 36 (Washington, DC: American Geographical Society, 1964); F.H. Eyre, ed., *Forest Cover Types of the United States and Canada* (Washington, DC: Society of American Foresters, 1980).

25. Gawler and Cutko, *Natural Landscapes of Maine.*

26. William B. Krohn, Randall B. Boone, and Stephanie L. Painton, "Quantitative delineation and characterization of hierarchical biophysical regions of Maine," *Northeastern Naturalist* 6 (1999): 139–164.

27. The discovery in the 1950s of such overlapping species distributions was another important argument against the prevailing "superorganism" theory of the first half of the twentieth century (e.g., Whittaker, *Communities and Ecosystems*). See sidebar 2.3.

28. David M. Olson et al., "Terrestrial ecoregions of the world: a new map of life on Earth," *BioScience* 51 (2001): 933–938; Terrestrial Ecoregions of the World website, National Geographic Society and World Wildlife Fund (retrieved on March 8, 2010, at www.nationalgeographic.com/wildworld/terrestrial.html).

29. Cleland et al, *Ecological Subregions.*

30. In older versions of ecoregional maps of North America, this province was combined with a similar province that continued west along the border all the way to Minnesota. Bailey, *Ecoregions of the United States* (Map).

31. See Olson et al., "Terrestrial ecoregions of the world"; Terrestrial Ecoregions of the World website (retrieved on March 8, 2010, at www.nationalgeographic.com/wildworld/terrestrial.html); for a Canadian Maritimes perspective, see A. Mosseler, J.A. Lynds, and J.E. Major, "Old-growth forests of the Acadian Forest Region," *Environ. Rev.* 11 (2003): S47–S77.

32. Lloyd C. Irland, "Maine's forest vegetation regions: selected maps 1858–1993," *Northeastern Naturalist* 4 (1997): 241–260.

33. On the north and west sides of the St. Lawrence Seaway, Deborah Elliot-Fisk, "The Taiga and Boreal Forest," in Michael G. Barbour and William Dwight Billings, eds., *North American Terrestrial Vegetation*, 2nd edition (Cambridge: Cambridge University Press, 2001), 41–74, cover photo and figure 2.1. Even the older Bailey ecoregional maps differ on whether northern Maine is temperate or in the boreal/subarctic zone, Bailey, *Ecoregions of the United States* (Map) vs. Bailey, *Ecoregions of the Continents* (Map).

34. Charles V. Cogbill, "Black growth and fiddlebutts: the nature of old-growth red spruce," in Mary B. Davis, ed., *Eastern Old-Growth Forests: Prospects of Rediscovery and Recovery* (Washington, DC: Island Press, 1996), 113–125.

35. Edward A. Johnson, *Fire and Vegetation Dynamics: Studies from the North*

American Boreal Forest (Cambridge: Cambridge University Press, 1992), quote from back cover; information on fire size and frequency from page 1.

36. An exception to this generalization might be the northwest corner of the state. As described in chapter 3, this area is dominated by spruce and fir and appears to be subject to regular fires.

37. Information in this paragraph from Bennett, *Maine's Natural Heritage,* 76–210; Janet S. McMahon, *The Biophysical Regions of Maine — Patterns in the Landscape and Vegetation* (master's thesis, University of Maine, 1990); Randall B. Boone and William B. Krohn, "Relationship between avian range limits and plant transition zones in Maine," *Journal of Biogeography* 27 (2000): 471–482.

38. We used those species listed and their range maps from the standard reference, Russell M. Burns and Barbara H. Honkala (technical coordinators), *Silvics of North America*, Agriculture Handbook 654 (Washington, DC: United States Department of Agriculture, Forest Service, 1990). For each relevant species in the volumes on conifers and hardwoods, we recorded whether or not its southern or northern range boundary occurred in each of the two areas. We excluded non-natives and species whose entire range is within either of the two analyzed areas. Using a chi-square test with Yates Correction for Continuity, we found that the probability that this difference is merely due to chance is 2.2 percent.

39. Animal range boundaries from Elizabeth C. Pierson, Jan Erik Pierson, and Peter D. Vickery, *A Birder's Guide to Maine* (Camden, ME: Down East Books, 1996); Malcolm L. Hunter Jr., Aram J.K. Calhoun, and Mark McCollough, *Maine Amphibians and Reptiles* (Orono: University of Maine Press, 1999); Maine Butterfly Survey, Maine Department of Inland Fisheries and Wildlife, and many other sponsors (retrieved at mbs.umf.maine.edu/ on May 18, 2010); Randall B. Boone and William B. Krohn, *Maine Gap Analysis Vertebrate Data — Parts I & II* (Orono: Maine Coop. Fish and Wild. Res. Unit, University of Maine, 1998).

40. R.F. Noss, "A regional landscape approach to maintain diversity," *BioScience* 33 (1983): 700–706.; P.S. Bourgeron, "Advantages and limitations of ecological classification for the protection of ecosystems," *Conservation Biology* 2 (1988): 218–220; M.L. Hunter Jr., G.L. Jacobson Jr., and T. Webb III, "Paleoecology and the coarse-filter approach to maintaining biological diversity," *Conservation Biology* 2 (1988): 375–385.

41. McMahon, *The Biophysical Regions of Maine*. It is important to recognize the precursor to this delineation, the first attempt at a comprehensive ecoregional analysis of Maine: Paul R Adamus, *The Natural Regions of Maine* (Augusta: Maine Critical Areas Program, Maine State Planning Office, 1976). This report is described in detail in Bennett, *Maine's Natural Heritage*.

42. Boone and Krohn, "Relationship between avian range limits."

43. The minor discrepancies between these two geographic demarcations stem from their different methods: whereas these four super-forests of presettlement

times are based on *abundance* of tree species, the modern ecoregions reflect *distributions* of tree species but also of the underlying climate and land.

44. See also Gawler et al., *Biological Diversity of Maine*, Appendices P and Q; Maine Natural Areas Program, Ecosystems in Maine website (retrieved March 17, 2010, at www.maine.gov/doc/nrimc/mnap/features/ecosystems.htm).

45. Part of the Department of Conservation, with a mission to "serve[s] Maine's citizens as the most comprehensive source on the State's important natural features." Maine Natural Areas Program website, "Overview and Mission" (retrieved on March 16, 2010, at www.maine.gov/doc/nrimc/mnap/index.htm). Gawler and Cutko have each also worked at NatureServe, a national non-profit conservation organization, whose "mission is to provide the scientific basis for effective conservation action."

46. Personal communication, Janet McMahon.

47. Information on variation across Maine in these factors and climate is from McMahon, *The Biophysical Regions of Maine*; Kendall, *Glaciers and Granite*, especially chapter 1; Gawler et al., *Biological Diversity in Maine*; Gregory A. Zielinski, *Conditions May Vary: A Guide to Maine Weather* (Camden, ME: Down East Books, 2009), especially chapters 4, 5, and 7.

48. Begon et al., *Ecology: From Individuals to Ecosystems*, 41.

49. G.L. Jacobson, I.J. Fernandez, P.A. Mayewski, and C.V. Schmitt, eds., *Maine's Climate Future: An Initial Assessment* (Orono: University of Maine, 2009), figure 4.

50. Lois Berg Stack, *Plant Hardiness Zone Map of Maine*, Bulletin No. 2242 (Orono: University of Maine Cooperative Extension, 2006).

51. Zielinski, *Conditions May Vary*, figure 4.3.

52. Moses Greenleaf, *A Survey of the State of Maine in Reference to Its Geographical Features, Statistics and Political Economy* (Portland, ME: Shirley and Hyde, 1829), 38.

53. A similar composite temperature index was the best discriminator among major vegetation zones in all of New England: Charles V. Cogbill, John Burk, and G. Motzkin, "The forests of presettlement New England, USA: spatial and compositional patterns based on town proprietor surveys," *Journal of Biogeography* 29 (2002): 1279–1304.

54. Craig G. Lorimer, "The presettlement forest and natural disturbance cycle of Northeastern Maine," *Ecology* 58 (1977): 139–148.

55. Information on disturbance and birds is from research on harvesting effects on bird species from John M. Hagan, and Stacie L. Grove, *Bird Abundance and Distribution in Managed and Old-Growth Forest in Maine*, Report No. MM-9901 (Brunswick, ME: Manomet Center for Conservation Sciences, 1999).

56. Ron Davis and Dennis Anderson argue similarly that Maine's uniquely steep temperature gradients account for the remarkable diversity of peatland types in

the state; Ronald B. Davis and Dennis S. Anderson, "Classification and distribution of freshwater peatlands in Maine," *Northeastern Naturalist* 8 (2001): 1–5.

57. Maine State Planning Office (chief author: Janet McMahon), *An Ecological Reserves System for Maine: Benchmarks in a Changing Landscape*, Report to the 116th Maine Legislature (Augusta: Maine State Planning Office, Natural Resources Planning Division, State of Maine, 1993), 67; J.S. McMahon, "Saving all the pieces: an ecological reserves proposal from Maine," *Maine Naturalist* 1 (1993): 213–222.

58. Public Laws of Maine, *An Act to Establish Standards and Conditions for Designation of Ecological Reserves on Lands Managed by the Bureau of Parks and Lands*, 2000, chapter 592 S.P. 157–L.D. 477 (retrieved March 8, 2010, at http://www.mainelegislature.org/ros/LOM/LOMDirectory.htm).

59. Ibid.

60. Maine Natural Areas Program, *Ecological Reserve Monitoring: Summary Report for the Maine Outdoor Heritage Fund and The Nature Conservancy* (Augusta: Maine Natural Areas Program, Department of Conservation, 2009), 4–5.

61. Andy Cutko and Rick Frisina, *Saving All the Parts: A Conservation Vision for Maine Using Ecological Land Units* (Augusta: Maine Natural Area Program, Department of Conservation, 2005), 11.

62. That value is now closer to 700,000.

63. David R. Foster et al., *Wildlands and Woodlands* (Petersham, MA: Harvard Forest, 2010) and www.wildlandsandwoodlands.org.

64. Using seven ecoregions and the ninety-eight community types designated at that time; Cutko and Frisina, *Saving All the Parts*, 27. This is also the approach developed in the original 1993 report, Maine State Planning Office, *An Ecological Reserves System for Maine*.

65. For studies that have this approach nationally, see J.M. Scott et al., "Nature reserves: do they capture the full range of America's biological diversity?" *Ecological Applications* 11 (2001): 999–1007; R.W. Dietz and B. Czech, "Conservation deficits for the continental United States: an ecosystem gap analysis," *Conservation Biology* 19 (2005):1478–1487.

66. Cutko and Frisina, *Saving All the Parts*, 25–32.

67. See also Christopher S. Cronan et al., "An assessment of land conservation patterns in Maine based on spatial analysis of ecological and socioeconomic indicators," *Environmental Management* 45 (2010): 1076–1095.

68. For a complete ecological and socioeconomic goals analysis of the Maine conservation land network, see Cronan et al., "An assessment of land conservation patterns in Maine."

69. The Nature Conservancy, *Determining the Size of Eastern Forest Reserves* (Boston: Eastern U.S. Conservation Region, 2004); see also R.F. Noss and A.Y. Cooperrider, *Saving Nature's Legacy: Protecting and Restoring Biodiversity* (Washington, DC: Island Press, 1994).

70. Cronan et al., "An assessment of land conservation patterns in Maine."

71. About 46 percent have conservation land and 89 percent have mature forest or water within a half-mile surrounding buffer. MNAP, *Ecological Reserve Monitoring* (2009), figure 3 and table 1.

72. Cutko and Frisina, *Saving All the Parts*, 32.

6. The Future of the Maine Woods

Epigraph from David R. Foster and Glenn Motzkin, "Ecology and conservation in the cultural landscape of New England: lessons from nature's history," *Northeastern Naturalist* 5 (1998): 111–126.

1. Arthur K. Ellis, *Teaching and Learning Elementary Social Studies* (Boston: Allyn and Bacon, 1970), 431

2. G.E. Likens, F.H. Bormann, and N.M. Johnson, "Acid rain," *Environment* 14 (1972): 33–40; G.E. Likens and F.H. Bormann, "Acid rain: a serious regional environmental problem," *Science* 184 (1974): 1176–1179; C.V. Cogbill and G.E. Likens, "Acid precipitation in the northeastern United States," *Water Resources Research* 10 (1974): 1133–1137.

3. Fundamentals of acid deposition from Gene Likens, "Acid rain," in *Encyclopedia of Earth* (2010) (retrieved on March 29, 2010, from www.eoearth.org/article/Acid_rain); Charles T. Driscoll, Kathy Fallon Lambert, and Limin Chen, "Acidic deposition: sources and ecological effects," in Gerald R. Visgilio and Diana W. Whitelaw, eds., *Acid in the Environment: Lessons and Future Prospects* (New York: Springer, 2007), 27–58; Charles T. Driscoll, Kathy Fallon Lambert, and Limin Chen, "Acidic deposition: sources and effects," in Malcolm G. Anderson, ed., *Encyclopedia of Hydrological Sciences* (Chichester, England: Wiley, 2007); Eric K. Miller, *Assessment of Forest Sensitivity to Nitrogen and Sulfur Deposition in Maine*, Conference of New England Governors and Eastern Canadian Premiers Forest Mapping Group (Augusta: Maine Department of Environmental Protection, 2006); C.T. Driscoll et al., *Acid Rain Revisited: Advances in Scientific Understanding since the Passage of the 1970 and 1990 Clean Air Act Amendments*, Science Links Publication 1 (Hanover, NH: Hubbard Brook Research Foundation, 2001).

4. J.D. Aber et al., "Nitrogen saturation in northern forest ecosystems," *BioScience* 39 (1989): 378–386; John Aber et al., "Nitrogen saturation in temperate forest ecosystems," *BioScience* 48 (1998): 921–934.

5. Information on calcium depletion: D.H. DeHayes et al., "Acid rain impacts calcium nutrition and forest health," *BioScience* 49 (1999): 789–800; T.G. Huntington, "Assessment of calcium status in Maine forests: review and future projection," *Canadian Journal of Forest Research* 35 (2005): 1109–1121; P.G. Schaberg et al., "Associations of calcium and aluminum with the growth and health of sugar maple trees in Vermont," *Forest Ecology and Management* 223 (2006): 159–169; P.G. Schaberg, D.H. DeHayes, and G.J. Hawley, "Anthropogenic calcium depletion:

a unique threat to forest ecosystem health?" *Ecosystem Health* 7 (2001): 214–228. Background on nitrogen saturation: Aber et al., "Nitrogen saturation in northern forest ecosystems"; Mark E. Fenn et al., "Nitrogen excess in North American ecosystems: predisposing factors, ecosystem responses, and management strategies," *Ecological Applications* 8 (1998): 706–733; John D. Aber et al., "Is nitrogen deposition altering the nitrogen status of northeastern forests?" *BioScience* 53 (2003): 375–389. Additional information: Steven P. Hamburg and Charles V. Cogbill, "Historical decline of red spruce populations and climatic warming," *Nature* (London) 331 (1988): 428–431; Likens, "Acid rain"; Driscoll et al., "Acidic deposition: sources and ecological effects," 27–58; Driscoll et al., "Acidic deposition: sources and effects"; Miller, *Assessment of Forest Sensitivity*; Driscoll et al, *Acid Rain Revisited.*

6. Sugar maple decline: P.G. Schaberg et al., "Associations of calcium and aluminum with the growth and health of sugar maple trees in Vermont"; B.A. Huggett et al., "Long-term calcium addition increases growth release, wound closure, and health of sugar maple (*Acer saccharum*) trees at the Hubbard Brook Experimental Forest," *Canadian Journal of Forest Research* 37 (2007): 1692–1700. Red spruce decline: J.M. Halman et al., "Calcium addition at the Hubbard Brook Experiment Forest increases sugar storage, antioxidant activity and cold tolerance in native red spruce (*Picea rubens*)," *Tree Physiology* 28 (2008): 855–862; Likens, "Acid Rain"; Driscoll et al., "Acidic deposition: sources and ecological effects," 27–58; Driscoll et al., "Acidic deposition: sources and effects"; Miller, *Assessment of Forest Sensitivity*; Driscoll et al., *Acid Rain Revisited*. Dieter Mueller-Dombois cautioned ecologists to tease apart the impacts of acid deposition from natural ones in "Natural dieback in forests," *BioScience* 37 (1987): 575–583.

7. National Atmospheric Deposition Program, *Hydrogen Ion Wet Deposition from Measurements Made at the Field Laboratories*, 2004, National Trends Network (retrieved on August 25, 2011, at nadp.sws.uiuc.edu); John D. Aber et al., "Is nitrogen deposition altering the nitrogen status?"; Huntington, "Assessment of calcium status in Maine forests."

8. Huntington, "Assessment of calcium status in Maine forests."

9. Stephen A. Norton and Ivan J. Fernandez, eds., *The Bear Brook Watershed in Maine: A Paired Watershed Experiment — the First Decade (1987–1997)* (Boston: Kluwer Academic, 1999); S. Norton et al., "The Bear Brook Watershed, Maine (BBWM), USA," *Environ. Monit. Assess.* 55 (1999): 7–51; Sultana Jefts et al., "Decadal responses in soil N dynamics at the Bear Brook Watershed in Maine, USA," *Forest Ecology and Management* 189 (2004): 189–205; J.A. Elvir, "Effects of enhanced nitrogen deposition on foliar chemistry and physiological processes of forest trees at the Bear Brook Watershed in Maine," *Forest Ecology and Management* 221 (2006): 207–214.

10. J.A. Lynch, V.C. Bowersox, and J.W. Grimm, "Acid rain reduced in the east-

ern United States," *Environ. Sci. Technol.* 34 (2000): 940–949; Huntington, "Assessment of calcium status in Maine forests"; Driscoll et al., "Acidic deposition: sources and effects."

11. For Hubbard Brook streams: Driscoll et al., "Acidic deposition: sources and effects," 1452. For New England lakes: J. Stoddard et al., *Response of Surface Water Chemistry to the Clean Air Act Amendments of 1990* (Washington, DC: U.S. Environmental Protection Agency, 2003), ix. There is also evidence for some recovery in Adirondack lakes: Johanna F. Polsenberg, "Adirondack lakes making a comeback?" *Frontiers of Ecology and the Environment* 8 (2010): 345.

12. Miller, *Assessment of Forest Sensitivity.*

13. R. Ouimet et al., "Critical loads of atmospheric S and N deposition and current exceedances for Northern temperate and boreal forests in Québec," *Water Air, and Soil Pollution: Focus* 1 (2001): 119–134.

14. Driscoll et al., "Acidic deposition: sources and effects," 1455.

15. Projections for the future: L. Chen and C.T. Driscoll, "Regional application of an integrated biogeochemical model to northern New England and Maine," *Ecological Applications* 15 (2005): 1783–1797; S.S. Gbondo-Tugbawa et al., "Validation of an integrated biogeochemical model (PnET-BGC) at a northern hardwood forest ecosystem," *Water Resources Research* 37 (2001): 1057–1070. Clean Air Transport Rule: U.S. Environmental Protection Agency. *Air Transport* (retrieved on October 12, 2010, at www.epa.gov/airtransport); Gabriel Nelson, "EPA unveils rules on smog-forming emissions from power plants," *New York Times*, October 12, 2010 (retrieved on October 12, 2010, at www.nytimes.com/gwire/2010/07/07/07greenwire-epa-uveils-rules-on-smog-forming-emissions-fr-27348.html).

16. Maine Forest Service, Press release: "New Hemlock Woolly Adelgid Infestation Discovered," May 13, 2010, Department of Conservation, Augusta, ME.

17. From Jennifer D'Appollonia, "Regeneration strategies of Japanese barberry (*Berberis thunbergii DC.*) in coastal forests of Maine" (master's thesis, University of Maine, 2006). Some information also from Mark Miller, *Monhegan Stewardship Management Plan* (Monhegan, ME: Monhegan Associates, 2005), 6–11; J.A. Silander Jr. and D.M. Klepeis, "The invasion ecology of Japanese barberry (*Berberis thunbergii*) in the New England landscape," *Biological Invasions* 1 (1999): 189–201; personal observation in summer 2011.

18. The official U.S. federal definition of "invasive species" in Executive Order 13112 signed by President William Clinton on February 3, 1999, is as follows: "*Invasive species* means an alien species whose introduction does or is likely to cause economic or environmental harm or harm to human health."

19. Thomas P. Holmes et al., "Economic impacts of invasive species in forests: past, present, and future," *Annals of the New York Academy of Sciences* 1162 (2009): 18–38; D.M. Richardson et al., "Naturalization and invasion of alien plants: con-

cepts and definitions," *Diversity and Distributions* 6 (2000): 93–107; R.N. Mack et al. *Biotic Invasions: Causes, Epidemiology, Global Consequences and Control,* Issues in Ecology No. 5 (Washington, DC: Ecological Society of America, 2000).

20. Massachusetts Introduced Pests Outreach Project (August 2008), a project of the MA Department of Agricultural Resources and University of Mass. Extension Agriculture and Landscape Program (retrieved on June 3, 2010, at www .massnrc.org/pests/alb); W.D. Morewood et al., "Oviposition preference and larval performance of *Anoplophora glabripennis* (Coleoptera: Cerambycidae) in four eastern North American hardwood tree species," *Environmental Entomology* 32 (2003): 1028–1034; Kelli Hoover et al., "Performance of Asian longhorned beetle among tree species," *Proceedings: 2002 U.S. Department of Agriculture Interagency Research Forum,* GTR-NE-300, abstract; K.J. Dodds and D.A. Orwig, "An invasive urban forest pest invades natural environments — Asian longhorned beetle in northeastern US hardwood forests," *Canadian Journal of Forest Research* 41: 1729–1742.

21. Information on emerald ash borer from the following presentation: Amy Stone, Robin M. Usborn, and Jodie Ellis, *Out of the Ashes — What We Know about EAB Workshop* (Dayton, OH: Michigan State University, Purdue University, and the Ohio State University, September 24–25, 2008) (retrieved on June 5, 2010, at www.emeraldashborer.info/educational.cfm). Response of Maine Indian tribes to emerald ash borer: Darren Ranco, Rob Lilieholm, John Daigle, and Maine Indian Basketmakers Alliance, *Mobilizing Researchers and Diverse Stakeholders to Address Emerging Environmental Threats: Brown Ash Sustainability as a Cultural Resource for Maine Indian Basketmakers,* Project of the Maine Sustainability Solutions Initiative (Orono: University of Maine (retrieved on November 10, 2010, at www.umaine.edu/sustainabilitysolutions/sustainability_science/ssi_summer_ 09/brown_ash.html); personal communication with Theresa Secord, Executive Director, Maine Indian Basketmakers Alliance.

22. State inventory of hemlock: W.H. McWilliams et al., *The Forests of Maine: 2003,* Resource Bulletin NE-164 (Newtown Square, PA: U.S. Department of Agriculture, Forest Service, Northeastern Research Station, 2005). Projection for future of hemlock woolly adelgid in Maine: David Struble and Dennis Souto, *Environmental Assessment Regarding Management of Hemlock Woolly Adelgid Impacts in Maine* (Augusta: Maine Forest Service, 2007), 12–17, 34; U.S. Forest Service, National Forest Health Monitoring, "New England insect and disease risk map: hemlock woolly adelgid" (retrieved on March 20, 2011, at fhm.fs.fed.us/ sp/na_riskmaps/new_england/hwa_rm.shtml); William Livingston, personal communication.

23. Based on the recent research of graduate student, Stacy Trosper, in William Livingston's lab at the University of Maine.

24. Maine Natural Areas Program, "Asiatic bittersweet," *Maine Invasive Plants,* Bulletin No. 2506 (Augusta: Maine Department of Conservation, 2001); Jil M.

Swearingen, *Oriental Bittersweet, Least Wanted: Alien Plant Invaders of Natural Areas* (Washington, DC: Plant Conservation Alliance's Alien Plant Working Group, 2009).

25. S.L. Webb and C.M.K. Kaunzinger, "Biological invasion of the Drew University Forest Preserve by Norway maple *(Acer platanoides* L)," *Bulletin of the Torrey Botanical Club* 120 (1993): 342–349; S.L. Webb et al., "The myth of the resilient forest: case study of the invasive Norway maple *(Acer platanoides),*" *Rhodora* 102 (2000): 332–354; P.H. Martin, C.D. Canham, and R.K. Kobe, "Divergence from the growth-survival trade-off and extreme high growth rates drive patterns of exotic tree invasions in closed-canopy forests," *Journal of Ecology* 98 (2010): 778–789.

26. A.M. Barton et al., "Nonindigenous invasive woody plants in a rural New England town," *Biological Invasions* 6 (2004): 205–211.

27. Mark Williamson and Alastair Fitter, "The varying success of invaders," *Ecology* 77 (1996): 1661–1666. Not surprisingly, deviations from the tens rule have been documented; e.g., Gregory M. Ruiz and James T. Carlton, eds., *Invasive Species: Vectors and Management Strategies* (Washington, DC: Island Press, 2003).

28. Anthony Standen, "Japanese beetle: voracious, libidinous, prolific, he is eating his way across the U.S., destroying $7,000,000 worth of plant life every year," *Life Magazine* 17 (July 17, 1944): 39–46.

29. In other parts of North America, plant invasives are playing a transformative role in ecosystems compared to that for New England. For example exotic grasses have taken over hundreds of thousands of acres in western North America, supplanting natives and altering fire regimes that further degrade native elements of these ecosystems. C.M. D'Antonio and P. M. Vitousek, "Biological invasions by exotic grasses, the grass/fire cycle and global change," *Annual Review of Ecology and Systematics* 23 (1992): 63–87; M.L. Brooks et al., "Effects of invasive alien plants on fire regimes," *BioScience* 54 (2004): 677–688.

30. Quotation from Adrian Higgins, "Harvard biologist comes to the defense of the much reviled tree of heaven," *Washington Post,* June 14, 2010. See recent article on this issue: Mark A. Davis et al., "Don't judge species on their origin," *Nature* 474 (2011): 153–154.

31. Ann Gibbs, personal communication.

32. See also R.J. Lilieholm, L.C. Irland, and J.M. Hagan, "Changing socioeconomic conditions for private woodland protection" in S.C. Trombulak, and R. Baldwin, eds., *Multi-scale Conservation Planning* (New York: Springer-Verlag, 2010). For an excellent review of issues in this section for all of New England, see David R. Foster et al., *Wildlands and Woodlands* (Petersham, MA: Harvard Forest, 2010).

33. Housing development in Maine: The Brookings Institution Metropolitan Policy Program, *Charting Maine's Future: An Action Plan for Promoting Sustainable Prosperity and Quality Places* (Washington, DC: The Brookings Institution, 2006), sections 2 and 3; Eric M. White, *Forests on the Edge: A Case Study of*

South-Central and Southwest Maine Watersheds, preliminary report (Portland, OR: Pacific Northwest Research Station, USDA Forest Service, in preparation). Maine population growth: United States Census Bureau, *Annual Estimates of the Resident Population for the United States, Regions, States, and Puerto Rico: April 1, 2000 to July 1, 2009* (retrieved on June 11, 2010, at www.census.gov/popest/states/ NST-ann-est.html). Plum Creek Plan: Maine Land Use Regulation Commission, *Concept Plan for the Moosehead Lake Region*, September 23, 2009 (retrieved on June 10, 2010, at www.maine.gov/doc/lurc/reference/resourceplans/moosehead .html); The Nature Conservancy in Maine, *Moosehead Forest Project* (retrieved on June 10, 2010, at www.nature.org/wherewework/northamerica/states/maine/ features/art19238.html#Changed_Dramatically); Bangor Daily News, "LURC approves Plum Creek Plan; protesters arrested," September 23, 2009 (retrieved on June 10, 2010, at www.bangordailynews.com/).

34. Susan M. Stein et al., *Forests on the Edge: Housing Development on America's Private Forests*, General Technical Report PNW-GTR-636 (Portland, OR: Pacific Northwest Research Station, USDA Forest Service, 2005); S.M. Stein et al., *Private Forests, Public Benefits: Increased Housing Density and Other Pressures on Private Forest Contributions*, General Technical Report No. PNW-GTR-795 (Portland, OR: Pacific Northwest Research Station, USDA Forest Service, 2010); Eric M. White and Rhonda Mazza, *A Closer Look at Forests on the Edge: Future Development of Private Forests in Three States*, General Technical Report PNW-GTR-758 (Portland, OR: Pacific Northwest Research Station, USDA Forest Service, 2008); Eric M. White et al., *A Sensitivity Analysis of "Forests on the Edge": Housing Development on America's Private Forests*, General Technical Report PNW-GTR-792 (Portland, OR: Pacific Northwest Research Station, USDA Forest Service, 2009); Eric M. White, *Forests on the Edge: A Case Study of South-Central and Southwest Maine Watersheds*, preliminary report (Portland, OR: Pacific Northwest Research Station, USDA Forest Service, in preparation). Quotation: Andrew Kekacs, "Forests on the edge," *Fresh from the Woods*, produced by *Forests for Maine's Future* (retrieved on March 1, 2010, at www.forestsformainesfuture .org/Default.aspx?tabid=73).

35. A.J. Plantinga, T. Mauldin, and R.J. Alig, *Land Use in Maine: Determinants of Past Trends and Projections of Future Changes*, Res. Pap. PNWRP- 511 (Portland, OR: U.S. Department of Agriculture, Forest Service, Pacific Northwest Research Station, 1999).

36. Maximum generating capacity is an upper limit that is rarely if ever reached. Actual electricity generated can be much lower than these values.

37. Wind projects in Maine: Natural Resources Council of Maine, *Wind Projects in Maine* (retrieved on November 30, 2011, at www.nrcm.org/maine_wind_ projects.asp). Wildlife impacts: Wing Goodale and Tim Divoll, *Birds, Bats and Coastal Wind Farm Development in Maine: A Literature Review*, Report BRI 2009– 18 (Gorham, ME: BioDiversity Research Institute, 2009). Kibby wind project

expansion: Maine Natural Areas Program, *Comments on Kibby Expansion Development Permit Application (DP 4860)*, February 24, 2010; Natural Resources Council of Maine, *Comments on Kibby Expansion Development Permit Application (DP 4860)*, April 28, 2010. For a striking contrast in perspective: Friends of the Boundary Mountains, *Comments on Kibby Expansion Development Permit Application (DP 4860)*, April 20, 2010, vs. Donald Hudson, *Comments on Kibby Expansion Development Permit Application (DP 4860)*, April 19, 2010.

38. Biomass energy plants in Maine: Tux Turkel, "Emission study undercuts biomass benefits," *Portland Press Herald*, June 12, 2010. Manomet report: T. Walker, ed., P. Cardellichio, A. Colnes, J. Gunn, B. Kittler, R. Perschel, C. Recchia, D. Saah, and T. Walker, contributors, *Massachusetts Biomass Sustainability and Carbon Policy Study: Report to the Commonwealth of Massachusetts Department of Energy Resources*, Natural Capital Initiative Report NCI-2010–03 (Brunswick, ME: Manomet Center for Conservation Sciences, 2010); Timothy D. Searchinger et al., "Fixing a Critical Climate Accounting Error," *Science* 326 (2009): 527–528.

39. Sommeraurer Forest Sector Advisory Services, "US producing wood pellets for Europe" (retrieved on September 12, 2011, at forestindustries.eu/content/us-producing-wood-pellets-europe).

40. William McComb and David Lindenmayer, "Dying, dead, and down trees," in Malcolm L. Hunter Jr., ed., *Maintaining Biodiversity in Forested Ecosystems* (Cambridge, UK: Cambridge University Press, 1999), 335–372.

41. Paul E. Sendak, Robert C. Abt, and Robert J. Turner, "Timber supply projections for northern New England and New York: integrating a market perspective," *Northern Journal of Applied Forestry* 20 (2003): 175–185.

42. Malcolm Hunter Jr., personal communication.

43. Maine Forest Service, *Assessment of Sustainable Biomass Availability: Absolute Supply Is Not the Issue. Improving Utilization and Silviculture While Keeping Costs Low Are* (Augusta: Maine Forest Service, Department of Conservation, July 17, 2008); Maine Forest Service, Department of Conservation, *Maine State Forest Assessment and Strategies* (Augusta: Maine Forest Service, Department of Conservation, 2010), 4; Michael Greenwood, Robert S. Seymour, and Marvin W. Blumenstock, *Productivity of Maine's Forest Underestimated — More Intensive Approaches Are Needed*, Maine Agricultural Experimental Station Miscellaneous Report No. 328 (Orono: Maine Agricultural Experimental Station, 1988).

44. Jared S. Nunery and William S. Keeton, "Forest carbon storage in the northeastern United States: net effects of harvesting frequency, post-harvest retention, and wood products," *Forest Ecology and Conservation* 259 (2010): 1363–1375; D.R. Foster and J. Aber, eds., *Forests in Time: The Environmental Consequences of 1,000 Years of Change in New England* (New Haven, CT: Yale University Press, 2004); Walker et al., *Massachusetts Biomass Sustainability and Carbon Policy Study*.

45. John M. Hagan, Lloyd C. Irland, and Andrew A. Whitman, *Changing Timberland Ownership in the Northern Forest and Implications for Biodiversity*, Report

No. MCCS-FCP-2005–1 (Brunswick, ME: Manomet Center for Conservation Sciences, 2005); 2005–2009 figures from Lilieholm et al. "Changing socioeconomic conditions for private woodland protection." Similar patterns for Maine, with a focus on the Mahoosuc region of western Maine, can be found in Abigail Weinberg and Chris Larson, *Forestland for Sale: Challenges and Opportunities for Conservation over the Next Ten Years* (New York: Open Space Institute, 2008). For a slice in time description of transactions just in the early part of 2004, see Phyllis Austin, "Continuing land sales in Maine's North Woods lead to conservation, fragmentation," *Maine Environmental News*, June 9, 2004 (retrieved on June 14, 2010, at www.meepi.com). For a specific example, see Misty Edgecomb, "IP sells 1.1 M acres to Boston firm," *Bangor Daily News*, November 10, 2004. Just before completion of this book, Huber Resources, longtime land owner in Maine, sold its 357,000 acres of timberland to Conservation Forestry, LLC, an investment group in New Hampshire, as described in Jym St. Pierre, "Huber sells Maine timberlands," *Maine Environmental News*, October 28, 2011. On the Malone purchase, see Beth Quimby, "Mogul's land buy prompts questions," *Portland Press Herald*, January 30, 2011.

46. Hagan et al, *Changing Timberland Ownership*, 18.

47. General impacts of habitat fragmentation: Ian F. Spellerberg, "Ecological effects of roads and traffic: a literature review," *Global Ecology and Biogeography Letters* 7 (1998): 317–333; Lenore Fahrig, "Effects of habitat fragmentation on biodiversity," *Annual Review of Ecology, Evolution, and Systematics* 34 (2003): 487–515; David B. Lindenmayer and Joern Fischer, *Habitat Fragmentation and Landscape Change* (Washington, DC: Island Press, 2006); R. Aguilar et al., "Plant reproductive susceptibility to habitat fragmentation: review and synthesis through a meta-analysis," *Ecology Letter* 9 (2006): 968–980.

48. For birds: M.F. Small and M.L. Hunter, "Forest fragmentation and avian nest predation in forested landscapes," *Oecologia* 76 (1988): 62–64; John M. Hagan et al., "The early development of forest fragmentation in birds," *Conservation Biology* 10 (1996): 188–202; John M. Hagan and Stacie L. Grove, "Bird abundance and distribution in managed and old-growth forest in Maine," Report No. MM-9901 (Brunswick, ME: Manomet Center for Conservation Sciences, 1999); Pierre Drapeau et al., Landscape-scale disturbances and changes in bird communities of boreal mixed-wood forests," *Ecological Monographs* 70 (2000): 423–444 (for forests in Quebec with similarities to those in Maine). For mammals: Theodore G. Chapin, Daniel J. Harrison, Donald D. Katnik, "Influence of landscape pattern on habitat use by American marten in an industrial forest," *Conservation Biology* 12 (1998): 1327–1337; D.C. Payer and D.J. Harrison, "Structural differences between forests regenerating following spruce budworm defoliation and clearcut harvesting: implications for marten," *Canadian Journal of Forest Research* 30 (2000): 1965–1972; Angela K. Fuller and Daniel J. Harrison, "Influence of partial timber harvesting on American martens in north-central Maine," *Journal of*

Wildlife Management 69 (2005): 710–722; Angela K. Fuller, Daniel J. Harrison, and Henry J. Lachowski, "Stand scale effects of partial harvesting and clearcutting on small mammals and forest structure," *Forest Ecology and Management* 191 (2004): 373–386. For reptiles and amphibians: D.A. Patrick, M.L. Hunter Jr., and A.J.K. Calhoun, "Effects of experimental forestry treatments on a Maine amphibian community," *Forest Ecology and Management* 234 (2006): 323–332; David A. Patrick, Aram J.K. Calhoun, Malcolm L. Hunter Jr., "The importance of understanding spatial population structure when evaluating the effects of silviculture on spotted salamanders (*Ambystoma maculatum*)," *Biological Conservation* 141 (2008): 807–814; F. Beaudry, P.G. deMaynadier, M.L. Hunter Jr., "Identifying road mortality threat at multiple spatial scales for semi-aquatic turtles," *Biological Conservation* 141 (2008): 2550–2563. For plants, bryophytes, lichens: Erika L. Rowland, Alan S. White, and William H. Livingston, *A Literature Review of the Effects of Intensive Forestry on Forest Structure and Plant Community Composition at the Stand and Landscape Levels*, Miscellaneous Publication 754 (Orono: Maine Agricultural and Forest Experiment Station, 2005); Andrew A. Whitman and John M. Hagan, "Herbaceous plant communities in upland and riparian forest remnants in western Maine," *Mosaic Science Notes* No. 2000–2003 (Brunswick, ME: Manomet Center for Conservation Sciences, 2000); John M. Hagan and Andrew A. Whitman, *Late-Successional Forest: A Disappearing Age Class and Implications for Biodiversity*, Report FMSN 2004–2 (Brunswick, ME: Manomet Center for Conservation Sciences, 2004); S.B. Selva, "Using lichens to assess ecological continuity in northeastern forests," in Mary Byrd Davis, ed., *Eastern Old-Growth Forests: Prospects for Rediscovery and Recovery* (Washington, D.C: Island Press, 1996), 35–48; J.M. Hagan and S.L. Grove, "Coarse woody debris: humans and nature competing for wood," *Journal of Forestry* 97 (1999): 6–11.

49. Hagan and Whitman, *Late-Successional Forest*, 2.

50. Ibid.

51. P.D. Vickery, M.L. Hunter Jr., and S.M. Melvin, "Effect of habitat area on the distribution of grassland birds in Maine," *Conservation Biology* 8 (1994): 1087–1097; U.S. Fish and Wildlife Service, *New England Cottontail* (Sylvilagus transitionalis) (Hadley, MA: U.S. Fish and Wildlife Service, Northeast Unit, 2006); Thomas P. Hodgman, *Grasshopper Sparrow Assessment* (Bangor: Maine Department of Inland Fisheries and Wildlife, 2009).

52. "On record" meaning recorded by thermometers, which has occurred dependably since the late 1800s. Global record: National Oceanic and Atmospheric Administration, "NOAA: June, April to June, and year-to-date global temperatures are warmest on record" (retrieved on July 30, 2010, at www.noaanews .noaa.gov/stories2010/20100715_globalstats.html); State record: Jonathan Erdman and Nick Wiltgen, "Records and notables: 2010 U.S. mid-term report," The Weather Channel Beta (retrieved on July 30, 2010, at www.weather.com/outlook/ weather-news/news/articles/records-us-jan-jun-2010_2010–07–09).

53. The result for simple linear regression is statistically significant, as follows: $t = 16.4$, $df = 128$, $P < 0.0001$, $r^2 = 0.68$. This means that, given these data, the chance that our conclusion of a rise in temperature over this period is incorrect is about one in ten thousand. A third-degree polynomial regression fits the data even more closely ($r^2 = 0.79$).

54. R.K. Pachauri and A. Reisinger, eds., *Climate Change 2007: Synthesis Report*, Fourth Assessment Report (Geneva, Switzerland: International Panel on Climate Change, 2007), chapter 1; S. Solomon, D. Qin, M. Manning, Z. Chen, M. Marquis, K.B. Averyt, M. Tignor and H.L. Miller, eds., *Contribution of Working Group I to the Fourth Assessment Report of the Intergovernmental Panel on Climate Change, 2007* (Cambridge, UK: Cambridge University Press, 2007).

55. Pachauri and Reisinger, *Climate Change 2007*, figure 1.2

56. Deborah Zabarenko, "Ten key indicators show global warming 'undeniable,'" *Reuters*, July 29, 2010; see also J. J. Kennedy et al., "We know the world has warmed?" in D.S. Arndt, M.O. Baringer, and M.R. Johnson, eds., *State of the Climate in 2009. Bull. Amer. Meteor. Soc.* 91 (2010): S26.

57. Sources for changes in biological systems are the following: Pachauri and Reisinger, *Climate Change 2007*, figure 1.2; C. Rosenzweig et al., "Assessment of observed changes and responses in natural and managed systems," in M.L. Parry et al., eds., *Climate Change 2007: Impacts, Adaptation and Vulnerability. Contribution of Working Group II to the Fourth Assessment Report of the Intergovernmental Panel on Climate Change* (Cambridge, UK: Cambridge University Press, 2007), chapter 1; T.L. Root et al., "Fingerprints of global warming on wild animals and plants," *Nature* 421 (2003): 57–60; C. Parmesan and G. Yohe, "A globally coherent fingerprint of climate change impacts across natural systems," *Nature* 421 (2003): 37–42; Allison L. Perry et al., "Climate change and distributional shifts in marine fishes," *Science* 308 (2005): 1912–1915; J.A. Pounds et al., "Widespread amphibian extinctions from epidemic disease driven by global warming," *Nature* 439 (2006):161–67; Camille Parmesan, "Ecological and evolutionary responses to recent climate change," *Annual Review of Ecology, Evolution, and Systematics* 37 (2006): 637–669; A. Menzel, "European phonological response to climate change matches the warming pattern," *Global Change Biology* 12 (2006): 1969–1976; Cynthia Rosenzweig et al., "Attributing physical and biological impacts to anthropogenic climate change," *Nature* 453 (2008): 353–357; Melanie A. Harsch et al., "Are treelines advancing? A global meta-analysis of treeline response to climate warming," *Ecology Letters* 12 (2009): 1040–1049; S.J. Thackeray et al., "Trophic level asynchrony in rates of phenological change for marine, freshwater and terrestrial environments," *Global Change Biology* 16 (2010): 3304–3313.

58. A recent article in the *New Yorker* describes widespread publication bias and selective reporting across the sciences, which tend to support accepted hypotheses. One wonders whether these phenomena also affect the "meta-values" reported here. Jonah Lehrer, "The truth wears off: is there something wrong with

the scientific method?" *New Yorker,* December 13, 2010 (retrieved on June 6, 2010, at www.newyorker.com).

59. The most recent comprehensive study shows migration rates two to three times faster than these earlier ones: I-C. Chen, J.K. Hill, R. Ohlemuller, D.B. Roy, and C.D. Thomas, "Rapid range shifts of species associated with high levels of climate warming," *Science* 333 (2011): 1024–1026

60. Rosenzweig et al., "Assessment of observed changes and responses in natural and managed systems," 106–107; L. Zhou et al. "Relation between interannual variations in satellite measures of northern forest greenness and climate between 1982 and 1999," *Journal of Geophysical Research* 108 (2003): 4004; S.J. Goetz et al., "Satellite-observed photosynthetic trends across boreal North America associated with climate and fire disturbance," *Proceedings of the National Academy of Sciences, USA* 102 (2005): 13521–13525; C. Boisvenue and S.W. Running, "Impacts of climate change on natural forest productivity — evidence since the middle of the 20th century," *Global Change Biology* 12 (2006): 862–882.

61. Maine temperature data: U.S. Historical Climatology Network, National Oceanic and Atmospheric Administration (NOAA), U.S. Department of Commerce (retrieved on August 2, 2010, at New England Integrated Sciences & Assessment, NEISA from inhale.unh.edu/Climate/index.html). Statistical result for Maine temperature: simple linear regression is as follows: for temperature, $t = 4.6$, $df = 104$, $P < 0.0001$, $r^2 = 0.17$; for growing season length, $t = 2.5$, $df = 51$, $P < 0.02$, $r^2 = 0.11$. This means that, given these data, the chance that our conclusion of a rise in temperature over this period is incorrect is about one in ten thousand and for growing season about two in one hundred. Similar warming results for the region: C.P. Wake and A. Markham, *Indicators of Climate Change in the Northeast* (Portsmouth, NH: Clean Air-Cool Planet, 2005); Cameron Wake et al., *Cross Border Indicators of Climate Change over the Past Century: Northeastern United States and Canadian Maritime Region* (Boscawen, NH: Gulf of Maine Council on the Marine Environment, 2006); T.G. Huntington et al., "Climate and hydrological changes in the northeastern United States: recent trends and implications for forested and aquatic ecosystems," *Canadian Journal of Forest Research* 39 (2009): 199–212. Other environmental changes consistent with warming in Maine: G.A. Hodgkins, R.W. Dudley, and T.G. Huntington, "Changes in the timing of high river flows in New England over the 20th century," *Journal of Hydrology* (Amst.) 278 (2003): 244–252; T.G. Huntington, G.A. Hodgkins, and R.W. Dudley, "Historical trend in river ice thickness and coherence in hydroclimatological trends in Maine," *Climate Change* 61: 217–236; T.G. Huntington et al., "Changes in the proportion of precipitation occurring as snow in New England (1949–2000)," *Journal of Climatology* 17 (2004): 2626–2636; Wake and Markham, *Indicators of Climate Change in the Northeast*; T.G. Huntington, "Evidence for intensification of the global water cycle: review and synthesis," *Journal of Hydrology* (Amst.) 319 (2006): 83–95; Cameron Wake et al., *Cross Border Indicators of Climate Change over the*

Past Century; T.G. Huntington, "CO_2-induced suppression of transpiration can not explain increasing runoff," *Hydrol. Process.* 22 (2008): 311–314; Huntington et al., "Climate and hydrological changes in the northeastern United States."

62. W. Herbert Wilson Jr., Daniel Kipervaser, and Scott A. Lilley, "Spring arrival dates of Maine migratory breeding birds: 1994–1997," *Northeastern Naturalist* 7 (2000): 1–6. Earlier arrival dates were found for New York and Massachusetts by Christopher J. Butler, "The disproportionate effect of global warming on the arrival dates of short-distance migratory birds in North America," *Ibis* 145 (2003): 484–495, and for coastal Massachusetts by Abraham J. Miller-Rushing et al., "Bird migration times, climate change, and changing population sizes," *Global Change Biology* 14 (2008): 1959–1972. For discussion about why such mixed results might occur, see W. Herbert Wilson Jr., "Spring arrival dates of migratory breeding birds in Maine: sensitivity to climate change," *Wilson Journal of Ornithology* 119 (2007): 665–677.

63. J.P. Gibbs and A.R. Breisch, "Climate warming and calling phenology of frogs near Ithaca, NY, 1900–1999," *Conservation Biology* 15 (2001): 1175–1178; D. Primack et al., "Herbarium specimens demonstrate earlier flowering times in response to warming in Boston," *American Journal of Botany* 91 (2004): 1260–1264; D.W. Wolfe et al., "Climate change and shifts in spring phenology of three horticultural woody perennials in northeastern USA," *International Journal of Biometeorology* 49 (2005): 303–309; A.D. Richardson et al., "Phenology of a northern hardwood forest canopy," *Global Change Biology* 12 (2006), 1174–1188; Leslie M. Smith, Sandra Whitehouse, and Candace A. Oviatt, "Impacts of climate change on Narragansett Bay," *Northeastern Naturalist* 17 (2010): 77–90; B. Beckage et al., "A rapid shift of a forest ecotone during 40 years of warming in the Green Mountains of Vermont," *Proceedings of the National Academy of Sciences USA* 105 (2008): 4197–4202.

64. A good example of these observations for insects is in Kevin O'Connor, "Is climate change bringing the state more bugs? Bitten by the bug," *Times Argus* (Barre-Monpelier, Vermont), August 26, 2007.

65. Daniel K. Niven, Gregory S. Butcher, and G. Thomas Bancroft, "Christmas bird counts and climate change: northward shifts in early winter abundance," *American Birds* 109 (2009) 10–15.

66. "Global warming changing birds' habits," *Associated Press*, February 10, 2009 (retrieved on August 3, 2010, at www.msnbc.msn.com/id/29104238).

67. Two species, the common raven and the merlin, have moved south. Malcolm Hunter Jr., personal communication.

68. Global climate change projections: Pachauri and Reisinger, *Climate Change 2007*, 44–47; G.A. Meehl et al., "Global Climate Projections," in S. Solomon et al., eds., *Climate Change 2007: The Physical Science Basis. Contribution of Working Group I to the Fourth Assessment Report of the Intergovernmental Panel on Climate Change* (Cambridge, UK: Cambridge University Press, 2007), chapter 10. Pro-

jections for Maine: Ivan Fernandez et al., "Chapter 2: Maine's climate yesterday, today, and tomorrow," in G.L. Jacobson et al., eds., *Maine's Climate Future: An Initial Assessment* (Orono: University of Maine, April 2009), 14–15; K. Hayhoe et al., "Past and future changes in climate and hydrological indicators in the U.S. Northeast," *Climate Dynamics* 28 (2007): 381–407; "Climate and hydrological changes in the northeastern United States," 199–212; Katharine Hayhoe et al., "Regional climate change projections for the Northeast U.S.," *Mitigation and Adaptation Strategies for Global Change* 13 (2008): 425–436; Bruce T. Anderson, Katharine Hayhoe, and Xin-Zhong Liang, "Anthropogenic-induced changes in twenty-first-century summertime hydroclimatology of the northeastern US," *Climatic Change* 99 (2010): 403–423.

69. Linda Joyce et al., "Potential consequences of climate variability and change for the forests of the United States," in National Assessment Synthesis Team, U.S. Global Change Research Program, *Climate Change Impacts on the United States: The Potential Consequences of Climate Variability and Change* (Cambridge, UK: Cambridge University Press, 2001), especially 502–506; Virginia H. Dale et al., "Climate change and forest disturbances," *BioScience* 51 (2001): 723–734; Dominique Bachelet et al., "Climate change effects on vegetation distribution and carbon budget in the United States," *Ecosystems* 4 (2001): 164–185; C.B. Field et al., "2007: North America," in M.L. Parry et al., eds., *Climate Change 2007: Impacts, Adaptation and Vulnerability. Contribution of Working Group II to the Fourth Assessment Report of the Intergovernmental Panel on Climate Change* (Cambridge, UK: Cambridge University Press, 2007), 629–631; S. V. Ollinger et al., "Potential effects of climate change and rising CO_2 on ecosystem processes in northeastern U.S. forests," *Mitigation and Adaptation Strategies for Global Change* 13 (2008): 467–485; Marcus Lindner et al., "Climate change impacts, adaptive capacity, and vulnerability of European forest ecosystems," *Forest Ecology and Management* 259 (2010): 698–709; Yongqiang Liu, John Stanturf, and Scott Goodrick, "Trends in global wildfire potential in a changing climate," *Forest Ecology and Management* 259 (2010): 685–697; C.D. Thomas et al., "Extinction risk from climate change," *Nature* 427 (2004): 145–148; W. Thuiller et al., "Biodiversity conservation: uncertainty in predictions of extinction risk," *Nature* 430 (2004): 1; Camille Parmesan, "Ecological and evolutionary responses to recent climate change," *Annual Review of Ecology, Evolution, and Systematics* 37 (2006): 637–669; Daniel B. Botkin et al., "Forecasting the effects of global warming on biodiversity," *BioScience* 57 (2007): 227–236.

70. These projections can be viewed interactively at the Climate Change Atlas website, maintained by the USDA Forest Service: www.nrs.fs.fed.us/atlas/.

71. George L. Jacobson Jr. and Ann Dieffenbacher-Krall, "White pine and climate change: insights from the past," *Journal of Forestry* 93 (1995): 39–42. Jacobson and Dieffenbacher-Krall do point out that a resurgence in white pine could be dampened in a warmer world by white pine blister rust and the white pine weevil.

72. Andrew J. Hansen et al., "Global change in forests: responses of species, communities and biomes," *BioScience* 51 (2001): 765–779; Guoping Tang and Brian Beckage, "Projecting the distribution of forests in New England in response to climate change," *Diversity and Distributions* 16 (2010), 144–158; USDA Forest Service, Northeastern Research Station, *Climate Change Atlas* (retrieved on September 10, 2010, at www.nrs.fs.fed.us/atlas).

73. W.H. Livingston, G. Granger, M. Fries, H. Trial, D. Struble, J. Steinman, and S. Howell, *White Pine Decline in Maine* (PowerPoint) (retrieved on September 15, 2010, at extension.unh.edu/forestry/Docs/livingsto.pdf).

74. Allison M. Kanoti and William H. Livingston, "Investigating the onset of radial growth reduction caused by balsam woolly adelgid damage on balsam fir in relation to climate using dendroecological methods," in Laura S. Kenefic and Mark J. Twery, eds., *Changing Forests — Challenging Times*, Proceedings of the New England Society of American Foresters 85th Winter Meeting, 2005 March 16–18, Gen. Tech. Rep. NE-325 (Newtown Square, PA: U.S. Department of Agriculture, Forest Service, Northeastern Research Station, 2005), 53.

75. D.K. Manter and W.H. Livingston, "Thawing rate and fungal infection by *Rhizosphaera kalkhoffi* Bubak influence freezing injury on red spruce (*Picea rubens* Sarg.) needles," *Canadian Journal of Forest Research* 26 (1996):918 — 927; W.H. Livingston and A.S. White, "May drought confirmed as likely cause of brown ash dieback in Maine," *Phytopathology* 87 (1997): S59; Matthew T. Kasson and William H. Livingston, "Spatial distribution of *Neonectria* species associated with beech bark disease in northern Maine," *Mycologia*, 101 (2009): 190–195; Allan N.D. Auclair, Warren E. Heilman, and Blondel Brinkman, "Predicting forest dieback in Maine, USA: a simple model based on soil frost and drought," *Canadian Journal of Forest Research* 40 (2010): 687–702.

76. Craig D. Allen et al., "A global overview of drought and heat-induced tree mortality reveals emerging climate change risks for forests," *Forest Ecology and Management* 259 (2010): 660–684. See also P.J. van Mantgem et al., "Widespread increase of tree mortality rates in the western United States," *Science* 323 (2009): 521–524; David D. Breshears et al., "Regional vegetation die-off in response to global-change-type drought," *Proceedings of the National Academies of Sciences* 102 (2005): 15144–15148.

77. P.D. Manion, *Tree Disease Concepts* (Upper Saddle River, NJ: Prentice-Hall, 1991); P.D. Manion and D. Lachance, *Forest Decline Concepts* (St. Paul, MN: APS Press, 1992).

78. From Robert Millikan, J.R. McNeill, *Something New Under the Sun: an Environmental History of the 20th-Century World* (New York: W.W. Norton, 2000), 357.

79. National Oceanic and Atmospheric Administration, *NOAA Celebrates 200 years of Science, Service, and Stewardship* (Washington, DC: U.S. Department of Commerce) (retrieved on October 20, 2010, at http://celebrating200years.noaa .gov/); James Hansen et al., "Global temperature change," *Proceedings of the Na-*

tional Academy of Sciences 103 (2006): 14288–14293; Jeffrey T. Richelson, U.S. Satellite Imagery, 1960–1999, National Security Archive Electronic Briefing Book No 13, April 14, 1999 (retrieved on October 25, 2010, at www.gwu.edu/~nsarchiv/NSAEBB/NSAEBB13).

80. Lloyd C. Irland et al., "Assessing socioeconomic impacts of climate change on US forests, wood-product markets, and forest recreation," BioScience 51 (2001): 753–764; Richard G. Pearson and Terence P. Dawson, "Predicting the impacts of climate change on the distribution of species: are bioclimate envelope models useful?" Global Ecology & Biogeography 12 (2003): 361–371; Ronald P. Neilson et al., "Forecasting regional to global plant migration in response to climate change: challenges and directions," BioScience 55 (2005): 749–759; Botkin et al., "Forecasting the effects of global warming on biodiversity."

81. Pearson and Dawson, "Predicting the impacts of climate change on the distribution of species."

82. John W. Williams, Stephen T. Jackson, and John E. Kutzbach, "Projected distributions of novel and disappearing climates by 2100 AD," Proceedings of the National Academy of Sciences 104 (2007): 5738–5742.

83. Chen, et al., "Rapid range shifts of species associated with high levels of climate warming"; Kai Zhu, Christopher W. Woodall, James S. Clark, "Failure to migrate: lack of tree range expansion in response to climate change," Global Change Biology (2011): doi: 10.1111/j.1365-2486.2011.02571.x.

84. Quotation: Gene E. Likens and Jerry F. Franklin, "Ecosystem thinking in the northern forest — and beyond," BioScience 59 (2009): 511–513. See also Irland et al., "Assessing socioeconomic impacts of climate change."

85. For a discussion of these issues for all of New England, see Foster et al., Wildlands and Woodlands.

Bibliography

Aber, J.D., K.J. Nadelhoffer, P. Steudler, and J.M. Mellilo. "Nitrogen saturation in northern forest ecosystems." *BioScience* 39 (1989): 378–386.

Aber, John D., Christine L. Goodale, Scott V. Ollinger, Marie-Loiuse Smith, Alison H. Magill, Mary E. Martin, Richard A. Hallett, and John L. Stoddard. "Is nitrogen deposition altering the nitrogen status of northeastern forests?" *BioScience* 53 (2003): 375–389.

Aber, John, William McDowell, Knute Nadelhoffer, Alison Magill, Glenn Berntson, Mark Kamakea, Steven McNulty, William Currie, Lindsey Rustad, and Ivan Fernandez. "Nitrogen saturation in temperate forest ecosystems." *BioScience* 48 (1998): 921–934.

Adams and Perham Field Notes. Historical Society of Pennsylvania. Philadelphia, PA.

Adamus, Paul R., *The Natural Regions of Maine*. Augusta, ME: Maine Critical Areas Program, Maine State Planning Office, 1976.

Aguilar, R., L. Ashworth, L. Galetto, and M.A. Aizen. "Plant reproductive susceptibility to habitat fragmentation: review and synthesis through a meta-analysis." *Ecology Letters* 9 (2006): 968–80.

Ahn, SoEun, William B. Krohn, Andrew J. Plantinga, Timothy J. Dalton, and Jeffrey A. Hepinstall. *Agricultural Land Changes in Maine: A Compilation and Brief Analysis of Census of Agriculture Data, 1850–1997*. Technical Bulletin 182. Orono: Maine Agricultural and Forest Experiment Station, 2002.

Albion, D. *Searsmont: the Old Towns of Quantabacook 1764–1796*. Camden, ME: Camden Herald Publ. Co., 1978.

Albion, R.G. *Forests and Sea Power: The Timber Problem of the Royal Navy, 1652–1862*. Cambridge, MA: Harvard University Press, 1926.

Allen, Craig D., Alison K. Macalady, Haroun Chenchouni, Dominique Bachelet, Nate McDowell, Michel Vennetier, Thomas Kitzberger, Andreas Rigling, David D. Breshears, E.H. Ted Hogg, Patrick Gonzalez, Rod Fensham, Zhen Zhang, Jorge Castro, Natalia Demidova, Jong-Hwan Lim, Gillian Allard, Steven W. Running, Akkin Semerci, and Neil Cobb. "A global overview of drought and heat-induced tree mortality reveals emerging climate change risks for forests." *Forest Ecology and Management* 259 (2010): 660–684.

Allen, Nancy. "Cutting out clearcutting." *Synthesis/Regeneration* 10, Spring 1996.

Alley, Richard B. *The Two-Mile Time Machine: Ice Cores, Abrupt Climate Change, and Our Future*. Princeton, NJ: Princeton University Press, 2000.

Almquist, Heather, Ann C. Dieffenbacher-Krall, Riley Flanagan-Brown, and

David Sanger. "The Holocene record of lake levels of Mansell Pond, central Maine, USA." *The Holocene* 11 (2001): 189–201.

Almquist-Jacobson, Heather, and David Sanger. "Holocene climate and vegetation in the Milford drainage basin, Maine, U.S.A., and their implications for human history." *Vegetation History and Archeobotany* 4 (1995): 211–222.

Alroy, John. "A multispecies overkill simulation of the end-Pleistocene megafaunal mass extinction." *Science* 292 (2001): 1893–1896.

American Association for the Advancement of Science. Board Statement on Climate Change, December 9, 2006.

American Friends Service Committee. *The Wabanakis of Maine and the Maritimes*. Philadelphia, PA: American Friends Service Committee, 1989.

Anderegga, William R. L., James W. Prall, Jacob Harold, and Stephen H. Schneider. "Expert credibility in climate change." *Proceedings of the National Academy of Sciences of the United States of America*, www.pnas.org/content/early/2010/06/04/1003187107.full.pdf+html.

Anderson, Bruce T., Katharine Hayhoe, and Xin-Zhong Liang. "Anthropogenic-induced changes in twenty-first century summertime hydroclimatology of the Northeastern US." *Climatic Change* 99 (2010): 403–423.

Anderson, R. Scott, George L. Jacobson Jr., Ronald B. Davis, and Robert Stuckenrath. "Gould Pond, Maine: late-glacial transitions from marine to upland environments." *Boreas* 21 (1992): 359–371.

Anderson, R. Scott, Ronald B. Davis, Norton G. Miller, and Robert Stuckenrath. "History of late- and post-glacial vegetation and disturbance around Upper South Branch Pond, northern Maine." *Canadian Journal of Botany* 64 (1986): 1977–1986.

Anderson, R.S., N.G. Miller, R.B. Davis, R.B., and R.E. Nelson. "Terrestrial fossils in the marine Presumpscot Formation: implications for late Wisconsinan paleoenvironments and isostatic rebound along the coast of Maine." *Canadian Journal of Earth Sciences* 27 (1990): 1241–1246.

Antonov, Levitus J., and T. Boyer. "Warming of the world ocean, 1955–2003." *Geophysical Research Letters* 32 (2005): L02604.

Arajo, M.B., R.G. Pearson, W. Thuiller, and M. Erhard. "Validation of species-climate impact models under climate change." *Global Change Biology* 11 (2005): 1504–1513.

Associated Press. "Burt's Bees founder Quimby wants to donate her land for national park." *Bangor Daily News*, March 28, 2011.

Associated Press. "Global warming changing birds' habits." February 10, 2009.

Atlantic Geoscience Society. *The Last Billion Years: A Geological History of the Maritime Provinces of Canada*. Halifax, NS: Nimbus Publishers, 2001.

Atwood, Stanley B. *The Length and Breadth of Maine*. Augusta, ME: Kennebec Journal Print Shop, 1946.

Auclair, Allan N.D., Warren E. Heilman, and Blondel Brinkman. "Predicting forest dieback in Maine, USA: a simple model based on soil frost and drought." *Canadian Journal of Forest Research* 40 (2010): 687–702.

Austin, Phyllis. "Continuing land sales in Maine's North Woods lead to conservation, fragmentation." *Maine Environmental News,* June 9, 2004, www.meepi .com.

Axtell, James. *Native and Newcomers: The Cultural Origins of North America.* New York: Oxford University Press, 2001.

Bachelet, Dominique, Ronald P. Neilson, James M. Lenihan, and Raymond J. Drapek. "Climate change effects on vegetation distribution and carbon budget in the United States." *Ecosystems* 4 (2001): 164–185.

Bailey, Roger G. *Ecoregions of the United States* (Map). Washington, DC: United States Department of Agriculture, Forest Service, 1994.

Bailey, Roger G. *Ecosystem Geography.* New York: Springer-Verlag, 1996.

Baldwin, Elizabeth Dennis, Laura S. Kenefic, and Will F. LaPage. "Alternative large-scale conservation visions for northern Maine: interviews with decision leaders in Maine." *Maine Policy Review* 16 (2007): 78–91.

Baldwin, Robert F., Stephen C. Trombulak, Karen Beazley, Conrad Reining, Gillian Woolmer, John R. Nordgren, and Mark Anderson. "Importance of Maine for ecoregional conservation planning." *Maine Policy Review* 16 (2007): 66–77.

Balee, William L. *Advances in Historical Ecology.* New York: Columbia University Press, 1998.

Bangor Daily News. "LURC approves Plum Creek Plan; Protesters Arrested." September 23, 2009.

Barrington, D.S., and C.A. Paris. "Refugia and migration in the Quaternary history of the New England flora." *Rhodora* 109 (2007): 369–386.

Barron, Hal S. *Those Who Stayed Behind: Rural Society in Nineteenth-Century New England.* Cambridge: Cambridge University Press, 1984.

Barton, A.M., L. Brewster, A. Cox, and N. Prentiss. "Nonindigenous invasive woody plants in a rural New England town." *Biological Invasions* 6 (2004): 205–211.

Barton, Andrew M., Andrea M. Nurse, Katelyn Michaud, and Sarah W. Hardy. "Use of CART analysis to differentiate pollen of red pine (*Pinus resinosa*) and jack pine (*P. banksiana*) in New England." *Quaternary Research* 75 (2011): 18–23.

Beale, Colin M., Jack J. Lennon, and Alessandro Gimona. "Opening the climate envelope reveals no macroscale associations with climate in European birds." *Proceedings of the National Academy of Sciences* 105 (2008): 14908–14912.

Beaudry, F., P.G. deMaynadier, and M.L. Hunter Jr. "Identifying road mortality threat at multiple spatial scales for semi-aquatic turtles." *Biological Conservation* 141 (2008): 2550–2563.

Beckage, B., B. Osborne, D.G. Gavin, C. Pucko, T. Siccama, and T. Perkins.

"A rapid shift of a forest ecotone during 40 years of warming in the Green Mountains of Vermont." *Proceedings of the National Academy of Sciences USA* 105 (2008): 4197–41202.

Begon, Michael, Colin R. Townsend, and John L. Harper. *Ecology: From Individuals to Ecosystems*, 4th edition. Oxford: Blackwell Publishing, 2006.

Bennett, Dean B. *Maine's Natural Heritage: Rare Species and Unique Natural Features* Camden, ME: Down East Books, 1988.

Bennett, Dean B. *The Forgotten Nature of New England*. Camden, ME: Down East Books, 1996.

Bhiry, Naja, and Louise Filion. "Mid-Holocene hemlock decline in eastern North America linked with phytophagous insect activity." *Quaternary Research* 45 (1996): 312–320.

Biggar, H.P., ed., W.F. Ganong, translator. *The Works of Samuel de Champlain, Volume 1* (and *Voyages du Sievr de Champlain*), 270–342. Toronto: The Champlain Society, 1922.

Birds of North America Online. bna.birds.cornell.edu/bna/species/.

Blasing, T.J. *Recent Greenhouse Gas Concentrations, July 2010*. Carbon Dioxide Information Analysis Center, Oak Ridge National Laboratory, U.S. Department of Energy (retrieved August 8, 2010, at cdiac.ornl.gov/pns/current_ghg .html).

Boisvenue, C., and S.W. Running. "Impacts of climate change on natural forest productivity — evidence since the middle of the 20th century." *Global Change Biology* 12 (2006): 862–882.

Bonnichsen, Robson. "The coming of the fluted-point people." *Habitat: Journal of the Maine Audubon Society* 5 (1988): 34–36.

Boone, Randall B., and William B. Krohn. "Relationship between avian range limits and plant transition zones in Maine." *Journal of Biogeography* 27 (2000): 471–482.

Boone, Randall B., and William B. Krohn. *Maine Gap Analysis Vertebrate Data — Parts I & II*. Orono: Maine Coop. Fish and Wild. Res. Unit, University of Maine, 1998.

Borns, H. "The making of Maine: The advance and retreat of the last great glacier." *Habitat: Journal of the Maine Audubon Society* 5(1988): 22–25.

Borns, H.W. Jr., L.A. Doner, C.C. Dorion, G.L. Jacobson Jr., M.R. Kaplan, K.J. Kreutz, T.V. Lowell, W.B. Thompson, and T.K. Weddle. "The deglaciation of Maine, U.S.A." In J. Ehlers and P.L. Gibbard, eds., *Quaternary Glaciations — Extent and Chronology, Part II: North America*, 89–109. Amsterdam, Netherlands: Elsevier, 2004.

Botkin, Daniel B., Henrik Saxe, Miguel B. Araujo, Richard Betts, Richard H.W. Bradshaw, Tomas Cedhagen, Peter Chesson, Terry P. Dawson, Julie R. Etterson, Daniel P. Faith, Simon Ferrier, Antoine Guisan, Anja Skjoldborg Hansen, David W. Hilbert, Craig Loehle, Chris Margules, Mark New, Matthew J.

Sobel, and David R.B. Stockwell. "Forecasting the effects of global warming on biodiversity." *BioScience* 57 (2007): 227–236.

Bourgeron, P.S. "Advantages and limitations of ecological classification for the protection of ecosystems." *Conservation Biology* 2 (1988): 218–220.

Bourque, Bruce J. *Twelve Thousand Years: American Indians in Maine.* Lincoln: University of Nebraska Press, 2001.

Brain, J.P. "The Popham Colony: an historical and archeological brief." *Maine Archeological Society Bulletin* 43 (2003): 1–28.

Braun, E. Lucy. *The Deciduous Forests of Eastern North America.* Philadelphia: Blakiston, 1950.

Breshears, David D., Neil S. Cobb, Paul M. Rich, Kevin P. Price, Craig D. Allen, Randy G. Balice, William H. Romme, Jude H. Kastens, M. Lisa Floyd, Jayne Belnap, Jesse J. Anderson, Orrin B. Myers, and Clifton W. Meyer. "Regional vegetation die-off in response to global-change-type drought." *Proceedings of the National Academies of Sciences* 102 (2005): 15144–15148.

Broecker, W.S., D. Peteet, and D. Rind. "Does the ocean-atmosphere system have more than one stable mode of operation?" *Nature* 315 (1985): 21–25.

Brooker, Rob W., and Ragan M. Callaway. "Facilitation in the conceptual melting pot." *Journal of Ecology* 97 (2009): 1117–1120.

Brookings Institution Metropolitan Policy Program. *Charting Maine's Future: An Action Plan for Promoting Sustainable Prosperity and Quality Places.* Washington, DC: Brookings Institution, 2006.

Brooks, M.L., C.M. D'Antonio, D.M. Richardson, J. Grace, J.J. Keeley, DiTomaso, R. Hobbs, M. Pellant, and D. Pyke. "Effects of invasive alien plants on fire regimes." *BioScience* 54 (2004): 677–688.

Brown, Richard D., and Jack Tager. *Massachusetts: A Concise History.* Amherst: University of Massachusetts Press, 2000.

Buckley, L.B., and W. Jetz. "Environmental and historical constraints on global patterns of amphibian richness." *Proc. R. Soc. Lond., B, Biol. Sci.* 274 (2007): 1167–1173.

Burns, Russell M., and Barbara H. Honkala, technical coordinators. *Silvics of North America.* Agriculture Handbook 654. Washington, DC: United States Department of Agriculture, Forest Service, 1990.

Burrage, Henry S., ed. *Early English and French Voyages, Chiefly from Hakluyt, 1534–1608.* New York: Charles Scribner's Sons, 1906.

Burrage, Henry S. *The Beginning of Colonial Maine, 1602–1658.* Portland, ME: Marks Printing House, 1914.

Butler, Christopher J. "The disproportionate effect of global warming on the arrival dates of short-distance migratory birds in North America." *Ibis* 145 (2003): 484–495.

Butler, Francis G. *A History of Farmington, Franklin County, 1776–1885.* Farmington, ME: Press of Knowlton, McLeary, and Co., 1885.

Caldwell, Dabney W. *Roadside Geology of Maine.* Missoula, MT: Mountain Press, 1998.

Carroll, Charles P. *The Timber Economy of Puritan New England.* Providence, RI: Brown University Press, 1973.

Cary, Austin. "On the growth of spruce." In *Second Annual Report, Maine Forest Commissioner,* 20–36. Augusta: Maine Forest Commission, 1894.

Cary, Austin. "Report." In *Third Annual Report of the Forest Commissioner of the State of Maine,* 15–203. Augusta, ME: Burleigh and Flint, 1896.

Catton, Theodore. *Land Reborn: A History of Administration and Visitor Use in Glacier Bay National Park and Preserve.* Seattle, WA: Government Printing, 1995.

Chapin, F.S., L.R. Walker, and L.C. Sharman. "Mechanisms of primary succession following deglaciation at Glacier Bay, Alaska." *Ecological Monographs* 64 (1994): 149–175.

Chapin, Theodore G., Daniel J. Harrison, and Donald D. Katnik. "Influence of landscape pattern on habitat use by American marten in an industrial forest." *Conservation Biology* 12 (1998): 1327–1337.

Chen, I-Ching, James K. Hill, Ralf Ohlemuller, David B. Roy, and Chris D. Thomas. "Rapid range shifts of species associated with high levels of climate warming," *Science* 333 (2011): 1024–1026

Chen, L., and C. T. Driscoll. "Regional application of an integrated biogeochemical model to northern New England and Maine." *Ecological Applications* 15 (2005): 1783–1797.

Chokkalingam, Unna. "Spatial and temporal patterns and dynamics in oldgrowth northern hardwood and mixed forests of northern Maine." PhD dissertation, University of Maine, 1998.

Chokkalingam, Unna, and Alan S. White. "Structure and spatial patterns of trees in old-growth northern hardwood and mixed forests of northern Maine." *Plant Ecology* 156 (2001): 139–160.

Clark, Charles E. *The Eastern Frontier: The Settlement of Northern New England, 1610–1763.* Hanover, NH: University of New England Press, 1970.

Clark, Sara A., and Peter Howell. "From Diamond International to Plum Creek: the era of large landscape conservation in the northern forest." *Maine Policy Review* 16 (2007): 56–65.

Cleland, David T., Jerry A. Freeouf, James E. Keys, Greg J. Nowacki, Constance A. Carpenter, and W. Henry McNabb. *Ecological Subregions: Sections and Subsections of the Conterminous United States.* Gen. Tech. Report WO-76. Washington, DC: United States Department of Agriculture, Forest Service, 2007.

Clinton, President William. Executive Order 13112, February 3, 1999.

Cogbill, Charles V. "Black growth and fiddlebutts: the nature of old-growth red spruce." In Mary B. Davis, ed., *Eastern Old-Growth Forests: Prospects of Redis-covery and Recovery*, 113–125. Washington, DC: Island Press, 1996.

Cogbill, Charles V. "Dynamics of the boreal forests of the Laurentian High-lands, Canada." *Canadian Journal of Forest Research* 15 (1985): 252–261.

Cogbill, C.V., John Burk, and G. Motzkin. "The forests of presettlement New England, USA: spatial and compositional patterns based on town proprietor surveys." *Journal of Biogeography* 29 (2002): 1279–1304.

Cogbill, Charles V., John Burk, and G. Motzkin. "The forests of presettlement New England, USA: spatial and compositional patterns based on town pro-prietor surveys." *Journal of Biogeography* 29 (2002): 1279–1304.

Conkling, Philip W. *Islands in Time: A Natural and Cultural History of the Islands of the Gulf of Maine*. Camden, ME: Down East Books, 1999.

Coolidge, Philip T. *History of the Maine Woods*. Bangor, ME: Furbish-Roberts, 1963.

Copenheaver, C.A., A.S. White, and W.A. Patterson III. "Vegetation develop-ment in a southern Maine pitch pine-scrub oak barren." *Journal of the Torrey Botanical Society* 127 (2000): 19–32.

Critical Areas Program. *Natural Old-Growth Forest Stands in Maine and Its Rel-evance to the Critical Areas Program*. Planning Report 77. Augusta, ME: State Planning Office, 1983.

Crocker, R.L., and J. Major. "Soil development in relation to vegetation and sur-face age at Glacier Bay, Alaska." *Journal of Ecology* 43 (1955): 427–448.

Croll, James. *Climate and Time in Their Geological Relations: A Theory of Secular Changes of the Earth's Climate*. New York: D. Appleton and Company, 1893; 1st edition, 1875.

Cronan, Christopher S., Robert J. Lilieholm, Jill Tremblay, and Timothy Glid-den. "An assessment of land conservation patterns in Maine based on spatial analysis of ecological and socioeconomic indicators." *Environmental Manage-ment* 45 (2010): 1076–1095.

Cronon, William. *Changes in the Land: Indians, Colonist, and the Ecology of New England*. New York: Hill and Wang, 1983.

Curtis, John T. *The Vegetation of Wisconsin*. Madison: University of Wisconsin Press, 1959.

Cutko, Andy, and Rick Frisina. *Saving All the Parts: A Conservation Vision for Maine Using Ecological Land Units*. Augusta: Maine Natural Area Program, Department of Conservation, 2005.

Dale, Virginia H., Linda A. Joyce, Steve McNulty, Ronald P. Neilson, Matthew P. Ayres, Michael D. Flannigan, Paul J. Hanson, Lloyd C. Irland, Ariel E. Lugo, Chris J. Peterson, Daniel Simberloff, Frederick J. Swanson, Brian J. Stocks, and B. Michael Wotton. "Climate change and forest disturbances." *BioScience* 51 (2001): 723–734.

Dame, Lorin L., and Henry Brooks. *Handbook of the Trees of New England*. Boston: Athenaeum Press, 1901.

Damman, A.N.H. "Geographic changes in the vegetation pattern of raised bogs in the Bay of Fundy region in Maine and New Brunswick." *Vegetatio* 35 (1977): 137–151.

D'Antonio, C.M., and P. M. Vitousek. "Biological invasions by exotic grasses, the grass/fire cycle and global change." *Annual Review of Ecology and Systematics* 23 (1992): 63–87.

D'Appollonia, Jennifer. "Regeneration strategies of Japanese barberry (*Berberis thunbergii DC.*) in coastal forests of Maine." Master's thesis, University of Maine, 2006.

Davies, R.G., C. David L. Orme, David Storch, Valerie A. Olson, Gavin H. Thomas, Simon G. Ross, Tzung-Su Ding, Pamela C. Rasmuseen, Peter M. Bennett, Ian P.F. Owens, Tim M. Blackburn, and Kevin J. Gaston. "Topography, energy and the global distribution of bird species richness." *Proc. R. Soc. Lond., B, Biol.Sci.* 274 (2007): 189–1197.

Davis, M.B. "Holocene vegetation of eastern United States." In H. E. Wright and S. Porter, eds., *The Late Quarternary of the United States*, Vol. 2, 166–181. Minneapolis: University of Minnesota Press, 1983.

Davis, M.B. "Palynology after Y2K — understanding the source area of pollen in sediments." *Annual Review of Earth and Planetary Sciences* 28 (2000): 1–18.

Davis, Margaret B. "Pleistocene biogeography of temperate deciduous forests." *Geoscience and Man* 13 (1976): 13–26.

Davis, Margaret B., Ray W. Spear, and Linda C.K. Shane. "Holocene climate of New England." *Quaternary Research* 14 (1980): 240–250.

Davis, Mark A., Matthew K. Chew, Richard J. Hobbs, Ariel E. Lugo, John J. Ewel, Geerat J. Vermeij, James H. Brown, Michael L. Rosenzweig, Mark R. Gardener, Scott P. Carroll, Ken Thompson, Steward T.A. Pickett, Juliet C. Stromberg, Peter Del Tredici, Katharine N. Suding, Joan G. Ehrenfeld, J. Philip Grime, Joseph Mascaro, and John C. Briggs. "Don't judge species on their origin." *Nature* 474 (2011): 153–154.

Davis, Mary Byrd. "Maine." In *Old Growth in the East: A Survey* (online ed., Copyright2008), www.primalnature.org.

Davis, Owen K. "Palynology: an important tool for discovering historic ecosystems." In Dave Egan and Evelyn A. Howell, eds., *The Historical Ecology Handbook*, 229–255. Washington, DC: Island Press, 2001.

Davis, Ronald B. "Spruce-fir forests of the coast of Maine." *Ecological Monographs* 36 (1966): 79–94.

Davis, Ronald B., and Dennis S. Anderson. "Classification and distribution of freshwater peatlands in Maine." *Northeastern Naturalist* 8 (2001): 1–50.

Davis, Ronald B., and George L. Jacobson Jr. "Late glacial and early Holocene landscapes in northern New England and adjacent areas of Canada." *Quaternary Research* 23 (1985): 341–368.

Day, Gordon M. "The Indian as an ecological factor in the northeastern forest." *Ecology* 34 (1953): 329–346.

de Champlain, Samuel. *The Voyages and Explorations of Samuel de Champlain.* Volume 1, chapters 3 and 4. New York: Allerton Book Co., 1904.

DeCosta, B.F. *Ancient Norumbega or the Voyages of Simon Ferinando and John Walker to the Penobscot River, 1579–1580.* Albany, NY: New England Historical Genealogical Society, John Munsell's Sons, 1890.

DeGraaf, R.M., and M. Yamasaki. *New England Wildlife, Habitats, Natural History and Distribution.* Hanover, NH: University Press of New England, 2001.

DeHayes, D.H., G.L. Jacobson Jr., P.G. Schaberg, B. Bongarten, L. Iverson, and A.C. Dieffenbacher-Krall. "Forest responses to changing climate: lessons from the past and uncertainty for the future." In R.A. Mickler, R.A. Birdsey, and J. Hom, eds., *Responses of Northern U.S. Forests to Environmental Change.* Ecological Studies 139, 495–540. New York: Springer, 2000.

DeHayes, D.H., P.G. Schaberg, G.J. Hawley, and G.R. Strimbeck. "Acid rain impacts calcium nutrition and forest health." *BioScience* 49 (1999): 789–800.

Delcourt, P.A., and H.R. Delcourt. *Long-Term Forest Dynamics of the Temperate Zone: A Case Study of Late-Quaternary Forests in Eastern North America.* New York: Springer-Verlag, 1987.

DeLorme. *Maine Atlas and Gazetteer,* 13th edition. Yarmouth, ME: DeLorme, 2007.

Dieffenbacher-Krall, Ann M., and Andrea M. Nurse. "Late-glacial and Holocene record of lake levels of Mathews Pond and Whitehead Lake, northern Maine, USA." *Journal of Paleolimnology* 34 (2005): 283–310.

Dietz, R.W., and B. Czech. "Conservation deficits for the continental United States: an ecosystem gap analysis." *Conservation Biology* 19 (2005): 1478–1487.

Dobbs, David, and Richard Ober. *The Northern Forest.* White River Junction, VT: Chelsea Green, 1995.

Dodds, K.J., and D.A. Orwig. "An invasive urban forest pest invades natural environments — Asian longhorned beetle in northeastern US hardwood forests." *Canadian Journal of Forest Research* 41: 1729–1742

Donlan, Josh, Harry W. Greene, Joel Berger, Carl E. Bock, Jane H. Bock, David A. Burney, James A. Estes, Dave Foreman, Paul S. Martin, Gary W. Roemer, Felisa A. Smith, and Michael E. Soulé. "Re-wilding North America." *Nature* 436 (2005): 913–914.

Drapeau, Pierre, Alain Leduc, Jean-François Giroux, Jean-Pierre L. Savard, Yves Bergeron, and William L. Vickery. "Landscape-scale disturbances and changes in bird communities of boreal mixed-wood forests." *Ecological Monographs* 70 (2000): 423–444.

Driscoll, Charles T., Kathy Fallon Lambert, and Limin Chen. "Acidic deposition: sources and ecological effects." In Gerald R. Visgilio and Diana W. Whitelaw, eds., *Acid in the Environment: Lessons and Future Prospects*, 27–58. New York: Springer, 2007.

Driscoll, Charles T., Kathy Fallon Lambert, and Limin Chen. "Acidic deposition: sources and effects." In Malcolm G. Anderson, ed., *Encyclopedia of Hydrological Sciences*. Chichester, England: Wiley, 2007.

Driscoll, C.T., G.B. Lawrence, A.J. Bulger, T.J. Butler, C.S. Cronan, C. Eagar, K.F. Lambert, G.E. Likens, J.L. Stoddard, and K.C. Weathers. *Acid Rain Revisited: Advances in Scientific Understanding since the Passage of the 1970 and 1990 Clean Air Act Amendments*. Science Links Publication 1. Hanover, NH: Hubbard Brook Research Foundation, 2001.

Dubos, Rene. *The Wooing of Earth*. New York: Charles Scribers' Sons, 1980.

Dullinger, Stefan, Thomas Dirnböck, and Georg Grabherr. "Modelling climate change–driven treeline shifts: relative effects of temperature increase, dispersal and invisibility." *Journal of Ecology* 92 (2004): 241–252.

Duncan, Richard P., Phillip Cassey, and Tim M. Blackburn. "Do climate envelope models transfer? A manipulative test using dung beetle introductions." *Proc. R. Soc. B* 276 (2009): 1449–1457.

Dunn, Robert R., Donat Agosti, Alan N. Andersen, Xavier Arnan, Carsten A. Bruhl, Xim Cerda, Aaron M. Ellison, Brian L. Fisher, Matthew C. Fitzpatrick, Heloise Gibb, Nicholas J. Gotelli, Aaron D. Gove, Benoit Guenard, Milan Janda, Michael Kaspari, Edward J. Laurent, Jean-Philippe Lessard, John T. Longino, Jonathan D. Majer, Sean B. Menke, Terrence P. McGlynn, Catherine L. Parr, Stacy M. Philpott, Martin Pfeiffer, Javier Retana, Andrew V. Suarez, Heraldo L.Vasconcelos, Michael D.Weiser, and Nathan J. Sanders. "Climatic drivers of hemispheric asymmetry in global patterns of ant species richness." *Ecology Letters* 12 (2009): 324–333.

Dyer, Frederick H. *A Compendium of the War of the Rebellion*. Des Moines, IA: The Dyer Publishing Company, 1908.

Eckstrom, Fannie H. "Champlain's visit to the Penobscot." *Sprague's Journal of Maine History* 1 (1913): 56–65.

Edgecomb, Misty. "IP sells 1.1 M acres to Boston firm." *Bangor Daily News,* November 10, 2004.

Egan, Dave, and Evelyn A. Howell. *The Historical Ecology Handbook: A Restorationist's Guide to Reference Ecosystems*. Washington, DC: Island Press, 2001.

Elkinton, J.S., and G.H. Boettner. "The effects of *Compsilura concinnata,* an introduced generalist tachinid, on non-target species in North America: a cautionary tale." In R.G. Van Driesche and R. Reardon, eds., *Assessing Host Ranges for Parasitoids and Predators Used for Classical Biological Control: A*

Guide to Best Practice, 4–14. Morgantown, WV: United States Department of Agriculture Forest Health Technology Enterprise Team, 2004.

Elkinton, Joseph S., Dylan Parry, and George H. Boettner. "Implicating an introduced generalist parasitoid in the invasive browntail moth's enigmatic demise." Ecology 87 (2006): 2664–2672.

Elkinton, Joseph S., Evan Preisser, George Boettner, and Dylan Parry. "Factors influencing larval survival of the invasive browntail moth (Lepidoptera: Lymantriidae) in relict North American populations." Environmental Entomology 37 (2008): 1429–1437.

Elliot-Fisk, Deborah. "The Taiga and Boreal Forest." In Michael G. Barbour and William Dwight Billings, eds., North American Terrestrial Vegetation, 2nd edition, 41–74. Cambridge: Cambridge University Press, 2001.

Ellis, Arthur K. Teaching and Learning Elementary Social Studies. Boston: Allyn and Bacon, 1970.

Elvir, J.A., G.B. Wiersma, M.E. Day, M.S. Greenwood, and I.J. Fernandez. "Effects of enhanced nitrogen deposition on foliar chemistry and physiological processes of forest trees at the Bear Brook Watershed in Maine." Forest Ecology and Management 221 (2006): 207–214.

Encyclopaedia Britannica. Volume 14, 11th edition. New York: Encyclopaedia Britannica Company, 1910.

Erdman, Jonathan, and Nick Wiltgen. "Records and notables: 2010 U.S. midterm report." The Weather Channel Beta.

Evelyn, John. Sylva: Or a Discourse of Forest-trees and Propagation of Timber in his Majesties Dominion. London: The Royal Society, 1664.

Eyre, F.H., ed. Forest Cover Types of the United States and Canada. Washington, DC: Society of American Foresters, 1980.

Faber-Langendoen, D., L. Master, J. Nichols, K. Snow, A. Tomaino, R. Bittman, G. Hammerson, B. Heidel, L. Ramsay, and B. Young. NatureServe Conservation Status Assessments: Methodology for Assigning Ranks. Arlington, VA: NatureServe, April 2009.

Fahrig, Lenore. "Effects of habitat fragmentation on biodiversity." Annual Review of Ecology, Evolution, and Systematics 34 (2003): 487–515.

Fastie, Christopher L. "Causes and ecosystem consequences of multiple pathways of primary succession at Glacier Bay, Alaska." Ecology 76 (1995): 1899–1916.

Felt, Ephraim Porter. "The gipsy and brown tail moths." New York State Museum Bulletin 103, Entomology 25 (1906): 6.

Fenn, Mark E., Mark A. Poth, John D. Aber, Jill S. Baron, Bernard T. Bormann, Dale W. Johnson, A. Dennis Lemly, Steven G. McNulty, Douglas F. Ryan, and Robert Stottlemyer. "Nitrogen excess in North American ecosystems: predisposing factors, ecosystem responses, and management strategies." Ecological Applications 8 (1998): 706–733.

Fesenmyer, Kurt A., and Norman L. Christensen Jr. "Reconstructing Holocene fire history in a southern Appalachian forest using soil charcoal." *Ecology* 91 (2010): 662–670.

Field, C.B., L.D. Mortsch, M. Brklacich, D.L. Forbes, P. Kovacs, J.A. Patz, S.W. Running, and M.J. Scott. "2007: North America." In M.L. Parry, O.F. Canziani, J.P. Palutikof, P.J. van der Linden, and C.E. Hanson, eds., *Climate Change 2007: Impacts, Adaptation and Vulnerability. Contribution of Working Group II to the Fourth Assessment Report of the Intergovernmental Panel on Climate Change.* Cambridge: Cambridge University Press, 2007.

Firestone, Richard, B.A. West, J.P Kennett, L. Beckere, T.E. Bunch, Z.S. Revay, P.H. Schultz, T. Belgyag, D.J. Kennett, J.M. Erlandson, O.J. Dickenson, A.C. Goodyear, R.S. Harris, G.A. Howard, J.B. Kloosterman, P. Lechlern, P.A. Mayewski, J. Montgomery, R. Poreda, T. Darrah, S.S. Que Hee, A.R. Smith, A. Stich, W. Topping, J.H. Wittke, and W.S. Wolbac. "Evidence for an extraterrestrial impact 12,900 years ago that contributed to the megafaunal extinctions and Younger Dryas cooling." *Proceedings of the National Academy of Sciences* 104 (2007): 16016–16021.

Fisher, R.T. "Second-growth white pine as related to the former uses of the land." *Journal of Forestry* 16 (1918): 253–254.

Flatebo, Gro, Carol R. Foss, and Steven K. Pelletier. *Biodiversity in the Forests of Maine: Guidelines for Land Management.* UMCE Bulletin No. 7147. Orono: University of Maine Cooperative Extension, 1999.

Flora of North America. Vol. 21. www.efloras.org.

Fogler Library, University of Maine. Maine census data. www.library.umaine.edu/census/default.htm.

Foster, David R., and John D. Aber, eds. *Forests in Time: The Environmental Consequences of 1,000 Years of Change in New England.* New Haven, CT: Yale University Press, 2004.

Foster, David R., Brian M. Donahue, David B. Kittredge, Kathleen F. Lambert, Malcolm L. Hunter, Brian R. Hall, Lloyd C. Irland, Robert J. Lilieholm, David A. Orwig, Anthony W. D'Amato, Elizabeth A. Colburn, Jonathan R. Thompson, James N. Levitt, Aaron M. Ellison, William S. Keeton, John D. Aber, Charles V. Cogbill, Charles T. Driscoll, Timothy J. Fahey, and Clarisse M. Hart. *Wildlands and Woodlands.* Petersham, MA: Harvard Forest, 2010.

Foster, David R., and Glenn Motzkin. "Ecology and conservation in the cultural landscape of New England: lessons from nature's history." *Northeastern Naturalist* 5 (1998): 111–126.

Foster, David R., Glenn Motzkin, Debra Bernardos, and James Cardoza. "Wildlife dynamics in the changing New England landscape." *Journal of Biogeography* 29 (2002): 1337–1357.

Foster, David R., Wyatt W. Oswald, Edward K. Faison, Elaine D. Doughty, and

Barbara C.S. Hansen. "A climatic driver for abrupt mid-Holocene vegetation dynamics and the hemlock decline in New England." *Ecology* 87 (2006):
2959–2966.

Foster, Michael K., and William Cowan, eds. *In Search of New England's Native
Past: Selected Essays by Gordon M. Day.* Amherst: University of Massachusetts
Press, 1998.

Fox, Steven. *The American Conservation Movement: John Muir and His Legacy.*
Madison: University of Wisconsin Press, 1985.

Fraver, Shawn. "Spatial and temporal patterns of natural disturbance in old-
growth forests of northern Maine, USA." PhD dissertation, University of
Maine, 2004.

Fraver, Shawn, and Alan S. White. "Disturbance dynamics of old-growth *Picea
rubens* forests of northern Maine." *Journal of Vegetation Science* 16 (2005):
597–610.

Fraver, Shawn, and Alan S. White. "Identifying growth releases in dendrochronological studies of forest disturbance." *Canadian Journal of Forest Research*
35 (2005): 1648–1656.

Fraver, Shawn, Alan S. White, and Robert S. Seymour. "Natural disturbance
in an old-growth landscape of northern Maine, USA." *Journal of Ecology* 97
(2009): 289–298.

Fraver, Shawn, Robert S. Seymour, James H. Speer, and Alan S. White.
"Dendrochronological reconstruction of spruce budworm outbreaks in
northern Maine, USA." *Canadian Journal of Forest Research* 37 (2007):
523–529.

Frederic, Paul B. "Public policy and land development: The Maine Land Use
Regulation Commission." *Land Use Policy* 8 (1991): 50–62.

Frederic, Paul B. "Going west! Leaving 19th century rural northern New England: diaspora from the town of Industry, Maine." In Dimitrios Konstadakopulos and Soterios Zoulas, *100 Years in America: Historical Determinants
and Images of the Identity and Culture of Diasporas from Southeastern Europe,*
85–96. Proceedings of the Conference held at the Hellenic College, Brookline, MA, October 11, 2008. Bristol, England: The University of the West of
England, 2010.

Freinkel, Susan. *American Chestnut: The Life, Death, and Rebirth of a Perfect
Tree.* Berkeley: University of California Press, 2007.

Friends of the Boundary Mountains. *Comments on Kibby Expansion Development Permit Application (DP 4860),* April 20, 2010.

Fuller, Angela K., and Daniel J. Harrison. "Influence of partial timber harvesting on American martens in north-central Maine." *Journal of Wildlife Management* 69 (2005): 710–722.

Fuller, Angela K., Daniel J. Harrison, and Henry J. Lachowski. "Stand scale effects of partial harvesting and clearcutting on small mammals and forest structure." *Forest Ecology and Management* 191 (2004): 373–386.

Fuller, Janice L. "Ecological impact of the mid-Holocene hemlock decline in southern Ontario, Canada." *Ecology* 79 (1998): 2337–2351.

Gajewski, K. "Late Holocene climate changes in eastern North America estimated from pollen data." *Quaternary Research* 29 (1987): 255–262.

Gajewski, K. "Late Holocene pollen stratigraphy in four northeastern United States lakes." *Geographie physique et quaternaire* 41 (1987): 377–386.

Gawler, S.C., J.J. Albright, P.D. Vickery, and F.C. Smith. *Biological Diversity in Maine: An Assessment of Status and Trends in the Terrestrial and Freshwater Landscape*. Report prepared for the Maine Forest Biodiversity Project. Augusta: Maine Natural Areas Program, Department of Conservation, 1996.

Gawler, Susan C., and Andrew R. Cutko. *Natural Landscapes of Maine: A Classification of Vegetated Natural Communities and Ecosystems*. Augusta: Maine Natural Areas Program, Department of Conservation, 2010.

Gawler, S.C., D.C. Waller, and E.S. Menges. "Environmental factors affecting establishment and growth of *Pedicularis furbishae*, a rare endemic of the St. John River Valley, Maine." *Bull. Torrey Bot. Club* 114 (1987): 280–292.

Gbondo-Tugbawa, S.S., C.T. Driscoll, J.D. Aber, and G.E. Likens. "Validation of an integrated biogeochemical model (PnET-BGC) at a northern hardwood forest ecosystem." *Water Resources Research* 37 (2001): 1057–1070.

Geological Survey of Maine and University of Maine Climate Change Institute. *Maine's Ice Age Trail*. www.maine.gov/doc/nrimc/mgs/explore/surficial/facts/jul07.htm.

Gibbs, J.P., and A.R. Breisch. "Climate warming and calling phenology of frogs near Ithaca, NY, 1900–1999." *Conservation Biology* 15 (2001): 1175–1178.

Gill, Jacquelyn L., John W. Williams, Stephen T. Jackson, Katherine B. Lininger, and Guy S. Robinson. "Pleistocene megafaunal collapse, novel plant communities, and enhanced fire regimes in North America." *Science* 326 (2009): 1100–1103.

Gledhill, David. *The Names of Plants*. 3rd edition. Cambridge, UK: Cambridge University Press, 2002.

Goetz, S.J., A.G. Bunn, G.J. Fiske, and R.A. Houghton. "Satellite-observed photosynthetic trends across boreal North America associated with climate and fire disturbance." *Proceedings of the National Academy of Sciences, USA* 102 (2005): 13521–13525.

Gonzalez, P., R.P. Neilson, K.S. McKelvey, J.M. Lenihan, and R.J. Drapek. *Potential Impacts of Climate Change on Habitat and Conservation Priority Areas for* Lynx canadensis *(Canada Lynx)*. Washington, DC: Watershed, Fish, Wildlife, Air, and Rare Plants Staff, National Forest System, Forest Service, U.S. Department of Agriculture, and Arlington, VA: NatureServe, 2007.

Goodale, Wing, and Tim Divoll. *Birds, Bats and Coastal Wind Farm Development* in Maine: A Literature Review. Report BRI 2009–18. Gorham, ME: BioDiversity Research Institute, 2009.

Gould, Stephen Jay. Bully for Brontosaurus. New York: W.W. Norton, 1991.

Graham, Ada, and Frank Graham Jr. Kate Furbish and the Flora of Maine. Gardiner, ME: Tilbury House, 1995.

Greenleaf, Moses. A Survey of the State of Maine in Reference to Its Geographical Features, Statistics and Political Economy. Portland, ME: Shirley and Hyde, 1829.

Greenwood, Michael, Robert S. Seymour, and Marvin W. Blumenstock. Productivity of Maine's Forest Underestimated — More Intensive Approaches Are Needed. Maine Agricultural Experimental Station Miscellaneous Report No. 328. Orono: Maine Agricultural Experimental Station, 1988.

Groom, Martha J., Gary K. Meffe, and C. Ronald Carroll. Principles of Conservation Biology. 3rd edition. Sunderland, MA: Sinauer, 2006.

Guilday, John E. "Differential extinction during late-Pleistocene and recent times." In P.S. Martin and H.E. Wright Jr., eds., Pleistocene Extinctions: The Search for a Cause, 121–140. New Haven, CT: Yale University Press, 1967.

Guthrie, R. Dale. "New carbon dates link climatic change with human colonization and Pleistocene extinctions." Nature 441 (2006): 207–209.

Haas, Jean Nicolas, and John H. McAndrews. "The summer drought related hemlock (Tsuga canadensis) decline in eastern North America 5,700 to 5,100 years ago." In Katherine A. McManus, Kathleen S. Shields, and Dennis R. Souto, eds., Proceedings: Symposium on Sustainable Management of Hemlock Ecosystems in Eastern North America (GTR-NE-267), 81–88. Newtown Square, PA: U.S. Department of Agriculture, Forest Service, Northeastern Research Station, 2000.

Hagan, J.M., and S.L. Grove. "Coarse woody debris: humans and nature competing for wood." Journal of Forestry 97 (1999): 6–11.

Hagan, John M., and Stacie L. Grove. Bird Abundance and Distribution in Managed and Old-Growth Forest in Maine. Report No. MM-9901. Brunswick, ME: Manomet Center for Conservation Sciences, 1999.

Hagan, John M., Lloyd C. Irland, and Andrew A. Whitman. Changing Timberland Ownership in the Northern Forest and Implications for Biodiversity. Report No. MCCS-FCP-2005-1. Brunswick, ME: Manomet Center for Conservation Sciences, 2005.

Hagan, John M., W. Matthew Vander Haegen, and Peter S. McKinley. "The early development of forest fragmentation in birds." Conservation Biology 10 (1996): 188–202.

Hagan, John M., and Andrew A. Whitman. Late-Successional Forest: A Disappearing Age Class and Implications for Biodiversity. Report No. FMSN 2004–2. Brunswick, ME: Manomet Center for Conservation Sciences, 2004.

Hall, Dorothy K., Carl S. Benson, and William O. Field. "Changes of glaciers in

Glacier Bay Alaska, using ground and satellite measurements." *Physical Geography* 16 (1995): 27–41.

Halman, J.M., P.G. Schaberg, G.J. Hawley, and C. Eagar. "Calcium addition at the Hubbard Brook Experiment Forest increases sugar storage, antioxidant activity and cold tolerance in native red spruce (*Picea rubens*)." *Tree Physiology* 28 (2008): 855–862.

Hamburg, Steven P., and Charles V. Cogbill. "Historical decline of red spruce populations and climatic warming." *Nature* (London) 331 (1988): 428–431.

Hansen, Andrew J., Ronald P. Neilson, Virginia H. Dale, Curtis H. Flather, Louis R. Iverson, David J. Currie, Sarah Shafer, Rosamonde Cook, and Patrick J. Bartlein. "Global change in forests: responses of species, communities and biomes." *BioScience* 51 (2001): 765–779.

Hansen, J., L. Nazarenko, R. Ruedy, M. Sato, J. Willis, A. Del Genio, D. Koch, A. Lacis, K. Lo, S. Menon, T. Novakov, J. Perlwitz, G. Russell, G.A. Schmidt, and N. Tausnev. "Earth's energy imbalance: confirmation and implications." *Science* 308 (2005): 1431–1435.

Hansen, James, Makiko Sato, Reto Ruedy, Ken Lo, David W. Lea, and Martin Medina-Elizade. "Global temperature change." *Proceedings of the National Academy of Sciences* 103 (2006): 14288–14293.

Harsch, Melanie A., Philip E. Hulme, Matt S. McGlone, and Richard P. Duncan. "Are treelines advancing? A global meta-analysis of treeline response to climate warming." *Ecology Letters* 12 (2009): 1040–1049.

Hayhoe, Katharine, Cameron Wake, Bruce Anderson, Xin-Zhong Liang, Edwin Maurer, Jinhong Zhu, James Bradbury, Art DeGaetano, Anne Hertel, and Donald Wuebbles. "Regional Climate Change Projections for the Northeast U.S." *Mitigation and Adaptation Strategies for Global Change* 13 (2008): 425–436.

Hayhoe, K., C.P. Wake, T.G. Huntington, L. Luo, M.D. Schwartz, J. Sheffield, E.F. Wood, B. Anderson, J.A. Bradbury, A.T. DeGaetano, and D.W. Wolfe. "Past and future changes in climate and hydrological indicators in the U.S. Northeast." *Climate Dynamics* 28 (2007): 381–407.

Hays, J.D., J. Imbrie, and N.J. Shackleton. "Variations in the Earth's orbit: pacemakers of the ice ages." *Science* 194 (1976): 1121–1132.

Heikkinen, Risto K., Miska Luoto, Miguel B. Araújo, Raimo Virkkala, Wilfried Thuiller, and Martin T. Sykes. "Methods and uncertainties in bioclimatic envelope modelling under climate change." *Progress in Physical Geography* 30 (2006): 1–27.

Heinlein, Robert. *Time Enough for Love.* New York: Putnam, 1973.

Higgins, Adrian. "Harvard biologist comes to the defense of the much reviled tree of heaven." *Washington Post*, June 14, 2010.

Hodgkins, A., R.W. Dudley, and T.G. Huntington. "Changes in the timing of high river flows in New England over the 20th century." *Journal of Hydrology* (Amst.) 278 (2003): 244–252.

Hodgman, Thomas P. *Grasshopper Sparrow Assessment.* Bangor: Maine Department of Inland Fisheries and Wildlife, 2009.

Hoffman, Kristen. "Farms to forests in Blue Hill Bay: Long Island, Maine, as a case study in reforestation." *Maine History* 44 (2008): 50–76.

Holmes, Thomas P., Juliann E. Aukema, Betsy Von Holle, Andrew Liebhold, and Erin Sills. "Economic impacts of invasive species in forests: past, present, and future." *Annals of the New York Academy of Sciences* 1162 (2009): 18–38.

Hoover, Kelli, Scott Ludwig, James Sellmer, Deborah McCullough, and Laura Lazarus. "Performance of Asian longhorned beetle among tree species." In *Proceedings: 2002 U.S. Department of Agriculture Interagency Research Forum* (GTR-NE-300, abstract).

Hoving, L.L., D.J. Harrison, W.B. Krohn, R.A. Joseph, and M. O'Brien. "Broadscale predictors of Canada lynx occurrence in eastern North America." *Journal of Wildlife Management* 69 (2005): 739–751.

Hoyle, B. Gary, Daniel C. Fisher, Harold W. Borns Jr., Lisa L. Churchill-Dickson, Christopher C. Dorion, and Thomas K. Weddlef. "Late Pleistocene mammoth remains from Coastal Maine, USA." *Quaternary Research* 61 (2004): 277–288.

Hudson, Donald. *Comments on Kibby Expansion Development Permit Application (DP 4860),* April 19, 2010.

Huggett, B.A., P.G. Schaberg, G.J. Hawley, and C. Eagar. "Long-term calcium addition increases growth release, wound closure, and health of sugar maple (*Acer saccharum*) trees at the Hubbard Brook Experimental Forest." *Canadian Journal of Forest Research* 37 (2007): 1692–1700.

Hughes, Patrick. "The meteorologist who started a revolution." *Weatherwise* 47 (1994): 1–29.

Hunt, John H. *River Otter Assessment.* Bangor: Maine Department of Inland Fisheries and Wildlife, 1986.

Hunter, Malcolm L. Jr. "The diversity of New England forest ecosystems." In J.A. Bissonette, ed., *Is Good Forestry Good Wildlife Management?* (Miscellaneous Publication No. 689), 35–47. Orono: Maine Agricultural Experimental Station, 1986.

Hunter, Malcolm L. Jr., ed. *Maintaining Biodiversity in Forested Ecosystems.* Cambridge, England: Cambridge University Press, 1999.

Hunter, Malcolm L. Jr., Aram J.K. Calhoun, and Mark McCollough. *Maine Amphibians and Reptiles.* Orono: University of Maine Press, 1999.

Hunter, Malcolm L. Jr. and James Gibbs. *Fundamentals of Conservation Biology.* 3rd edition. Oxford, England: Blackwell Publishing, 2007.

Hunter, M.L. Jr., G.L. Jacobson Jr., and T. Webb III. "Paleoecology and the coarse-filter approach to maintaining biological diversity." *Conservation Biology* 2 (1988): 375–385.

Huntington, T.G. "Assessment of calcium status in Maine forests: review and future projection." *Canadian Journal of Forest Research* 35 (2005): 1109–1121.

Huntington, T.G. "CO_2-induced suppression of transpiration can not explain increasing runoff." *Hydrol. Process.* 22 (2008): 311–314.

Huntington, T.G. "Evidence for intensification of the global water cycle: review and synthesis." *Journal of Hydrology* (Amst.) 319 (2006): 83–95.

Huntington, T.G., G.A. Hodgkins, and R.W. Dudley. "Historical trend in river ice thickness and coherence in hydroclimatological trends in Maine." *Climate Change* 61: 217–236.

Huntington, T.G., G.A. Hodgkins, B.D. Kleim, and R.W. Dudley. "Changes in the proportion of precipitation occurring as snow in New England (1949–2000)." *Journal of Climatology* 17 (2004): 2626–2636.

Huntington, T.G., A.D. Richardson, K.J. McGuire, and K. Hayhoe. "Climate and hydrological changes in the northeastern United States: recent trends and implications for forested and aquatic ecosystems." *Canadian Journal of Forest Research* 39 (2009): 199–212.

Hyland, Fay, and Ferdinand H. Steinmetz. *The Woody Plants of Maine, Their Occurrence and Distribution.* University of Maine Studies, 2nd Ser., No. 59. Orono: University of Maine, 1944.

Hyland, F., W.B. Thompson, and R. Stuckenrath Jr. "Late Wisconsinan wood and other tree remains in the Presumpscot Formation, Portland, Maine." *Maritime Sediments* 14 (1978): 103–120.

Imbrie, John, and Katherine P. Imbrie. *Ice Ages: Solving the Mystery.* Cambridge, MA: Harvard University Press, 1986.

International Commission on Stratigraphy. *International Stratigraphic Chart.* www.stratigraphy.org/upload/ISChart2009.pdf.

Irland, Lloyd C. "Maine forests: a century of change, 1900–2000." *Maine Policy Review* 9 (2000): 66–77.

Irland, Lloyd C. "Maine's forest industry: from one era to another." In Richard Barringer, ed., *Changing Maine: 1960–2010,* 363–387. Gardiner, ME: Tilbury House, 2004.

Irland, Lloyd C. *Maine's Forest Area, 1600–1995: Review of Available Estimates.* Miscellaneous Publication 736. Augusta: Maine Agricultural and Forest Experiment Station, University of Maine, 1998.

Irland, Lloyd C. "Maine's forest vegetation regions: selected maps 1858–1993." *Northeastern Naturalist* 4 (1997): 241–260.

Irland, Lloyd C. *The Northeast's Changing Forest.* Petersham, MA: Harvard University Press and Harvard Forest, 1999.

Irland, Lloyd C. *Wildlands and Woodlots: The Story of New England's Forests.* Hanover, NH: University Press of New England, 1982.

Irland, Lloyd C., Darius Adams, Ralph Alig, Carter J. Betz, Chi-Chung Chen, Mark Hutchins, Bruce A. McCarl, Ken Skog, and Brent L. Sohngen. "Assessing socioeconomic impacts of climate change on US forests, wood-product markets, and forest recreation." *BioScience* 51 (2001): 753–764.

Iverson, Louis R., Anantha M. Prasad, Stephen N. Matthews, and Matthew Peters. "Estimating potential habitat for 134 eastern U.S. tree species under six climate scenarios." *Forest Ecology and Management* 254 (2008): 390–406.

Iverson, Louis, Anantha Prasad, and Stephen Matthews. "Modeling potential climate change impacts on the trees of the northeastern United States." *Mitigation and Adaptation Strategies for Global Change* 13 (2008): 487–516.

Jackson, S.T. "Pollen source area and representation in small lakes in the northeastern United States." *Review of Palaeobotany and Palynology* 63 (1990): 53–76.

Jackson, Stephen T., Jonathan T. Overpeck, Thompson Webb III, Sheran E. Keattch, and Katherine H. Anderson. "Mapped plant-macrofossil and pollen records of late Quarternary vegetation change in eastern North America." *Quaternary Science Reviews* 16 (1997): 1–70.

Jacobson, George Jr. and Ronald B. Davis. "The real forest primeval: the evolution of Maine's forests over 14,000 years." *Habitat: Journal of the Maine Audubon Society* 5 (1988): 26–29.

Jacobson, G.L. Jr., R.B. Davis, R.S. Anderson, M. Tolonen, and R. Stuckenrath Jr. "Post-glacial vegetation of the coastal lowlands of Maine." In *Global Pollen Database, NOAA Paleoclimatology,* www.ncdc.noaa.gov/paleo/metadata.

Jacobson, George Jr., and Ann Dieffenbacher-Krall. "White pine and climate change: insights from the past." *Journal of Forestry* 93 (1995): 39–42.

Jacobson, G.L., I.J. Fernandez, P.A. Mayewski, and C.V. Schmitt, eds. *Maine's Climate Future: An Initial Assessment.* Orono: University of Maine, April 2009.

Jacobson, G.L. Jr., T. Webb III, and E.C. Grimm. "Patterns and rates of vegetation change during the deglaciation of eastern North America." In W.F. Ruddiman and H.W. Wright Jr., eds., *North America and Adjacent Oceans during the Last Deglaciation* (DNAG v. K-3), 277–288. Boulder, CO: The Geology Society of North America, 1987.

James, Sydney V. Jr., ed. *Three Visitors to Early Plymouth.* Bedford, MA: Applewood Books, 1963.

Jardine, Nicholas, James A. Secord, and Emma C. Spary. *Cultures of Natural History.* Cambridge: Cambridge University Press, 1996.

Jefts, Sultana, Ivan J. Fernandez, Lindsey E. Rustad, and D. Bryan Dail. "Decadal responses in soil N dynamics at the Bear Brook Watershed in Maine, USA." *Forest Ecology and Management* 189 (2004): 189–205.

Jin, Suming, and Steven A. Sader. "Effects of forest ownership and change on forest harvest rates, types and trends in northern Maine." *Forest Ecology and Management* 228 (2006): 177–186.

Johnson, Edward A. *Fire and Vegetation Dynamics: Studies from the North American Boreal Forest.* Cambridge: Cambridge University Press, 1992.

Josselyn, John. *New-Englands Rarities Discovered.* Bedford, MA: Applewood Books, 1992 (original publication, London: G. Widdowes at the Green Dragon in St. Pauls Church-yard, 1672).

Jouzel, J., V. Masson-Delmotte, O. Cattani, G. Dreyfus, S. Falourd, G. Hoffmann, B. Minster, J. Nouet, J.M. Barnola, J. Chappellaz, H. Fischer, J.C. Gallet, S. Johnsen, M. Leuenberger, L. Loulergue, D. Luethi, H. Oerter, F. Parrenin, G. Raisbeck, D. Raynaud, A. Schilt, J. Schwander, E. Selmo, R. Souchez, R. Spahni, B. Stauffer, J.P. Steffensen, B. Stenni, T.F. Stocker, J.L. Tison, M. Werner, and E.W. Wolff. "Orbital and millennial Antarctic climate variability over the past 800,000 years." *Science* 317 (2007): 793–797

Joyce, Linda, John Aber, Steve McNulty, Virginia Dale, Andrew Hansen, Lloyd Irland, Ron Neilson, and Kenneth Skog. "Potential consequences of climate variability and change for the forests of the United States." In National Assessment Synthesis Team, U.S. Global Change Research Program, *Climate Change Impacts on the United States: The Potential Consequences of Climate Variability and Change,* 489–521. Cambridge: Cambridge University Press, 2001.

Judd, Richard W. "The Maine Woods: a legacy of controversy." The Margaret Chase Smith Essay, *Maine Policy Review* 16 (2007): 8–10.

Judd, Richard W. *Aroostook: A Century of Logging in Northern Maine.* Orono: University of Maine Press, 1989.

Judd, Richard W. *Common Lands, Common People: The Origins of Conservation in Northern New England.* Cambridge, MA: Harvard University Press, 1997.

Judd, Richard W., Edwin A. Churchill, and Joel W. Eastman, eds. *Maine: The Pine Tree State from Prehistory to the Present.* Orono: University of Maine Press, 1995.

Kane, Ailene, and Johanna Schmitt. *New England Plant Conservation Program Conservation and Research Plan:* Liatris borealis Nuttall ex MacNab (*Northern Blazing Star*). Framingham, MA: New England Wildflower Society, 2001.

Kanoti, Allison M., and William H. Livingston. "Investigating the onset of radial growth reduction caused by balsam woolly adelgid damage on balsam fir in relation to climate using dendroecological methods." In Laura S. Kenefic and Mark J. Twery, eds., *Changing Forests — Challenging Times.* Proceedings of the New England Society of American Foresters 85th Winter Meeting, March 16–18, 2005 (Gen. Tech. Rep. NE-325). Newtown Square, PA: U.S. Department of Agriculture, Forest Service, Northeastern Research Station, 2005.

Kasson, Matthew T., and William H. Livingston. "Spatial distribution of Neonectria species associated with beech bark disease in northern Maine." *Mycologia,* 101 (2009): 190–195.

Kearney, Michael, and Warren Porter. "Mechanistic niche modelling: combining physiological and spatial data to predict species ranges." *Ecology Letters* 12 (2009): 334–350.

Kekacs, Andrew. "Forests on the edge." *Fresh from the Woods* (retrieved on March 1, 2010, at www.forestsformainesfuture.org/Default.aspx?tabid=73).

Kendall, David L. *Glaciers and Granite: A Guide to Maine's Landscape and Geology*. Camden, ME: Down East Books, 1987.

Kenefic, Laura S., Alan S. White, Andrew R. Cutko, and Shawn Fraver. "Reference stands for silvicultural research: a Maine perspective." *Journal of Forestry* 103 (2005): 363–367.

Kennedy, J.J., P.W. Thorne, T.C. Peterson, R.A. Ruedy, P.A. Stott, D.E. Parker, S.A. Good, H.A. Titchner, and K.M. Willett. "We know the world has warmed? In D.S. Arndt, M.O. Baringer, and M.R. Johnson, eds., *State of the Climate in 2009. Bull. Amer. Meteor. Soc.* 91 (2010): S26.

Kerchner, Theresa. "The improved acre: the Besse farm as a case study in land-clearing, abandonment, and reforestation." *Maine History* 44 (2008): 77–102.

Kerchner, Theresa. *The Improved Acre: The Besse Farm, Maine, as a Case Study in Land Clearing, Farm Abandonment, and Reforestation in Northern New England* (unpublished report, 2009).

Kiltie, Richard A. "Seasonality, gestation time, and large mammal extinctions." In Paul S. Martin and Richard G. Klein, eds., *Quaternary Extinctions: A Prehistoric Revolution*, 299–314. Tucson, AZ: University of Arizona Press, 1984.

Klyza, Christopher M., and Stephen C. Trombulak, eds. *The Future of the Northern Forest*. Hanover, NH: University Press of New England, 1994.

Koch, Paul L., and Anthony D. Barnosky. "Late Quaternary extinctions: state of the debate." *Annual Reviews of Ecology, Evolution, and Systematics* 37 (2006): 215–250.

Köster, Dorte, John Lichter, Peter D. Lea, and Andrea Nurse. "Historical eutrophication in a river-estuary complex in mid-coast Maine." *Ecological Applications* 17 (2007): 765–778.

Kreft, H., and W. Jetz. "Global patterns and determinants of vascular plant diversity." *Proc. Natl Acad. Sci. U.S.A.* 104 (2007): 5925–5930.

Krohn, William B., Randall B. Boone, and Stephanie L. Painton. "Quantitative delineation and characterization of hierarchical biophysical regions of Maine." *Northeastern Naturalist* 6 (1999): 139–164.

Krohn, W.B., K.D. Elowe, and R.B. Boone. "Relations among fishers, snow, and martens: development and evaluation of two hypotheses." *The Forestry Chronicle* 71 (1995): 97–105.

Krohn, William B., and Christopher L. Hovin. *Early Maine Wildlife: Historical Accounts of Canada Lynx, Moose, Mountain Lion, White-Tailed Deer, Wolverine, Wolves, and Woodland Caribou, 1603–1930*. Orono: University of Maine Press, 2010.

Küchler, A.W. *Potential Natural Vegetation of the Conterminous United States.* American Geographical Society, Special Publication No. 36. Washington, DC: American Geographical Society, 1964.

Langenheim, Jean. H. "Early history and progress of women ecologists: emphasis upon research contributions." *Annual Review of Ecology and Systematics* 27 (1996): 1–53.

Lansky, Mitch. *Beyond the Beauty Strip: Saving What Is Left of Our Forests.* Gardiner, ME: Tilbury House, 1992.

Lastovicka, J., R.A. Akmaev, G. Beig, J. Bremer, and J.T. Emmert. "Global change in the upper atmosphere." *Science* 314 (2006): 1253–1254.

Lavoie, Martin, Louise Filion, and Elisabeth C. Robert. "Boreal peatland margins as repository sites of long-term natural disturbances of balsam fir/spruce forests." *Quaternary Research* 71 (2009): 295–306.

Lawrence, D.B. "Glaciers and vegetation in south-eastern Alaska." *American Scientist* 46 (1958): 89–122.

Lear, Alex. "Forest service warns of browntail moth caterpillar infestation in mid-coast." *The Forecaster* (Portland, ME), May 12, 2010.

Lehrer, Jonah. "The truth wears off: is there something wrong with the scientific method?" *New Yorker,* December 13, 2010.

LeVert, Mike, Charles S. Colgan, and Charles Lawton. "Are the economics of a sustainable maine forest sustainable." *Maine Policy Review* 16 (2007): 26–37.

Likens, Gene. "Acid Rain." In *Encyclopedia of Earth* (2010), www.eoearth.org/article/Acid_rain.

Likens, G.E., F.H. Bormann, and N.M. Johnson. "Acid rain." *Environment* 14 (1972): 33–40.

Likens, Gene E., and Jerry F. Franklin. "Ecosystem thinking in the northern forest — and beyond." *BioScience* 59 (2009): 511–513.

Lilieholm, Robert J. "Forging a common vision for Maine's North Woods. *Maine Policy Review* 16 (2007): 12–25.

Lilieholm, R.J., L.C. Irland, and J.M. Hagan. "Changing socio-economic conditions for private woodland protection." In S.C. Trombulak, and R. Baldwin, eds., *Multi-scale Conservation Planning.* New York: Springer-Verlag, 2010.

Lindbladh, Matts, W. Wyatt Oswald, David R. Foster, Edward K. Faison, Juzhi Hou, and Yongsong Huang. "A late-glacial transition from *Picea glauca* to *Picea mariana* in southern New England." *Quaternary Research* 67 (2007): 502–508.

Lindenmayer, David B., and Joern Fischer. *Habitat Fragmentation and Landscape Change.* Washington, DC: Island Press, 2006.

Lindner, Marcus, Michael Maroschek, Sigrid Netherer, Antoine Kremer, Anna Barbati, Jordi Garcia-Gonzalo, Rupert Seidl, Sylvain Delzon, Piermaria Corona, Marja Kolstro, Manfred J. Lexer, and Marco Marchetti. "Climate change impacts, adaptive capacity, and vulnerability of European forest ecosystems." *Forest Ecology and Management* 259 (2010): 698–709.

Little, Elbert Jr. *Digital Representations of Tree Species Range Maps from "Atlas of United States Trees."* Washington, DC: U.S. Geological Survey, U.S. Department of the Interior (esp.cr.usgs.gov/data/atlas/little).

Liu, Yongqiang, John Stanturf, and Scott Goodrick. "Trends in global wildfire potential in a changing climate." *Forest Ecology and Management* 259 (2010): 685–697.

Livingston, W.H., and A.S. White. "May drought confirmed as likely cause of brown ash dieback in Maine." *Phytopathology* 87 (1997): S59.

Livingston, W.H., G. Granger, M. Fries, H. Trial, D. Struble, J. Steinman, and S. Howell. *White Pine Decline in Maine.* PowerPoint presentation (retrieved on September 15, 2010, at extension.unh.edu/forestry/Docs/livingsto.pdf).

Longfellow, Henry Wadsworth. *Evangeline, a Tale of Acadie.* London: David Bogue, 6th edition (originally 1847).

Lorimer, Craig G. "The presettlement forest and natural disturbance cycle of Northeastern Maine." *Ecology* 58 (1977): 139–148.

Lorimer, Craig G., and Alan S. White. "Scale and frequency of natural disturbances in the northeastern US: implications for early successional forest habitats and regional age distributions." *Forest Ecology and Management* 185 (2003): 41–64.

Luoto, M., J. Pöyry, R. K. Heikkinen, and K. Saarinen. "Uncertainty of bioclimate envelope models based on the geographical distribution of species." *Global Ecology and Biogeography* 14 (2005): 575–584.

Lynch, J.A., V.C. Bowersox, and J.W. Grimm. "Acid rain reduced in the eastern United States." *Environ. Sci. Technol.* 34 (2000): 940–949.

Mack, R.N., D. Simberloff, W.M. Lonsdale, H. Evans, M. Clout, and F. Bazzazz. *Biotic Invasions: Causes, Epidemiology, Global Consequences and Control.* Issues in Ecology No. 5. Washington, DC: Ecological Society of America, 2000.

Maine Bureau of Parks and Lands. *Flagstaff Region Management Plan.* Augusta: Maine Bureau of Parks and Lands, Department of Conservation, June 12, 2007.

Maine Butterfly Survey. Maine Department of Inland Fisheries and Wildlife et al., mbs.umf.maine.edu.

Maine Department of Inland Fisheries and Wildlife. "Moose." www.maine.gov/ifw/wildlife/species/moose/index.htm.

Maine Department of Inland Fisheries and Wildlife. "Katahdin Arctic." www.maine.gov/IFW/wildlife/species/endangered_species/katahdin_arctic/index.htm.

Maine Department of Inland Fisheries and Wildlife. *Maine's Comprehensive Wildlife Conservation Strategy.* Augusta: Maine Department of Inland Fisheries and Wildlife, 2005.

Maine Forest Certification Initiative. *The Final Report of the Maine Forest Certification Advisory Committee,* January 28, 2005.

Maine Forest Service. *An Evaluation of the Effects of the Forest Practices Act.* Augusta: Maine Forest Service, Department of Conservation, 1995.

Maine Forest Service. *The State of the Forest and Recommendations for Forest Sustainability Standards.* Augusta: Maine Forest Service, Department of Conservation, 1999.

Maine Forest Service. *Assessment of Sustainable Biomass Availability: Absolute Supply Is Not the Issue. Improving Utilization and Silviculture While Keeping Costs Low Are.* Augusta: Maine Forest Service, Department of Conservation, July 17, 2008.

Maine Forest Service. *Browntail Moth,* Euproctis chrysorrhoea *(L.).* Augusta: Insect and Disease Laboratory, Maine Forest Service, Department of Conservation, 2008.

Maine Forest Service. *Forest Trees of Maine, Centennial Edition.* Augusta: Maine Forest Service, Department of Conservation, 2008.

Maine Forest Service. Maine Forest Practices Act Rule Promulgation: MFS Rule, chapter 23, "Timber harvesting standards to substantially eliminate liquidation harvesting," 2005, and MFS Rule, chapter 20, "Forest regeneration and clearcutting standards," 1999. Maine Department of Conservation, Maine Forest Service.

Maine Forest Service. Press release: "New hemlock woolly adelgid infestation discovered." May 13, 2010, Maine Department of Conservation.

Maine Forest Service. *Silvicultural Activities Report.* Augusta: Maine Forest Service, Department of Conservation, 1988 and 2009.

Maine Forest Service. *Wood Processor Reports.* Augusta: Maine Forest Service, Department of Conservation, 2000–2009.

Maine Forest Service, Department of Conservation. *Maine State Forest Assessment and Strategies.* Augusta: Maine Forest Service, Department of Conservation, 2010.

Maine Land Use Regulation Commission. *Concept Plan for the Moosehead Lake Region.* Augusta: Maine Department of Conservation, September 23, 2009.

Maine Natural Areas Program. "Overview and Mission." www.maine.gov/doc/nrimc/mnap/index.htm.

Maine Natural Areas Program. "Ecosystems in Maine." www.maine.gov/doc/nrimc/mnap/features/ecosystems.htm.

Maine Natural Areas Program. "Areas of Statewide Ecological Significance." www.maine.gov/doc/nrimc/mnap/focusarea/.

Maine Natural Areas Program. "Asiatic bittersweet," "Shrubby honeysuckles," "Purple loosestrife," *Maine Invasive Plants,* Bulletin Nos. 2506–2508. Augusta: Maine Department of Conservation, 2001.

Maine Natural Areas Program. *Comments on Kibby Expansion Development Permit Application DP 4860).* Augusta, ME: Maine Natural Areas Program, Department of Conservation, February 24, 2010.

Maine Natural Areas Program. *Ecological Reserve Monitoring: Summary Report for the Maine Outdoor Heritage Fund and The Nature Conservancy.* Augusta: Maine Natural Areas Program, Department of Conservation, 2009.

Maine Public Broadcasting. "A love for the land," Episode 201. In *Home: The Story of Maine.*

Maine State Archives. Land Office Records, Field Notes.

Maine State Planning Office (chief author: Janet McMahon). *An Ecological Reserves System for Maine: Benchmarks in a Changing Landscape.* Report to the 116th Maine Legislature. Augusta: Maine State Planning Office, Natural Resources Planning Division, 1993.

Manion, P.D. *Tree Disease Concepts.* Upper Saddle River, NJ: Prentice-Hall, 1991.

Manion, P.D., and D. Lachance. *Forest Decline Concepts.* St. Paul, MN: APS Press, 1992.

Manter, D.K., and W.H. Livingston. "Thawing rate and fungal infection by *Rhizosphaera kalkhoffi* Bubak influence freezing injury on red spruce (*Picea rubens* Sarg.) needles." *Canadian Journal of Forest Research* 26 (1996): 918–927.

Marsh, George Perkins. Address delivered before the Agricultural Society of Rutland County, Sept. 30, 1847. In *The Evolution of the Conservation Movement, 1850–1920.* Washington, DC: Library of Congress, American Memory (http://memory.loc.gov/ammem/amrvhtml/conshome.html).

Marsh, George Perkins. *Man and Nature.* New York: C. Scribner, 1864.

Marshak, Stephen. *Earth: Portrait of a Planet.* 2nd edition. New York: W.W. Norton, 2005.

Marshall, Humphrey. *Arbustrum Americanum: The American Grove.* Philadelphia: J. Cruickshank, 1785.

Martin, P.H., C.D. Canham, and R.K. Kobe. "Divergence from the growth-survival trade-off and extreme high growth rates drive patterns of exotic tree invasions in closed-canopy forests." *Journal of Ecology* 98 (2010): 778–789.

Martin, Paul S. "Prehistoric overkill." In P.S. Martin and H.W. Wright Jr., eds., *Pleistocene Extinctions: The Search for a Cause,* 75–120. New Haven, CT: Yale University Press, 1967.

Marvinney, Robert G. *Simplified Bedrock Geologic Map of Maine.* Augusta: Maine Geological Survey, Department of Conservation, 2002.

Massachusetts Introduced Pests Outreach Project. MA Department of Agricultural Resources and UMass Extension Agriculture and Landscape Program, August, 2008.

Master, L., D. Faber-Langendoen, R. Bittman, G.A. Hammerson, B. Heidel, J. Nichols, L. Ramsay, and A. Tomaino. *NatureServe Conservation Status Assessments: Factors for Assessing Extinction Risk.* Arlington, VA: NatureServe, April 2009.

May, Diane Ebert, and Ronald B. Davis. *Alpine Tundra Vegetation of Maine*

Mountains and Its Relevance to the Critical Areas Program. Planning Report 36. Augusta: Maine Critical Areas Program, State Planning Office, 1978.

McCann, Paul K. Timber! The Fall of Maine's Paper Giant. Ellsworth, ME: Ellsworth American, 1994.

McCaskill, G.L., W.H. McWilliams, B.J. Butler, D.M. Meneguzzo, C.J. Barnett, and M.H. Hansen. Maine's Forest Resources, 2008. Res. Note NRS-53. Newtown Square, PA: U.S. Department of Agriculture, Forest Service, Northern Research Station, 2010.

McComb, William, and David Lindenmayer. "Dying, dead, and down trees." In Malcolm L. Hunter Jr., ed., Maintaining Biodiversity in Forested Ecosystems, 335–372. Cambridge: Cambridge University Press, 1999.

McIntosh, Robert P. The Background of Ecology, Concept and Theory. New York: Cambridge University Press, 1985.

McKelvey, Kevin S., Keith B. Aubry, and Yvette K. Ortega. "History and distribution of lynx in the contiguous United States." In Leonard F. Ruggiero, Keith B. Aubry, Steven W. Buskirk, Gary M. Koehler, Charles J. Krebs, Kevin S. McKelvey, and John R. Squires, eds., Ecology and Conservation of Lynx in the United States. General Technical Report RMRS-GTR-30WWW, chapter 8. Fort Collins, CO: U.S. Department of Agriculture, Forest Service, Rocky Mountain Research Station, 1999.

McKibben, Bill. "An Explosion of Green." Atlantic Monthly 275 (April 1995): 61–83.

McLachlan, Jason S., James S. Clark, and Paul S. Manos. "Molecular indicators of tree migration capacity under rapid climate change." Ecology 86 (2005): 2088–2098.

McLaughlin, Craig R. Black Bear Assessment and Strategic Plan 1999. Bangor: Maine Department of Inland Fisheries and Wildlife, 1999.

McMahon, J.S. "Saving all the pieces: an ecological reserves proposal from Maine." Maine Naturalist 1 (1993): 213–222.

McMahon, Janet S. "The biophysical regions of Maine — patterns in the landscape and vegetation." Master's thesis, University of Maine, 1990.

McNab, W.H., D.T. Cleland, J.A. Freeouf, J.E. Keys Jr., G.J. Nowacki, and C.A. Carpenter, compilers. Description of Ecological Subregions: Sections of the Conterminous United States. Gen. Tech. Report WO-76B. Washington, DC: U.S. Department of Agriculture, Forest Service, 2007.

McNeill, J.R. Something New Under the Sun: An Environmental History of the 20th-Century World. New York: W.W. Norton, 2000.

McWilliams, W.H., Brett J. Butler, Laurence E. Caldwell, Douglas M. Griffith, Michael L. Hoppus, Kenneth M. Laustsen, Andrew J. Lister, Tonya W. Lister, Jacob W. Metzler, Randall S. Morin, Steven A. Sader, Lucretia B. Stewart, James R. Steinman, James A. Westfall, David A. Williams, Andrew Whitman, and Christopher Woodall. The Forests of Maine: 2003. Resource Bulletin

NE-164. Newtown Square, PA: U.S. Department of Agriculture, Forest Service, Northeastern Research Station, 2005.

Meehl, G.A., T.F. Stocker, W.D. Collins, P. Friedlingstein, A.T. Gaye, J.M. Gregory, A. Kitoh, R. Knutti, J.M. Murphy, A. Noda, S.C.B. Raper, I.G. Watterson, A.J. Weaver and Z.-C. Zhao. "Global Climate Projections." In S. Solomon, D. Qin, M. Manning, Z. Chen, M. Marquis, K.B. Averyt, M. Tignor, and H.L. Miller, eds., *Climate Change 2007: The Physical Science Basis. Contribution of Working Group I to the Fourth Assessment Report of the Intergovernmental Panel on Climate Change,* chapter 10. Cambridge: Cambridge University Press, 2007.

Meehl, G.A., W.M. Washington, C.M. Ammann, J.M. Arblaster, T.M.L. Wigley, and C. Tebaldi. "Combinations of natural and anthropogenic forcings in twentieth-century climate." *Journal of Climate* 17 (2004): 3721–3727.

Mehrhoff, L.J., J.A. Silander Jr., S.A. Leicht, E.S. Mosher, and N.M. Tabak. *Invasive Plant Atlas of New England.* Storrs, CT: Department of Ecology & Evolutionary Biology, University of Connecticut, 2003 (nbii-nin.ciesin.columbia. edu/ipane).

Menges, Eric S. "Population viability analysis for an endangered plant." *Conservation Biology* 4 (1990): 52–62.

Menzel, A., T.H. Sparks, N. Estrella, E. Koch, A. Aasa, R. Ahas, K. Alm-Kübler, P. Bissolli, O. Braslavská, A. Briede, F.M. Chmielewski, Z. Crepinsek,Y. Curnel, Å. Dahl, C.Defila, A. Donnelly, Y. Filella, K. Jatczak, F. Måge, A. Mestre, Ø. Nordli, J. Peñuelas, P. Pirinen, V. Remišová, H. Scheifinger, M. Striz, A. Susnik, A.J.H. vanVliet, F.-E.Wielgolaski, S. Zach, and A. Zust. "European phonological response to climate change matches the warming pattern." *Global Change Biology* 12 (2006): 1969–1976

Merriam, C. Hart. "The geographic distribution of animals and plants in North America." In *Yearbook of the United States Department of Agriculture 1894,* 203–214. Washington, DC: U.S. Department of Agriculture, 1895.

Merriam, C.H., and L. Steineger. *Results of a Biological Survey of the San Francisco Mountain Region and the Desert of the Little Colorado, Arizona.* North American Fauna Report 3. Washington, DC: Division of Ornithology and Mammalia, U.S. Department of Agriculture, 1890.

Miller, Char. *Gifford Pinchot and the Making of Modern Environmentalism.* Washington, DC: Island Press, 2001.

Miller, Eric K. *Assessment of Forest Sensitivity to Nitrogen and Sulfur Deposition in Maine.* Conference of New England Governors and Eastern Canadian Premiers Forest Mapping Group. Augusta: Maine Department of Environmental Protection, 2006.

Miller, Kevin. "Report: Maine woods can support heavier logging." *Bangor Daily News,* August 19, 2011.

Miller, Mark. *Monhegan Stewardship Management Plan.* Monhegan, ME: Monhegan Associates, Inc., 2005.

Miller, Phillip. *The Gardeners Dictionary*. 4th edition. London: John Rivington, 1754.

Miller-Rushing, Abraham J., Trevor L. Lloyd-Evans, Richard B. Primack, and Paul Satzinger. "Bird migration times, climate change, and changing population sizes." *Global Change Biology* 14 (2008): 1959–1972.

Moesswilde, M.J. "Age structure, disturbance, and development of old growth red spruce stands in northern Maine." Master's thesis, University of Maine, 1995.

Moore, Elizabeth H., and Jack W. Witham. "From forest to farm and back again: land use history as a dimension of ecological research in coastal Maine." *Environmental History* 1 (1996): 50–69.

Morewood, W.D., P.R. Neiner, J.R. McNeil, J.C. Sellmer, and K. Hoover. "Oviposition preference and larval performance of *Anoplophora glabripennis* (Coleoptera: Cerambycidae) in four eastern North American hardwood tree species." *Environmental Entomology* 32 (2003): 1028–1034.

Morey, David C., editor and annotator. *The Voyage of Archangell, James Rosier's Account of the Waymouth Voyage of 1605, A True Relation*. Gardiner, ME: Tilbury House, 2005.

Morin, Hubert, Yves Jardon, and Réjean Gagnon. "Relationship between spruce budworm outbreaks and forest dynamics in eastern North America." In Edward A. Johnson and Kiyoko Miyanishi, eds., *Plant Disturbance Ecology: The Process and the Response* (Amsterdam, Netherlands: Elsevier, 2007), 555–577.

Morin, Xavier, Carol Augsburger, and Isabelle Chuine. "Process-based modeling of species' distributions: what limits temperate tree species' range boundaries?" *Ecology* 88 (2007): 2280–2291.

Morin, Xavier, David Viner, and Isabelle Chuine. "Tree species range shifts at a continental scale: new predictive insights from a process-based model." *Journal of Ecology* 96 (2008): 784–794.

Mueller-Dombois, Dieter. "Natural dieback in forests." *BioScience* 37 (1987): 575–583.

Muller, R.A., and G. J. MacDonald. "Glacial cycles and astronomical forcing." *Science* (1997) 277: 215–218.

Naiman, R.J. and H. Décamps. "The ecology of interfaces: riparian zones." *Annual Review of Ecology and Systematics* 28 (1997): 621–658.

National Big Tree Register. www.americanforests.org/resources/bigtrees/index.php.

National Geographic Society and World Wildlife Fund. *Terrestrial Ecoregions of the World*. www.nationalgeographic.com/wildworld/terrestrial.html.

National Oceanic and Atmospheric Administration. "NOAA: June, April to June, and year-to-date global temperatures are warmest on record." July 30, 2010.

National Oceanic and Atmospheric Administration. *NOAA Celebrates 200 years of Science, Service, and Stewardship.* Washington, DC: U.S. Department of Commerce (http://celebrating200years.noaa.gov/).

Natural Resources Council of Maine. *Comments on Kibby Expansion Development Permit Application DP 4860).* Augusta: Maine Natural Areas Program, Department of Conservation, April 28, 2010.

Natural Resources Council of Maine. *Wind Projects in Maine* (www.nrcm.org/maine_wind_projects.asp).

The Nature Conservancy in Maine. "Mount Agamenticus Conservation Region," "Moosehead Forest Project," and "Kennebunk Plains." www.nature.org/ourinitiatives/regions/northamerica/unitedstates/maine/.

The Nature Conservancy in Maine. *Upper St. John River Forest Management Plan.* Brunswick: The Nature Conservancy in Maine, April 2003, updated September 2009.

The Nature Conservancy. *Determining the Size of Eastern Forest Reserves.* Boston: Eastern U.S. Conservation Region, 2004.

NatureServe. *Digital Distribution Maps of the Birds of the Western Hemisphere Version 3.0* (www.natureserve.org/getData/birdMaps.jsp).

Neilson, Ronald P., Louis F. Pitelka, Allen M. Solomon, Ran Nathan, Guy F. Midgley, Jóse M. Fragoso, Heike Lischke, and Ken Thompson. "Forecasting regional to global plant migration in response to climate change: Challenges and directions." *BioScience* 55 (2005): 749–759.

Nelson, Gabriel. "EPA unveils rules on smog-forming emissions from power plants." *New York Times,* October 12, 2010.

New England Integrated Sciences & Assessment (NEISA). *Indicators of Climate Change in the Northeastern United States.* inhale.unh.edu/Climate/index.html.

Newman, W.A., A.N. Genes, T. Brewer. "Pleistocene geology of north-eastern Maine." In H.W. Borns Jr., P. LaSalle, and W.B. Thompson, eds., *Late Pleistocene History of Northern New England and Adjacent Quebec.* Special Paper 197, 59–70. Boulder, CO: Geological Society of America, 1985.

Nilsson, S.G., U. Arup, R. Baranowski, and S. Ekman. "Tree-dependent lichens and beetles as indicators in conservation forests." *Conservation Biology* 9 (1995): 1208–1215.

Niven, Daniel K., Gregory S. Butcher, and G. Thomas Bancroft. "Christmas bird counts and climate change: northward shifts in early winter abundance." *American Birds* 109 (2009) 10–15.

NOAA Paleoclimatology. "North American Pollen Database." www.ncdc.noaa.gov/paleo/napd.html.

Northern Forest Lands Council. *Finding Common Ground: Conserving the Northern Forest.* Concord, NH: Northern Forest Lands Council, 1994.

Norton, Stephen A., and Ivan J. Fernandez, eds. *The Bear Brook Watershed in Maine: A Paired Watershed Experiment — the First Decade (1987–1997).* Boston: Kluwer Academic, 1999.

Norton, S., J. Kahl, I. Fernandez, T. Haines, L. Rustad, S. Nodvin, J. Scofield, T. Strickland, H. Erickson, P. Wigington Jr., and J. Lee. "The Bear Brook Watershed, Maine (BBWM), USA." *Environ. Monit. Assess.* 55 (1999): 7–51.

Noss, R.F. "A regional landscape approach to maintain diversity." *BioScience* 33 (1983): 700–706.

Noss, R.F., and A.Y. Cooperrider. *Saving Nature's Legacy: Protecting and Restoring Biodiversity.* Washington, DC: Island Press, 1994.

Nowell, K. *"Lynx Canadensis."* In *IUCN Red List of Threatened Species.* Version 2010.3. Washington, DC: International Union for Conservation of Nature and Natural Resources, 2010 (www.iucnredlist.org).

Nunery, Jared S., and William S. Keeton. "Forest carbon storage in the northeastern United States: net effects of harvesting frequency, post-harvest retention, and wood products." *Forest Ecology and Conservation* 259 (2010): 1363–1375.

O'Connor, Kevin. "Is climate change bringing the state more bugs? Bitten by the bug." *Times Argus* (Barre-Monpelier, Vermont), August 26, 2007.

Ollinger, S.V., C.L. Goodale, K. Hayhoe, and J.P. Jenkins. "Potential effects of climate change and rising CO_2 on ecosystem processes in northeastern U.S. forests." *Mitigation and Adaptation Strategies for Global Change* 13 (2008): 467–485.

Olson, David M., Eric Dinnerstein, Eric D. Wikramanayake, Neil D. Burgess, George V.N. Powell, Emma C. Underwood, Jennifer A. D'Amico, Illanga Itoua, Holly E. Strand, John C. Morrison, Colby J. Loucks, Thomas F. Allnut, Taylor H. Ricketts, Yumiko Kura, John F. Lamoreux, Wesley W. Wettengel, Prashant Hedao, and Kenneth R. Kassem. "Terrestrial ecoregions of the world: a new map of life on Earth." *BioScience* 51 (2001): 933–938.

Oreskes, Naomi, ed. *Plate Tectonics: An Insider's History of the Modern Theory of the Earth.* Boulder, CO: Westview Press, 2003.

Orono Bog Boardwalk. www.oronobogwalk.org/index.htm.

Otis, Charles P., translator. *Voyages of Samuel de Champlain.* Boston: The Prince Society, 1880.

Ouimet, R., L. Duchesne, D. Houle, and P.A. Arp. "Critical loads of atmospheric S and N deposition and current exceedances for northern temperate and boreal forests in Québec." *Water Air, and Soil Pollution: Focus* 1 (2001): 119–134.

Pachauri, R.K., and A. Reisinger, eds. *Climate Change 2007: Synthesis Report.* Fourth Assessment Report. Geneva, Switzerland: International Panel on Climate Change, 2007.

Parker, Everett L. *Kineo: Moosehead Sentinel from Native Americans to Hotel Grandeur.* Greenville, ME: Moosehead Communications, 2004.

Parker, Thomas. *History of Farmington, Maine, from Its Settlement to the Year 1846.* Farmington, ME: J.S. Swift, 1846.

Parkman, Francis. *Francis Parkman's Works: A Half-century of Conflict*. Boston: Little, Brown, 1907.

Parmesan, Camille. "Ecological and evolutionary responses to recent climate change." *Annual Review of Ecology, Evolution, and Systematics* 37 (2006): 637–669.

Parmesan, C., and G. Yohe. "A globally coherent fingerprint of climate change impacts across natural systems." *Nature* 421 (2003): 37–42.

Patrick, David A., Aram J.K. Calhoun, and Malcolm L. Hunter Jr. "The importance of understanding spatial population structure when evaluating the effects of silviculture on spotted salamanders (*Ambystoma maculatum*)." *Biological Conservation* 141 (2008): 807–814.

Patrick, D.A., M.L. Hunter Jr., and A.J.K. Calhoun. "Effects of experimental forestry treatments on a Maine amphibian community." *Forest Ecology and Management* 234 (2006): 323–332.

Patterson, W.A., III, and K. E. Sassaman. "Indian fires in the prehistory of New England." In G.P. Nicholas, ed., *Holocene Human Ecology in Northeastern North America*, 107–135. New York: Plenum, 1988.

Payer, D.C., and D.J. Harrison. "Structural differences between forests regenerating following spruce budworm defoliation and clear-cut harvesting: implications for marten." *Canadian Journal of Forest Research* 30 (2000): 1965–1972.

Pearson, Richard. "Climate change and the migration capacity of species." *Trends in Ecology and Evolution* 21 (2006): 111–113.

Pearson, Richard G., and Terence P. Dawson. "Predicting the impacts of climate change on the distribution of species: are bioclimate envelope models useful?" *Global Ecology & Biogeography* 12 (2003): 361–371.

Perry, Allison L., Paula J. Low, Jim R. Ellis, and John D. Reynolds. "Climate change and distributional shifts in marine fishes." *Science* 308 (2005): 1912–1915.

Pew Center on Global Climate Change. *The Causes of Global Climate Change*. Science Brief 1. Arlington, VA: Pew Center on Global Climate Change, 2008.

Philipona, Rolf, Bruno Dürr, Christoph Marty, Atsumu Ohmura, and Martin Wild. "Radiative forcing — measured at Earth's surface — corroborate the increasing greenhouse effect." *Geophysical Research Letters* 31 (2004): LO3202.

Pierson, Elizabeth C., Jan Erik Pierson, and Peter D. Vickery. *A Birder's Guide to Maine*. Camden, ME: Down East Books, 1996.

Pinchot, Gifford. *Breaking New Ground*. Washington, DC: Island Press, 1998, from original 1947.

Plantinga, A.J., T. Mauldin, and R.J. Alig. *Land Use in Maine: Determinants of Past Trends and Projections of Future Changes*. Res. Pap. PNWRP- 511. Portland, OR: U.S. Department of Agriculture, Forest Service, Pacific Northwest Research Station, 1999.

Pollock, Stephen, and Jan E. Pierson. "Foundation of the past: an introduction to Maine's bedrock geology." *Habitat: Journal of the Maine Audubon Society* 5 (1988): 18–21.

Polsenberg, Johanna F. "Adirondack lakes making a comeback?" *Frontiers of Ecology and the Environment* 8 (2010): 345.

Pounds, J.A., Martin R. Bustamante, Luis A. Coloma, Jamir A. Consugra, Michael P.L. Fogden, Pru N. Foster, Enrique La Marca, Karen L. Masters, Andres Merino-Viteri, Robert Puschendorf, Santiago R. Ron, G. Arturo Sanches-Azofeifa, Christopher J. Still, and Bruce E. Young. "Widespread amphibian extinctions from epidemic disease driven by global warming." *Nature* 439 (2006): 161–67.

Pöyryl, Juha, Miska Luota, Risto K. Heikkinen, and Kimmo Saarinen. "Species traits are associated with the quality of bioclimatic models." *Global Ecology and Biogeography* 17 (2008): 403–414.

Primack, D., C. Imbres, R.B. Primack, A. Miller-Rushing, and P. del Tredici. "Herbarium specimens demonstrate earlier flowering times in response to warming in Boston." *American Journal of Botany* 91 (2004): 1260–1264.

Public Laws of Maine. *An Act to Establish Standards and Conditions for Designation of Ecological Reserves on Lands Managed by the Bureau of Parks and Lands*, 2000. Chapter 592, S.P. 157–L.D. 477.

Purvis, William. *Lichens*. Washington, DC: Smithsonian Institution Press, 2000.

Quimby, Beth. "Mogul's land buy prompts questions." *Portland Press Herald*, January 30, 2011.

Ramankutty, Navin, Elizabeth Heller, and Jeanine Rhemtulla. "Prevailing myths about agricultural abandonment and forest regrowth in the United States." *Annals of the Association of American Geographers* 100 (2010): 502–512.

Ranco, Darren, Rob Lilieholm, John Daigle, and Maine Indian Basketmakers Alliance. *Mobilizing Researchers and Diverse Stakeholders to Address Emerging Environmental Threats: Brown Ash Sustainability as a Cultural Resource for Maine Indian Basketmakers*. Project of the Maine Sustainability Solutions Initiative. Orono: University of Maine (http://www.umaine.edu/brownash).

Reidel, Carl. "The political process of the Northern Forest Lands Study." In Christopher M. Klyza and Stephen C. Tromulak, eds., *The Future of the Northern Forest*, 93–111. Hanover, NH: University Press of New England, 1994.

Reimer, P.J., M.G.L. Baillie, E. Bard, A. Bayliss, J.W. Beck, P.G. Blackwell, Ramsey C. Bronk, C.E. Buck, G.S. Burr, R.L. Edwards, M. Friedrich, P.M. Grootes, T.P. Guilderson, I. Hajdas, T.J. Heaton, A.G. Hogg, K.A. Hughen, K.F. Kaiser, B. Kromer, F.G. McCormac, S.W. Manning, R.W. Reimer, D.A. Richards, J.R. Southon, S. Talamo, C.S.M. Turney, J. van der Plicht, and C.E. Weyhenmeyer. "IntCal09 and Marine09 radiocarbon age calibration curves, 0–50,000 years cal BP." *Radiocarbon* 51 (2009): 1111–50.

Report of the Commissioner appointed to superintend the Survey of the North American Railway (Augusta, ME, 1851), 6, quoted in Woods, *A History of Lumbering in Maine, 1820–1861*.

Reusch, Douglas N. "New Caledonian carbon sinks at the onset of Antarctic glaciation." *Geology* 39 (2011): 807–810.

Richardson, A.D., A.S. Bailey, E.G. Denny, C.W. Martin, and J. O'Keefe. "Phenology of a northern hardwood forest canopy." *Global Change Biology* 12 (2006), 1174–1188.

Richardson, D.M., P. Pyšek, M. Rejmánek, M.G. Barbour, F.D. Panetta, and C.J. West. "Naturalization and invasion of alien plants: concepts and definitions." *Diversity and Distributions* 6 (2000): 93–107.

Richelson, Jeffrey T. *U.S. Satellite Imagery, 1960–1999*. National Security Archive Electronic Briefing Book No 13, April 14, 1999 (www.gwu.edu/~nsarchiv/NSAEBB/NSAEBB13).

Rodenhouse, N.L., S.N. Matthews, K.P. McFarland, J.D. Lambert, L.R. Iverson, A. Prasad, T.S. Sillett, and R.T. Holmes. "Potential effects of climate change on birds of the Northeast." *Mitigation and Adaptation Strategies for Global Change* 13 (2008): 517–540.

Rolde, Neil. *The Interrupted Forest: A History of Maine's Wildlands*. Gardiner, ME: Tilbury House, 2001.

Root, T.L., J.T. Price, K.R. Hall, S.H. Schneider, C. Rosenzweig, and J.A. Pounds. "Fingerprints of global warming on wild animals and plants." *Nature* 421 (2003): 57–60.

Rosenzweig, C., G. Casassa, D.J. Karoly, A. Imeson, C. Liu, A. Menzel, S. Rawlins, T.L. Root, B. Seguin, and P. Tryjanowski. "Assessment of observed changes and responses in natural and managed systems." In M.L. Parry, O.F. Canziani, J.P. Palutikof, P.J. van der Linden, and C.E. Hanson, eds., *Climate Change 2007: Impacts, Adaptation and Vulnerability. Contribution of Working Group II to the Fourth Assessment Report of the Intergovernmental Panel on Climate Change*, chapter 1. Cambridge, UK: Cambridge University Press, 2007.

Rosenzweig, Cynthia, David Karoly, Marta Vicarelli, Peter Neofotis, Qigang Wu, Gino Casassa, Annette Menzel, Terry L. Root, Nicole Estrella, Bernard Seguin, Piotr Tryjanowski, Chunzhen Liu, Samuel Rawlins, and Anton Imeson. "Attributing physical and biological impacts to anthropogenic climate change." *Nature* 453 (2008): 353–357.

Rowland, Erika. "Disturbance pattern of multiple temporal scales in old-growth conifer–northern hardwood stands in northern Maine, USA." PhD dissertation, University of Maine, 2006.

Rowland, Erika L., and Alan S. White. "Topographic and compositional influences on disturbance patterns in a northern Maine old-growth landscape." *Forest Ecology and Management* 259 (2010): 2399–2409.

Rowland, Erika L., Alan S. White, and William H. Livingston. *A Literature Review of the Effects of Intensive Forestry on Forest Structure and Plant Community Composition at the Stand and Landscape Levels.* Miscellaneous Publication 754. Orono: Maine Agricultural and Forest Experiment Station, 2005.

Ruiz, Gregory M., and James T. Carlton, eds. *Invasive Species: Vectors and Management Strategies.* Washington, DC: Island Press, 2003.

Runkle, James R. "Patterns of disturbance in some old-growth mesic forests of eastern North America." *Ecology* 63 (1982): 1533–1546.

Russell, Emily W. B., Ronald B. Davis, R. Scott Anderson, Thomas E. Rhodes, and Dennis S. Anderson. "Recent centuries of vegetational change in the glaciated north-eastern United States." *Journal of Ecology* 81 (1993): 647–664.

Sader, Steven A., Matthew Bertrand, and Emily H. Wilson. "Satellite change detection of forest harvest patterns on an industrial forest landscape." *Forest Science* 49 (2003): 341–353.

Sanger, David. "The original native Mainers." *Habitat: Journal of the Maine Audubon Society* 5 (1988): 37–41.

Sanger, D., H. Almquist, and A.C. Dieffenbacher-Krall. "Mid-Holocene cultural adaptations to central Maine." In D.H. Sandweiss and K.A. Maasch, eds., *Climate Change and Cultural Dynamics: A Global Perspective on Holocene Transitions,* 435–456. London: Academic Press, 2006.

Santer, B.D., T.M.L. Wigley, A.J. Simmons, P.W. Kallberg, G.A. Kelly, S.M. Uppala, C. Ammann, J.S. Boyle, W. Bruggemann, C. Doutriaux, M. Fiorino, C. Mears, G.A. Meehl, R. Sausen, K.E. Taylor, W.M. Washington, M.F. Wehner, and F.J. Wentz. "Identification of anthropogenic climate change using a second-generation reanalysis." *Journal of Geophysical Research-Atmospheres* 109 (2004): D21104.

Sayen, Jamie. "Northern Appalachian wilderness: the key to sustainable natural and human communities in the northern forest." In Christopher M. Klyza and Stephen C. Tromulak, eds., *The Future of the Northern Forest,* 177–197. Hanover, NH: University Press of New England, 1994.

Schaberg, P.G., D.H. DeHayes, and G.J. Hawley. "Anthropogenic calcium depletion: a unique threat to forest ecosystem health?" *Ecosystem Health* 7 (2001): 214–228.

Schaberg, P.G., J.W. Tilley, G.J. Hawley, D.H. DeHayes, and S.W. Bailey. "Associations of calcium and aluminum with the growth and health of sugar maple trees in Vermont." *Forest Ecology and Management* 223 (2006): 159–169.

Schauffler, Molly, and George Jacobson. "Persistence of coastal spruce refugia during the Holocene in northern New England, USA, detected by stand-scale pollen stratigraphies." *Journal of Ecology* 90 (2002): 235–250.

Scott, J.M., F.W. Davis, R.G. McGhie, R.G. Wright, C. Groves, and J. Estes. "Nature reserves: do they capture the full range of America's biological diversity?" *Ecological Applications* 11 (2001): 999–1007.

Searchinger, Timothy D., Steven P. Hamburg, Jerry Melillo, William Chameides, Petr Havlik, Daniel M. Kammen, Gene E. Likens, Ruben N. Lubowski, Michael Obersteiner, Michael Oppenheimer, G. Philip Robertson, William H. Schlesinger, and G. David Tilman. "Fixing a critical climate accounting error." *Science* 326 (2009): 527–528.

Selva, S.B. "Using lichens to assess ecological continuity in northeastern forests." In Mary Byrd Davis, ed., *Eastern Old-Growth Forests: Prospects for Rediscovery and Recovery*, 35–48. Washington, DC: Island Press, 1996.

Sendak, Paul E., Robert C. Abt, and Robert J. Turner. "Timber supply projections for northern New England and New York: integrating a market perspective." *Northern Journal of Applied Forestry* 20 (2003): 175–185.

Seymour, R.S. "The red spruce–balsam fir forest of Maine: evolution of silvicultural practice in response to stand development patterns and disturbances." In M.J. Kelty, B.C. Larson, and C.D. Oliver, eds., *The Ecology and Silviculture of Mixed-Species Forests*, 217–244. Norwell, MA: Kluwer, 1992.

Seymour, Robert S., Alan S. White, and Philip deMaynadier. "Natural disturbance regimes in northeastern North America: evaluating silvicultural systems using natural scales and frequencies." *Forest Ecology and Management* 155 (2002): 357–367.

Shugart, Herman, Roger Sedjo, and Brent Sohngen. *Forests and Global Climate Change: Potential Impacts on U.S. Forest Resources.* Arlington, VA: Pew Center on Global Climate Change, 2003.

Shuman, Bryan, Paige Newby, Yongsong Huang, and Thompson Webb III. "Evidence for the close climatic control of New England vegetation history." *Ecology* 85 (2004): 1297–1310.

Silander, J.A. Jr., and D.M. Klepeis. "The invasion ecology of Japanese barberry (*Berberis thunbergii*) in the New England landscape." *Biological Invasions* 1 (1999): 189–201.

Simberloff, Daniel. "A succession of paradigms in ecology: essentialism to materialism and probabalism." *Synthese* 43 (1980): 3–39.

Small, M.F., and M.L. Hunter. "Forest fragmentation and avian nest predation in forested landscapes." *Oecologia* 76 (1988): 62–64.

Smith, Captain John. *A Description of New England* (1616). Electronic version, Paul Royster, ed. DigitalCommons@University of Nebraska–Lincoln, 2006 (digitalcommons.unl.edu).

Smith, Captain John. Map of New England (1614). DigitalCommons@University of Nebraska–Lincoln (digitalcommons.unl.edu).

Smith, David C. *A History of Lumbering in Maine, 1861–1960.* Maine Studies No. 93. Orono: University of Maine Press, 1972.

Smith, Leslie M., Sandra Whitehouse, and Candace A. Oviatt. "Impacts of climate change on Narragansett Bay." *Northeastern Naturalist* 17 (2010): 77–90.

Solomon, S., D. Qin, M. Manning, Z. Chen, M. Marquis, K.B. Averyt, M. Tignor and H.L. Miller, eds. *Contribution of Working Group I to the Fourth Assessment Report of the Intergovernmental Panel on Climate Change, 2007.* Cambridge: Cambridge University Press, 2007.

Spear, Ray W., Margaret B. Davis, and Linda C.K. Shane. "Late Quaternary history of low- and mid-elevation vegetation in the White Mountains of New Hampshire." *Ecological Monographs* 64 (1994): 85–109.

Spellerberg, Ian F. "Ecological effects of roads and traffic: a literature review." *Global Ecology and Biogeography Letters* 7 (1998): 317–333.

Spiess, Arthur. "Comings and goings: Maine's prehistoric wildlife." *Habitat: Journal of the Maine Audubon Society* 5 (1988): 30–33.

Springer, John S. *Forest Life and Forest Trees: Comprising Winter Camp-life among the Loggers.* New York: Harper and Brothers, Publishers, 1851.

Stack, Lois Berg. *Plant Hardiness Zone Map of Maine*, Bulletin #2242. Orono: University of Maine Cooperative Extension, 2006.

Standen, Anthony. "Japanese beetle: Voracious. libidinous, prolific, he is eating his way across the U.S., destroying $7,000,000 worth of plant life every year." *Life Magazine* 17 (July 17, 1944): 39–46.

Stein, Susan M., Ronald E. McRoberts, Ralph J. Alig, Mark D. Nelson, David M. Theobald, Mike Eley, Mike Dechter, and Mary Carr. *Forests on the Edge: Housing Development on America's Private Forests.* General Technical Report PNW-GTR-636. Portland, OR: Pacific Northwest Research Station, USDA Forest Service, 2005.

Stein, S.M., R.E. McRoberts, L.G. Mahal, M.A. Carr, R.J. Alig, S.J. Comas, D.M. Theobald, and A. Cundiff. *Private Forests, Public Benefits: Increased Housing Density and other Pressures on Private Forest Contributions.* General Technical Report No. PNW-GTR-795. Portland, OR: Pacific Northwest Research Station, USDA Forest Service, 2010.

Stinchcomb, G.E., T.C. Messner, S.G. Driese, L.C. Nordt, and R.M. Stewart. "Pre-colonial (A.D. 1100–1600) sedimentation related to prehistoric maize agriculture and climate change in eastern North America." *Geology* 39 (2011): 363–366.

Stoddard, J., J.S. Kahl, F. Deviney, D. DeWalle, C. Driscoll, A. Herlihy, J. Kellogg, P. Murdoch, J. Webb, and K. Webster. *Response of Surface Water Chemistry to the Clean Air Act Amendments of 1990.* Washington, DC: US Environmental Protection Agency, 2003.

Stolzenburg, William. "Where the wild things were." *Conservation Magazine* 7 (2006): 28–34.

Stone, Amy, Robin M. Usborn, and Jodie Ellis. *Out of the Ashes — What We Know about EAB Workshop.* Dayton, OH: Michigan State University, Purdue University, and the Ohio State University, September 24–25, 2008.

St. Pierre, Jym. "Huber sells Maine timberlands," *Maine Environmental News*, October 28, 2011.

Struble, David, and Dennis Souto. *Environmental Assessment Regarding Management of Hemlock Woolly Adelgid Impacts in Maine*. Augusta: Maine Forest Service, 2007.

Swearingen, Jil M. *Least Wanted: Alien Plant Invaders of Natural Areas*. Washington, DC: Plant Conservation Alliance's Alien Plant Working Group, 2009.

Takemura, T., Y. Tsushima, T. Yokohata, T. Nozawa, T. Nagashima, and T. Nakajima. "Time evolutions of various radiative forcings for the past 150 years estimated by a general circulation model." *Geophysical Research Letters* 33 (2006): L19705.

Tang, Guoping, and Brian Beckage. "Projecting the distribution of forests in New England in response to climate change." *Diversity and Distributions* 16 (2010): 144–158.

Taylor, Alan. *Liberty Men and the Great Proprietors: The Revolutionary Settlement on the Maine Frontier, 1760–1820*. Chapel Hill: University of North Carolina Press, 1990.

Thackeray, S.J., T.H. Sparks, M. Frederiksen, et al. "Trophic level asynchrony in rates of phenological change for marine, freshwater and terrestrial environments." *Global Change Biology* 16 (2010): 3304–3313.

Thomas, C.D., A. Cameron, R.E. Green, M. Bakkenes, L.J. Beaumont, Y.C. Collingham, B.F.N. Erasmus, M.F.D. Siqueira, A. Grainger, and L. Hannah. "Extinction risk from climate change." *Nature* 427 (2004): 145–148.

Thompson, Robert S., Sarah L. Shafer, Katherine H. Anderson, Laura E. Strickland, Richard T. Pelltier, Patrick J. Bartlein, and Michael W. Kerwin. "Topographic, bioclimatic, and vegetation characteristics of three ecoregion classification systems in North America: comparisons along continent-wide transects." *Environmental Management* 34 (2005), Suppl. 1: S125–S148.

Thompson, W.B. and H.W. Borns Jr. *Surficial Geologic Map of Maine*. Augusta: Maine Geological Survey, 1985 (updated by Robert G. Marvinney, 2003).

Thompson, Woodrow B., Carol B. Griggs, Norton G. Miller, Robert E. Nelson, Thomas K. Weddle, and Taylor M. Kilian. "Associated terrestrial and marine fossils in the late-glacial Presumpscot Formation, southern Maine, USA, and the marine reservoir effect on radiocarbon ages." *Quaternary Research* 75 (2011): 552–565.

Thoreau, Henry David. *The Maine Woods*. New York: Penguin Books, 1988; first published by Ticknor and Fields, 1864.

Thoreau, Henry David. *Walden; or Life in the Woods*. Boston: Ticknor and Fields, 1854).

Thorson, Robert M. *Stone by Stone: The Magnificent History in New England's Stone Walls*. New York: Walker and Company, 2002.

Thuiller, W., M.B. Araújo, R.G. Pearson, R.J. Whittaker, L. Brotons, and S. Lavorel. "Biodiversity conservation: Uncertainty in predictions of extinction risk." *Nature* 430 (2004): 1.

Trombulak, Stephen C. "Ecological health and the Northern Forest." *Vermont Law Review* 19 (1995): 283–333.

Turkel, Tux. "Emission study undercuts biomass benefits." *Portland Press Herald,* June 12, 2010.

U.S. Bureau of the Census (Charles Sargent). "Report on the Forests of North America." In *Tenth Census Report* Washington, DC: Government Printing Office, 1884.

U.S. Census Bureau. *Annual Estimates of the Resident Population for the United States, Regions, States, and Puerto Rico: April 1, 2000 to July 1, 2009.* Washington, DC: U.S. Census Bureau.

U.S. Census Bureau. U.S. Population Census Records, 1790–1880.

U.S. Department of Agriculture. *Natural Resources Conservation Service Plants Database.* plants.usda.gov.

U.S. Department of Agriculture. *USDA Plant Hardiness Zone Map.* Miscellaneous Publication 1475. Washington, DC: United States Department of Agriculture, 1990.

U.S. Environmental Protection Agency. *Air Transport.* www.epa.gov/airtransport.

U.S. Fish and Wildlife Service. *New England Cottontail* (Sylvilagus transitionalis). Hadley, MA: U.S. Fish and Wildlife Service, Northeast Unit, 2006.

U.S. Forest Service. "New England insect and disease risk map: hemlock woolly adelgid." http://fhm.fs.fed.us/sp/na_riskmaps/new_england/hwa_rm.shtml.

U.S. Forest Service. Northeastern Area, *Forest Health Protection — Dutch Elm Disease,* www.na.fs.fed.us/fhp/ded.

U.S. Forest Service. *Pest Alert: Browntail Moth.* Durham, NH: United States Department of Agriculture Forest Service, Northeastern Area, NA–PR–04–02.

U.S. Geological Survey. Water Science for Schools Webpage, ga.water.usgs.gov/edu/waterproperties.html.

U.S. Productions of Industry Census, 1850–1880.

Ulrich, Laurel Thatcher. *A Midwife's Tale: The Life of Martha Ballard, Based on Her Diary, 1785–1812.* New York: Vintage Books, 1990.

van Mantgem, P.J., N.L. Stephenson, J.C. Byrne, L.D. Daniels, J.F. Franklin, P.Z. Fulé, M.E. Harmon, A.J. Larson, J.M. Smith, A.H. Taylor, and T.T. Veblen. "Widespread increase of tree mortality rates in the western United States." *Science* 323 (2009): 521–524.

van Staal, Cees R. "Pre-Carboniferous metallogeny of the Canadian Appalachians." In W.D. Goodfellow, ed., *Mineral Resources of Canada: A Synthesis of Major Deposit-Types, District Metallogeny, the Evolution of Geological Provinces, and Exploration Method,* 793–818. Ottawa, ON: Mineral Deposit Division, Geological Association of Canada, and the Geological Survey of Canada. (Canada, 2007).

van Staal, Cees. "The northern Appalachians." In R.C. Selley, L. Robin, M. Cocks, and I.R. Plimer, eds., *Encyclopedia of Geology*. Volume 4, 81–91. Oxford, England: Elsevier, 2005.

van Staal, Cees R. and Robert D. Hatcher Jr. "Global setting of Ordovician orogenesis." *The Geological Society of America* Special Paper 466 (2010).

Van Wagner, C.E. "Age-class distribution and the forest fire cycle." *Canadian Journal of Forest Research* 2 (1978): 220–227.

Varekamp, Johan C. "The historic fur trade and climate change." *Eos* 87 (2006): 593, 596–597.

Varney, George J. "History of Greenville, Maine." In *A Gazetteer of the State of Maine*. Boston: B.B. Russell, 1886.

Vickery, Peter. "Effects of prescribed fire on the reproductive ecology of northern blazing star *Liatris scariosa* var. *novae-angliae*." *American Midland Naturalist* 148 (2002): 20–27.

Vickery, P.D., M. L. Hunter Jr., and S. M. Melvin. "Effect of habitat area on the distribution of grassland birds in Maine." *Conservation Biology* 8 (1994): 1087–1097.

Wahll, Andrew J., ed. *Sabino: Popham Colony Reader, 1602–2000*. Bowie, MD: Heritage Books, 2000, 72–102.

Wake, Cameron, Liz Burakowski, Gary Lines, Kyle McKenzie, and Thomas Huntington. *Cross Border Indicators of Climate Change over the Past Century: Northeastern United States and Canadian Maritime Region*. Boscawen, NH: Gulf of Maine Council on the Marine Environment, 2006.

Wake, C.P., and A. Markham. *Indicators of Climate Change in the Northeast*. Portsmouth, NH: Clean Air-Cool Planet, 2005.

Walker, T., editor, and P. Cardellichio, A. Colnes, J. Gunn, B. Kittler, R. Perschel, C. Recchia, D. Saah, and T. Walker, contributors. *Massachusetts Biomass Sustainability and Carbon Policy Study: Report to the Commonwealth of Massachusetts Department of Energy Resources*. Natural Capital Initiative Report NCI-2010–03. Brunswick, ME: Manomet Center for Conservation Sciences, 2010.

Washuk, Bonnie. "Persistent beavers cause problems on Stetson Road." *Lewiston Sun Journal*, October 1, 2010.

Webb, S.L., M.E. Dwyer, C.K. Kaunzinger, and P.H. Wyckoff. "The myth of the resilient forest: case study of the invasive Norway maple (*Acer platanoides*)." *Rhodora* 102 (2000): 332–354.

Webb, S.L., and C.M.K. Kaunzinger. "Biological invasion of the Drew University Forest Preserve by Norway maple (*Acer platanoides* L.)." *Bulletin of the Torrey Botanical Club* 120 (1993): 342–349.

Webb, T. III, P.J. Bartlein, S.P. Harrison, and K.H Anderson. "Vegetation, lake levels, and climate in eastern North America for the past 18,000 years." In H.E. Wright Jr., J.E. Kutzbach, T. Webb III, and P.J. Bartlein, eds., *Global Climates since the Last Glacial Maximum*, 415–467. Minneapolis: University of Minnesota Press, 1993.

Weddle, Thomas K., and Michael J. Retelle. "Deglacial history and relative sea-level changes, Northern New England and adjacent Canada." *Geological Society of America*, Special Paper 351 (2001).

Weinberg, Abigail, and Chris Larson. *Forestland for Sale: Challenges and Opportunities for Conservation over the Next Ten Years*. New York: Open Space Institute, 2008.

Wendel, G.W., and H. Clay Smith. "White Pine (*Pinus strobus* L.)." In Russell M. Burns and Barbara H. Honkala, technical coordinators. *Silvics of Forest Trees of the United States*. Volume 1, Agriculture Handbook 654. Washington, DC: U.S. Department of Agriculture, Forest Service, 1990.

Westveld, M. "Observations on cutover pulpwood lands in the Northeast," *Journal of Forestry* 26 (1928): 649–664.

Whelan, Robert J. *The Ecology of Fire*. Cambridge: Cambridge University Press, 1995.

White, Eric M., and Rhonda Mazza. *A Closer Look at Forests on the Edge: Future Development of Private Forests in Three States*. General Technical Report PNW-GTR-758. Portland, OR: Pacific Northwest Research Station, USDA Forest Service, 2008.

White, Eric M. *Forests on the Edge: A Case Study of South-Central and Southwest Maine Watersheds* (preliminary report). Portland, OR: Pacific Northwest Research Station, USDA Forest Service, in preparation.

White, Eric M., Ralph J. Alig, Susan M. Stein, Lisa G. Mahal, and David M. Theobald. *A Sensitivity Analysis of "Forests on the Edge": Housing Development on America's Private Forests*. General Technical Report PNW-GTR-792. Portland, OR: Pacific Northwest Research Station, USDA Forest Service, 2009.

Whitman, Andrew A., and John M. Hagan. "An index to identify late-successional forest in temperate and boreal zones." *Forest Ecology and Management* 246 (2006): 144–154.

Whitman, Andrew A., and John M. Hagan. "Herbaceous plant communities in upland and riparian forest remnants in western Maine." *Mosaic Science Notes* No. 2000–2003. Brunswick, ME: Manomet Center for Conservation Sciences, 2000.

Whitney, Gordon G. *From Coastal Wilderness to Fruited Plain: A History of Environmental Change in Temperate North America from 1500 to the Present*. Cambridge: Cambridge University Press, 1994.

Whittaker, Robert H. "A consideration of climax theory, the climax as a population and pattern." *Ecological Monographs* 23 (1953): 41–78.

Whittaker, Robert H. *Communities and Ecosystems*. 2nd edition. New York: Macmillan, 1975.

Williams, John W., Bryan N. Shuman, and Thompson Webb III. "Dissimilarity analyses of late-Quaternary vegetation and climate in eastern North America." *Ecology* 82 (2001): 3346–3362.

Williams, John W., Stephen T. Jackson, and John E. Kutzbach. "Projected distributions of novel and disappearing climates by 2100 AD." *Proceedings of the National Academy of Sciences* 104 (2007): 5738–5742.

Williamson, Mark, and Alastair Fitter. "The varying success of invaders." *Ecology* 77 (1996): 1661–1666.

Williamson, William D. *The History of the State of Maine.* Volume 1. Hallowell, ME: Glazier, Masters and Company, 1832.

Wilson, W. Herbert Jr. "Spring arrival dates of migratory breeding birds in Maine: sensitivity to climate change." *The Wilson Journal of Ornithology* 119 (2007): 665–677.

Wilson, W. Herbert Jr., Daniel Kipervaser, and Scott A. Lilley. "Spring arrival dates of Maine migratory breeding birds: 1994–1997." *Northeastern Naturalist* 7 (2000): 1–6.

Wolfe, D.W., M.D. Schwartz, A.N. Lasko, Y. Otsuki, R.M. Pool, and N.J. Shaulis. "Climate change and shifts in spring phenology of three horticultural woody perennials in northeastern USA." *International Journal of Biometeorology* 49 (2005): 303–309.

Wood, Richard G. *A History of Lumbering in Maine, 1820–1861.* Maine Studies No. 33 Orono: University of Maine Press, 1935.

Wroth, L.C., ed. *The Voyages of Giovanni da Verrazzano, 1524–1528.* New Haven, CT: Yale University Press, 1970.

Wunsch, Carl. "Quantitative estimate of the Milankovitch-forced contribution to observed Quaternary climate change." *Quaternary Science Reviews* 23 (2004): 1001–1012.

Young, Aaron. "The forests of York County." *The Maine Farmer and Mechanic,* May 5, 1848.

Zabarenko, Deborah. "Ten key indicators show global warming 'undeniable.'" *Reuters,* July 29, 2010.

Zachos, James, Mark Pagani, Lisa Sloan, Ellen Thomas, and Katharina Billups. "Trends, rhythms, and aberrations in global climate 65 Ma to present." *Science* 292 (2001): 686–693.

Zhou, L., R.K. Kaufmann, Y. Tian, R.B. Myneni, and C.J. Tucker. "Relation between interannual variations in satellite measures of northern forest greenness and climate between 1982 and 1999." *Journal of Geophysical Research* 108 (2003): 4004.

Zhu, Kai, Christopher W. Woodall, and James S. Clark. "Failure to migrate: lack of tree range expansion in response to climate change," *Global Change Biology* (2011): doi: 10.1111/j.1365-2486.2011.02571.x

Zielinski, Gregory A. *Conditions May Vary: A Guide to Maine Weather.* Camden, ME: Down East Books, 2009.

Illustration Credits

Frontispiece. Community GIS, Farmington, ME.

Figure 1.1. Photo by Woodrow Thompson.

Figure 1.2. Basemap: Maine Office of GIS.

Figure 1.3. Photo by Andrew Barton.

Figure 1.4. Photo by Dana Moos.

Figure 2.1. Adapted by permission from Figures 1 and 2, C.R. van Staal and R.D. Hatcher Jr., "Global setting of Ordovician orogenesis," *The Geological Society of America, Special Paper 466* (2010).

Figure 2.2. *Source*: Jouzel, J. et al., "Orbital and Millennial Antarctic Climate Variability over the Past 800,000 Years," *Science* 317 (2007): 793–797; R.B. Alley, 2004, GISP2 Ice Core Temperature and Accumulation Data. IGBP PAGES/World Data Center for Paleoclimatology Data Contribution Series #2004–013. NOAA/NGDC Paleoclimatology Program, Boulder, CO, USA.

Figure 2.3. Diagram by Andrew M. Barton.

Figure 2.4. Adapted by permission from Eric Selz, *Bangor Daily News*, Bangor, ME, January 12, 2006; original data from Borns et al., "The deglaciation of Maine, U.S.A," modified by Woodrow Thompson, personal communication. *Source*: H.W. Borns Jr. et al., "The deglaciation of Maine, U.S.A.," in J. Ehlers and P.L. Gibbard, eds., *Quaternary Glaciations — Extent and Chronology, Part II: North America* (Amsterdam, Netherlands: Elsevier, 2004), 89–109; Woodrow B. Thompson et al., "Associated terrestrial and marine fossils in the late-glacial Presumpscot Formation, southern Maine, USA, and the marine reservoir effect on radiocarbon ages," *Quaternary Research* 75 (2011): 552–565.

Figure 2.5. Photos: plucked cliff by Tkessler at http://en.wikipedia, used under GNU Free Documentation License 1.2; all others by Andrew Barton.

Figure 2.6. Photo by Andrew Barton.

Figure 2.7. Adapted by permission from George Jacobson Jr. and Ronald B. Davis, "The real forest primeval: the evolution of Maine's forests over 14,000 years," *Habitat: Journal of the Maine Audubon Society* 5 (1988): 26–29. Courtesy Maine Audubon.

Figure 2.8. Photo by Charles V. Cogbill.

Figure 2.9. Photo by Andrew Barton.

Figure 2.10. Adapted from Herman Shugart, Roger Sedjo, and Brent Sohngen, *Forests and Global Climate Change: Potential Impacts on U.S. Forest Resources* (Arlington, VA: Pew Center on Global Climate Change, 2003). Original illustrations from M.B. Davis, "Holocene vegetation of eastern United States," in H.E. Wright and S. Porter, eds., *The Late Quarternary of the United States*, vol. 2 (Minneapolis: University of Minnesota Press, 1983), 166–181.

Figure 2.11. Photo by Andrew Barton.

Figure 2.12. Photo by Andrew Barton.

Figure 2.13. Painting by Gary Hoyle.

Figure 3.1. Sources in text. Basemap: Maine Office of GIS.

Figure 3.2. Photograph courtesy of Molly O'Guinness Carlson; artifact now in the permanent collection of the Maine State Museum.

Figure 3.3. Thomas Wentworth Higginson, *A Book of American Explorers* (Boston: Lee and Shepard, 1877), 220.

Figure 3.4. Used by permission of the Penobscot County Registry of Deeds and Susan Bulay, Registrar. Photo by Charles Cogbill.

Figure 3.5. Survey by Philip Greeley and Pelham Sturtevant in 1802. Map from the Maine State Archives Land Records Office, Field Notes 24:27.

Figure 3.6. *Source:* Town-wide species abundances from personal collection of Charles V. Cogbill. See Charles V. Cogbill, John Burk, and G. Motzkin, "The forests of presettlement New England, USA: spatial and compositional patterns based on proprietor surveys," *Journal of Biogeography* 29 (2002): 1279–1304.

Figure 3.7. *Source:* Town-wide species abundances from the collection of Charles Cogbill.

Figure 3.8. Photo by Bill Silliker; courtesy of Maryellen Chiasson and The Nature Conservancy in Maine.

Figure 3.9. Photo by Charles V. Cogbill.

Figure 3.10. *Source:* Shawn Fraver, Morton Mossewilde, Unna Chokkalingam, Erika Rowland, and Charles Cogbill.

Figure 3.11. By permission, Unna Chokkalingam and Alan S. White, "Structure and spatial patterns of trees in old-growth northern hardwood and mixed forests of northern Maine," *Plant Ecology* 156 (2001): 139–160.

Figure 3.12. *Source:* Big Reed from Shawn Fraver, Morton Mossewilde, Unna Chokkalingam, Erika Rowland, and Charles Cogbill; Cobbosseecontee from Alan White and Charles Cogbill.

Figure 3.13. Adapted from Shawn Fraver, "Spatial and temporal patterns of natural disturbance in old-growth forests of northern Maine, USA (PhD dissertation, University of Maine, 2004). Coring and dating by Laura Conkey and Charles Cogbill.

Figure 3.14. Adapted by permission from Molly Schauffler and George Jacobson, "Persistence of coastal spruce refugia during the Holocene in northern New England, USA, detected by stand-scale pollen stratigraphies," *Journal of Ecology* 90 (2002): 235–250.

Figure 3.15. Drawing by Andrea Sulzer, courtesy of the *Natural Areas Journal.*

Figure 3.16. Photo by Charles V. Cogbill.

Figure 3.17. Photo by Charles V. Cogbill.

Figure 4.1. Photo by Andrew Barton.

Figure 4.2. *Source:* U.S. Natural Resources Conservation Service, Farmington, ME.

Figure 4.3 *Source:* Stanley B. Atwood, *The Length and Breadth of Maine* (Augusta, ME: Kennebec Journal Print Shop, 1946), 20–22; Philip T. Coolidge, *History of the Maine Woods* (Bangor, ME: Furbish-Roberts, 1963), 38–40; Wikipedia sites for some towns (accessed March 15, 2011). Basemap: Maine Office of GIS.

Figure 4.4. Photo by Andrew Barton.

Figure 4.5. *Source:* U.S. Population Censuses.

Figure 4.6. Photo by Andrew Barton.

Figure 4.7. *Source:* U.S. Agricultural Censuses and SoEun Ahn et al., *Agricultural Land Changes in Maine: A Compilation and Brief Analysis of Census of Agriculture Data, 1850–1997*, Technical Bulletin 182 (Orono: Maine Agricultural and Forest Experiment Station, 2002).

Figure 4.8. *Source:* 1850 and 1880: Philip T. Coolidge, *History of the Maine Woods* (Bangor, ME: Furbish-Roberts, 1963); 2008: G.L. McCaskill et al., *Maine's Forest Resources*, 2008, Res. Note NRS-53 (Newtown Square, PA: U.S. Department of Agriculture, Forest Service, Northern Research Station, 2010); all others: Lloyd C. Irland, *Maine's Forest Area, 1600–1995: Review of Available Estimates*, Miscellaneous Publication 736 (Augusta: Maine Agricultural and Forest Experiment Station, University of Maine, 1998). Data from Irland were modified in the following ways: 1872 estimate, clearly an outlier, was not used; estimates given for years within two years of each other were averaged. Total land area used in calculation of forest percentages was 19,751,400 acres from the second source listed in the figure caption. Some estimates from the 1800s do not jibe with the agricultural censuses for cleared land shown in Figure 4.7, likely because of uncertainties associated with estimated forest area.

Figure 4.9. Basemap: Maine Office of GIS.

Figure 4.10. *Source*: Philip T. Coolidge, *History of the Maine Woods* (Bangor, ME: Furbish-Roberts, 1963); W.H. McWilliams et al., *The Forests of Maine: 2003*, Resource Bulletin NE-164 (Newtown Square, PA: U.S. Department of Agriculture, Forest Service, Northeastern Research Station, 2005); Maine Forest Service Wood Processor Reports, 2000–2009.

Figure 4.11. *Source:* David. C. Smith, *A History of Lumbering in Maine, 1861–1960*, Maine Studies No. 93 (Orono: University of Maine Press, 1972), 12–13.

Figure 4.12. *Source:* Richard G. Wood, *A History of Lumbering in Maine, 1820–1861*, Maine Studies No. 33 (Orono: University of Maine Press, 1935), 135.

Figure 4.13. Photo by Andrew Barton.

Figure 4.14. *Source:* Philip T. Coolidge, *History of the Maine Woods* (Bangor, ME: Furbish-Roberts, 1963); W.H. McWilliams et al., *The Forests of Maine: 2003*, Resource Bulletin NE-164 (Newtown Square, PA: U.S. Department of Agriculture, Forest Service, Northeastern Research Station, 2005); Maine Forest Service Wood Processor Reports, 2000–2009.

Figure 4.15. *Source:* Craig G. Lorimer, "The presettlement forest and natural disturbance cycle of northeastern Maine," *Ecology* 58 (1977): 139–148; Gro Flatebo, Carol R. Foss, and Steven K. Pelletier, *Biodiversity in the Forests of Maine: Guidelines for Land Management,* Bulletin 7147 (Orono: University of Maine Cooperative Extension, 1999), 102.

Figure 5.1. Basemap: Maine Office of GIS.

Figure 5.2. Photo by Parker Schuerman.

Figure 5.3. Photo by Parker Schuerman.

Figure 5.4. Photo by Andrew Barton.

Figure 5.5. Photo by Andrew Barton.

Figure 5.6. Photo by Andrew Barton.

Figure 5.7. Photo by Andrew Barton.

Figure 5.8. Photo by Andrew Barton.

Figure 5.9. *Sources:* Michael Begon, Colin R. Townsend, and John L. Harper, *Ecology: from Individuals to Ecosystems,* 4th edition (Oxford, England: Blackwell, 2006), 20–24 and C. Barry Cox and Peter D. Moore, *Biogeography: An Ecological and Evolutionary Approach*, 5th edition (Oxford: Blackwell Scientific Publications, 1993), 88.

Figure 5.10. Adapted from fig. 4.10, p. 167, from *Communities and Ecosystems,* 2nd ed., by Robert H. Whittaker. Copyright © 1975 by Robert H. Whittaker. Reprinted by permission of Pearson Education, Inc.

Figure 5.11. Adapted by permission from David T. Cleland, Jerry A. Freeouf, James E. Keys, Greg J. Nowacki, Constance A. Carpenter, and W. Henry McNabb, *Ecological Subregions: Sections and Subsections of the Conterminous United States,* Gen. Tech. Report WO-76 (Washington, DC: U.S. Department of Agriculture, Forest Service, 2007).

Figure 5.12. *Source:* GIS data for trees from Elbert Little Jr., *Digital Representations of Tree Species Range Maps from "Atlas of United States Trees,"* U.S. Geological Survey, U.S. Department of the Interior (esp.cr.usgs.gov/data/atlas/little/) and for birds from NatureServe, *Digital Distribution Maps of the Birds of the Western Hemisphere Version 3.0* (www.natureserve.org/getData/birdMaps.jsp).

Figure 5.13. Top photo by Andrew Barton; bottom photo by Trevor Persons.

Figure 5.14. Maps are updates of Janet S. McMahon, "The biophysical regions of Maine — patterns in the landscape and vegetation" (master's thesis, University of Maine, 1990) by Randall B. Boone and William B. Krohn, "Relationship between avian range limits and plant transition zones in Maine," *Journal of Biogeography* 27 [2000]: 471–482, from which this was adapted with permission.

Figure 5.15. Adapted by permission from Andy Cutko and Rick Frisina, *Saving All the Parts: A Conservation Vision for Maine Using Ecological Land Units* (Augusta: Maine Natural Area Program, Department of Conservation, 2005), 13. Basemap: Maine Office of GIS.

Figure 5.16. Photo by Andrew Barton. *Source:* Susan Gawler and Andrew Cutko, *Natural Landscapes of Maine: A Guide to Natural Communities and Ecosystems* (Augusta: Maine Natural Areas Program, 2010).

Figure 5.17. *GIS data source:* Maine Office of GIS.

Fig 6.1. *Source:* Charles T. Driscoll, Kathy Fallon Lambert, and Limin Chen, "Acidic deposition: sources and ecological effects," in Gerald R. Visgilio and Diana W. Whitelaw, eds., *Acid in the Environment: Lessons and Future Prospects* (New York: Springer, 2007), 27–58.

Fig 6.2. Adapted from Eric K. Miller, *Assessment of Forest Sensitivity to Nitrogen and Sulfur Deposition in Maine*, Conference of New England Governors and Eastern Canadian Premiers Forest Mapping Group (Augusta: Maine Department of Environmental Protection, 2006). Basemap: Maine Office of GIS.

Fig 6.3. Photos: adelgid — USDA Forest Service Region 8, Image 1520082 at Insect Images, The Bugwood Network, used under Creative Commons Attribution 3.0 U.S.; borer — by permission from David Cappaert, Michigan State University, Image 2106098 at Insect Images, The Bugwood Network; ooze — Joseph O'Brien, USDA Forest Service, Image 5044024 at Insect Images, The Bugwood Network, used under Creative Commons Attribution 3.0 U.S.; beetle grub — Thomas B. Denholm, New Jersey Department of Agriculture, Image 1253027 at Insect Images, The Bugwood Network, used under Creative Commons Attribution 3.0 U.S.; beetle adult — Jennifer Forman Orth, MA Department of Agricultural Resources.

Fig 6.4. Photos: Loosestrife from Leo Michels, www.imagines-plantarum.de; all others by Andrew Barton.

Fig 6.5. Figure produced by S.M. Stein and the U.S. Forest Service (Stein et al. 2005, 2010) for Foster et al., *Wildlands and Woodlands* (2010).

Fig 6.6. *Source:* John M. Hagan, Lloyd C. Irland, and Andrew A. Whitman, *Changing Timberland Ownership in the Northern Forest and Implications for Biodiversity*, Report #MCCS-FCP-2005–1 (Brunswick, ME: Manomet Center for Conservation Sciences, 2005), Table 2.

Fig 6.7. *Source:* John M. Hagan, Lloyd C. Irland, and Andrew A. Whitman, *Changing Timberland Ownership in the Northern Forest and Implications for Biodiversity*, Report No. MCCS-FCP-2005–1 (Brunswick, ME: Manomet Center for Conservation Sciences, 2005), table 4, figure 6.

Fig 6.8. *Source:* J.E. Hansen, R. Ruedy, M. Sato, and K. Lo, *Global Annual Tem-*

perature Anomalies (Land Meteorological Stations), National Aeronautics and Space Administration. Retrieved August 2, 2010, from cdiac.ornl.gov/trends/temp/hansen/data.html.

Fig 6.9. Climate Change Research Center, University of New Hampshire–Durham), *Cross Border Indicators of Climate Change in the Northeast US and Eastern Canada; Data source:* U.S. Historical Climatology Network, lwf.ncdc.noaa.gov.

Fig 6.10. Adapted by permission from G.L. Jacobson, I.J. Fernandez, P.A. Mayewski, and C.V. Schmitt, eds., *Maine's Climate Future: An Initial Assessment* (Orono: University of Maine, April 2009), 40; originals from Arbor Day Foundation, "New arborday.org Hardiness Zone Map reflects warmer climate," press release (Nebraska City, NE: Arbor Day Foundation, 2006).

Fig 6.11. Adapted by permission from T.G. Huntington et al., "Climate and hydrological changes in the northeastern United States: recent trends and implications for forested and aquatic ecosystems," *Canadian Journal of Forest Research* 39 (2009): 205 and G.A. Hodgkins, R.W. Dudley, and T.G. Huntington, "Changes in the timing of high river flows in New England over the 20th century," *Journal of Hydrology* (Amst.) 278 (2003): 250.

Fig 6.12. Adapted from R.K. Pachauri and A. Reisinger, eds., *Climate Change 2007: Synthesis Report*, Fourth Assessment Report (Geneva, Switzerland: International Panel on Climate Change, 2007), 45, fig. 3.2 left panel. Updated global emissions data from International Energy Agency, "Latest Information" (retrieved on December 1, 2011, at www.iea.org/index_info.asp?id=1959) and SkepticalScience, "IEA CO_2 Emissions Update 2010 — Bad News" (retrieved on December 1, 2011 at www.skepticalscience.com/news.php?n=779)

Fig 6.13. Adapted by permission from G.L. Jacobson, I.J. Fernandez, P.A. Mayewski, and C.V. Schmitt, eds., *Maine's Climate Future: An Initial Assessment* (Orono: University of Maine, April 2009), 15.

Fig 6.14. Maps by Community GIS, Farmington, ME, using results from *Climate Change Atlas*, United States Department of Agriculture, Forest Service, Northern Research Station. Retrieved on August 15, 2010, from www.nrs.fs.fed.us/atlas.

Fig 6.15. Adapted by permission from Guoping Tang and Brian Beckage, "Projecting the distribution of forests in New England in response to climate change," *Diversity and Distributions* 16 (2010), 144–158.

Fig 6.16. Photos: lynx — Art G Wikimedia Commons, used under Creative Commons Attribution 2.0 U.S; marten — Yukon Flats National Wildlife Refuge, U.S. Fish and Wildlife Service; bobcat — Dave Menke, Yukon Flats National Wildlife Refuge, U.S. Fish and Wildlife Service; fisher — Tom Murray, by permission. Maps: B.D. Patterson, G. Ceballos, W. Sechrest, M.F. Tognelli, T. Brooks, L. Luna, P. Ortega, I. Salazar, and B.E. Young, *Digital Distribution*

Maps of the Mammals of the Western Hemisphere, version 1.0. (Arlington, VA: NatureServe, 2003). Data for maps: NatureServe in collaboration with Bruce Patterson, Wes Sechrest, Marcelo Tognelli, Gerardo Ceballos, The Nature Conservancy–Migratory Bird Program, Conservation International-CABS, World Wildlife Fund-US, and Environment Canada–WILDSPACE. Copyright © 2010 NatureServe, 1101 Wilson Boulevard, 15th Floor, Arlington Virginia 22209, U.S.A. All Rights Reserved.

Index

Page numbers in italics refer to figures and tables. Trees and animals are indexed under their common names.

Acadia (Acadie), 55, 84–85, 106, 160
Acadian Forest, 37, 70, 72, 85, 160–161
acid rain: causes, 191–193, *194*; and the Clean Air Act, 195–196; in the future, 196–197; and Maine forests, 127, 194–197, *195*; recent recovery from, 194–196
Adams, Solomon, 56
Adirondack–New England Mixed Forest Province, 160
albedo, 20–21, 22, 23, 218. *See also* climate change
alder, 29, 67
Allagash River and Wilderness Waterway (State of Maine), 115, 125, 129
Allen, Nancy, 127
Alley, Richard, 45, 231
Almquist-Jacobson, Heather, 44
alpine communities, 7, 145, 147, *150*, 169, 206, 228
American Revolution, and settlement of Maine, 101, *105*, 106, 108, 110, 111
Anderson, Dennis, 171
Androscoggin River, 106, *107*, 110, 127
animals: changes in response to past climate change, 43, 45; climate change impacts now and in the future, 220–221, *221*, 223, 228–234, *231*; development impacts, 206, 210–211; establishment of modern species after deglaciation, 43–44;

extirpation during settlement, 104, 117–119; mixing of boreal and temperate species, *162*, 163, *163*, 165; post-glacial megafauna, 40–44; 20th-century return of extirpated and diminished species, 126, 133. *See also* names of specific animals
apple tree, 102–103
Aroostook River, 26, 112
ash: 29, 55–56, 103, 111, *135*, 189, 199, 226; black (brown), 61, *62*, 72, 199; white, 61, *62*, 72, 132. *See also* emerald ash borer
Asian longhorned beetle, 198–199, *198*, 204, 232
Asiatic bittersweet, 199
aspen (popple): in current forests, 6, 73, 103, 134, 171; in future forests, 233; harvest and use, 121; in post-glacial forests, 29, 36; in presettlement forests, 58, 61, *62*, 73, 88
Augusta, ME, 55, 110, 194
auk, great, 43, 118

Bailey-McMahon ecoregion system, 158–160, *159*, 164–169
Ballard, Ephraim, 56–58, 101, 108, 132
balm-of-Gilead. *See* poplar, balsam
Bangor, ME, 6, 55, 58–59, *59*, 63, 115, *116*, *150*
barberry, Japanese, 197–198
Barton, Andrew, 103, 152
Basin Preserve (Nature Conservancy), 4, 15, *31*, 67, 88
basswood, 48, 55, 57, 59, *62*, 67, 101

Baxter State Park (State of Maine), 84, 88, 125, 145, 147, *150*

Baxter, Percival P., 125

Bear Brook Watershed, ME, 192

bear: black, 118, 126, 152; brown, 42; grizzly, 42; polar, 42; short-faced, 42

beaver, American, 42, 43, 44, 60, 104, 117–119, 126, 133

beech bark-scale insect-*Neonectria* disease, 125, 134, 152, 198, 232

beech, American: in current forests, 48, 55, 88, 103, 125, 132, 134, *135*, 152, 158, 168–169, 171, 226; in future forests, 232; harvest and use of, 111; in old-growth forests, 72, *72*, *80*, 82; in post-glacial forests, 16, 36–37, 39, 44, 67–68, 132; in presettlement forests, 6, 46, 54–58, 60, 61–65, *62*, *65*, 101, 132, 134, *135*, 168. See beech bark-scale insect-*Neonectria* disease

Belmont, ME, 56

Bennett, Dean, 142

Besse farm, 132, 174

Bhiry, Najat, 37

Big Reed Forest Reserve (The Nature Conservancy), 68–81, *69*, *70*, *71*, *72*, *73*, *74*, *77*, *79*, 204

Big Twenty Township, ME, 66

Bigelow Mountains and Preserve (State of Maine), 101, 129, 152

Bingham Purchase, 101, 108

Bingham, ME, 25

biological diversity, 6–8, 52, 128, 139, 142–143; assessment in Maine, 134, 169, 177–179; causes in Maine, 172–175; of communities, 20, 29, 151, 157; of ecoregions, 155–160, *157*, *159*; of old-growth, 71–72, 82; of species, 4, 128, 134, 152

biomass energy in Maine, 207–208

biomes, 37, 156, *157*, 158, 220; and Maine, 37, 160–163, *162*, *163*

biophysical regions. *See* ecoregions

birch, gray, 138, 174

birch, heartleaf, 71, 72, 206

birch, mountain paper. *See* heartleaf birch

birch, paper. *See* white birch

birch, sweet, 154–155, 161, 221

birch, white: in current forests, 55, 88, 103, *135*, 170, 226; in future forests, 221; in old-growth forests, 71, 72; in post-glacial forests, 68; in presettlement forests, 61–62, *62*, 66, *135*

birch, yellow: in current forests, 48, 134, *135*, 152; in old-growth forests, 70–72, *72*; in post-glacial forests, 36, 68; in presettlement forests, 60, 61, *62*, 64, 78, 134, *135*

birch: in current forests, 5–6, 48, 55, 88, 103, 134, *135*, 138, 152–154, 169–171, 174, 226; in future forests, 199, 206, 221; in old-growth forests, 70–72, *72*; in post-glacial forests, 16, 29, 36–37, 67, 68; in presettlement forests, 50, 54–61, *60*, *62*, 64, 66–67, 78, 101, 132, 134, *135*

black gum. *See* tupelo

black racer, 7, 144, 163, *163*

bobcat, 43, 118, 221, 228, 229

bogs and fens, 28, 30–31, 79, 150–151, *151*, 167

Bohr, Niels, 191

Boothbay Harbor, ME, 173

boreal forest biome, 37, 66, 70, 88, 147, 156, *157*, 160–163, *162*, *163*, 228

Borns, Harold, 24

Boundary Mountains, 174

Brain, Jeffrey, 50–51

Bramhall Hill, 1–2, 6

Braun, E. Lucy, 39, 48, 50, 63

Brewer, ME, 55

broad arrow mark and English policy, 50, 106

Brookings Institution report, *Charting Maine's Future*, 204

browntail moth, 200–201

Brunswick, ME, 110, 200

Buckfield, ME, 56, 117

Bulay, Susan, 58

butternut, 55, 62

Cabot, John, 53

Camden Hills, 26, 54–55, 63, 82

Camden, ME, 53–54

Cape Elizabeth, ME, 55

carbon credits from Maine forests, 208

caribou, 42–43, 45, 118, 126, 134

Carrabassett River, 113

Cary, Austin, 86, 121–122

Casco Bay, 64, 106, 167, 197

cedar, 53–54, 57–58, 61, 64, 66–67, 78, *80*, 101, 168; Atlantic white, *62*, 155, 171; northern white, 5, *62*, *62*, 72, 76, 88, *135*, *221*

Champlain, Samuel de, 54–55

cherry, 56, 66, 72, 174

chestnut blight, 37, 124–125, 198

chestnut, American, 37, *62*, 124–125. *See also* chestnut blight

Childs, Craig, 15

Chisholm, Hugh, 120

clear-cutting, 127, 129, *135*, 136–137

Clements, Frederic, 40

climate change (human-caused): bioclimatic vs. mechanistic projection models, 224–225; evidence for, 219–220; during the past century, 211–213, *212*; impacts on Earth's biological systems, 213; impacts on Maine's biological systems, 216; in Maine, 213–215, *214*, *215*; climate projections for 21st century, 216, *217*, 220, 231–232; biological impact projections for 21st century, 220–221, *221*, *222*, *223*, 224–225, *226*, 228–229, *229*, 231–232

climate change, (long-term, not human-caused): causes of, 20–23, *22*; and greenhouse gases, 20–21, *22*; in Maine in the past, *21*, 29–37, 44–47; role of albedo, 20–21, *22*; role of insolation, 20–21, *22*; temperature and glacial cycles over time, 20, *21*; weather vs. climate, 20

coarse filter approach in conservation, 164

Coburn Gore, ME, 173

Cogbill, Charles V., 61, 81

Communities (natural), 40, 42–44, 46–48, 143, 156, 158, 160; in Maine, 7, 56, 82, 85, 144–155, 158, 166–179, *170*, 189–190, 206–207

Conkling, Phillip, 111

conservation of Maine forests: assessment of success, 175–179; importance of ecoregions and ecosystem classification, 164–169, 175–179; initiatives and projects, 4, 6, 122–123, 125, 128–132, *130*, 148, 175–179, *176*, *177*; landscape connectivity and habitat fragmentation, 128, 178–179, 210–211; Maine's rarity ranking system, 146–147; minimum reserve size considerations, 178–179; role of conservation biology, 128

continental drift. *See* plate tectonics

Coolidge, Phillip, 104

Coplin Center Plantation, ME, 151–152, *153*

cottontail, eastern, 126, 163, 211

crab-apple, *62*

Cummings, Bob, 151
Curtis, John, 40
Cutko, Andy, 169, 175, 177–179

Damariscove Island, 104
Davis, Chris, 189–190, 216
Davis, Margaret B., 40
Davis, Ron, 150, 171
deer, white-tailed, 43–45, 118, 122, 133, 221
deglaciation. *See* glaciation and degla-ciation of Maine
DeGraaf, Richard, 119
Dexter, ME, 67
disease introduced to Indians by Europeans, 100–101, 105
diseases of trees, 124–125, 134, 154, 190, 197–198, 220, 226–227, 232
distributions of species and communi-ties, 61–65, 158–165, *162–164*, 169, 200–201; changes caused by climate change, 213, 216, 222–223, 224–226, *226*, 232
District of Maine. *See* Province of Maine
Dodds, Kevin, 199
dogwood, flowering, 7, 155, 161, 221
Driscoll, Charles, 192
Dubos, Rene, 100
Dunstable, MA, 100
Dutch elm disease, 125, 198

East Kennebago Mountain, 152
East Machias, ME, 122
East Millinocket, ME, 128
Eastern Broadleaf Forest Province, 160
eastern racer. *See* black racer
Ecological Reserves Program of Maine, 175–177
ecoregions, 143, 156, 158, *159*; in Maine, 160–169, *162*, *163*, *164*, *167*, 175–179

Elkinton, Joseph, 200–201
elm, 29, 55, 62, 71, 125. *See also* Dutch elm disease
emerald ash borer, *198*, 199, 204, 232
English colonies in Maine, 3, 48–50, 55, 84–85, 88, 104–107. *See also* land clearing for farms; land surveys in Maine; logging in Maine: during Colonial times
exotic species. *See* non-native invasive organisms
explorers of Maine forests, European, 48–51, *49*, 53–56, *54*. *See also* English colonies in Maine

F.W. Ayer mill (South Brewer), 113
Fairfield, ME, 110
farm abandonment, 4, 38, *113*, *114*, 117–120, 124, 132–133
Farmington, ME, 25, 56, 58, 100–103, 113, 118–119, 201
Felt, Ephraim Porter, 200
fens. *See* bogs and fens
Filial, Louise, 37
fir, balsam: in current forests, 5, *5*, 152, 154, 161, 166–169, *170*, 171, 189, 206; and dieback, 192–193, 225–226, 232; in future forests, 206, 216, 221, 225, 232; harvest and use of, 103, 129, 134, 135, 138; increased abundance as a result of spruce harvesting, 134, *135*, 138; in old-growth forests, 72, *80*, 82; in post-glacial forests, 29, 34, 35–36, 44; in presettlement forests, 46, 51, 53–54, 56, 61, 62, 64, 66–68, 78; and spruce budworm, 76–77
fire. *See* wildfire
Firestone, Richard, 43
firewood, 3, 110, 127, 136, 208
Fore River, 1–2
forest certification, 129, 131
forest dieback: projections for the

future, 232–233; role of interacting stresses and Manion's Spiral, 193, *194*, 226–227, 230; syndrome and examples, 221, 225–227, 230

Forest Practices law, 129, 131

forests of Maine: changes due to climate change, 29–40; old growth, 68–84; modern (summary), 131–139, 160–163, 166–167; post-glacial, 29–40; presettlement (summary), 84–89; presettlement vs. today, 134, *135*, 138–139, *139*; projecting the future, 191, *206*, *217*, 220–221, *221*–223, *226*, 228–234, *229*

Foster, David R., 189

Franklin, Jerry, 233

Frederic, Paul, 138

French and Indian Wars, 3, 55, 84–85, 100–101; effect on settlement of Maine, 106

French colony in Maine. *See* Acadia

Frisina, Rick, 177–179

Ft. Kent, ME, 174

fuelwood. *See* firewood

fur trading impact on forests, 44, 50, 104–105, 117, 118–119, 126, 133

Furbish, Kate, 153

Gawler, Susan, 169

geography and ecological diversity in Maine, 172–175

geological formation of Maine: 17–20; role of Gandaria and Avalonia, 18, *19*

Georges Bank, 25

Gibbs, Ann, 204

Gilbert, Raleigh, 48–50

Gill, Jacquelyn, 43

glaciation and deglaciation of Maine, 7, 24–26, *24*; impacts on the modern Maine landscape, 26–29, *27*. *See also* Younger Dryas

Goldsmith, Sir James, 128

Gorham, ME, 4

Gould Pond, 67

Grafton Notch, ME, 27, *27*

grasshopper sparrows, 144, 211

grassland, 7, 144–147, *145*, 148–149, *148*, 156, *157*, 169, 204, 211

Great Northern Paper Company, 120–121, 127

"great proprietors," 57, 107

greenhouse gases, 20–23, *22*, 206–207, 216, *217*, 218–219. *See also* climate change

Greenleaf, Moses, 173

Grenier, Dan, 150

habitat fragmentation and habitat loss, 128, 178–179, 210–211. *See also* conservation of Maine forests

Hagan, John, 208

Hallowell, ME, 55

Harpswell, ME, 197

Hartford, ME, 56

Harvard Forest, MA, 192

Haslam, Bill, 152

Hawthorn Elementary School, 200

hawthorn, 62, 71

Hayden, Charles, 66

hemlock woolly adelgid, 197–199, *198*, 204, 227

hemlock, eastern: in current forests, 48, 88, 134, *135*; harvest and use of, 110, 115, *116*, 152, 168, 171, 174; in old-growth forests, 72, 81; population collapse 5400 years ago, 36–37; in post-glacial forests, 16, 36–37, 44, 46, 79; in presettlement forests, 46, 50, 55–61, *60*, 62, 64–68, 85, 101, 132–134, *135*. *See* hemlock woolly adelgid

Hess, Harry, 17

hickory, 6, 62, 154–155, 161, 167, 220–221, *221*

hornbeam, hop, 72, 111. *See* ironwood

Houlton, ME, 174

Hubbard Brook Experimental Forest, NH, 191–192, 196

Humboldt, Alexander Von, 143

Hunter, Malcolm (Mac), 68

Huntington, Tom, 194

Hurricane Island, 18

Indian Island, 110

Indians: Abenaki, 3, 34, 51–52, 55, 89, 100–101, 105–106, 108, 110; Paleo-indians, 2–3, 43, 45; and Younger Dryas climate change, 45

Industry, ME, 56

International Paper Company in Rumford, 120, *120*

invasive species. *See* non-native invasive organisms

Irland, Lloyd, 124, 208

ironwood, *62, 67. See* hornbeam

Iverson, Louis, 224–225

Jacobson, George, 37–38, 67

Jay, ME, 25

Josselyn, John, 50

Judd, Richard, 122

juniper, common, 66

Kanoti, Allison, 197, 225

Kennebec River, 3, 5, 48, 55, 57, 63, 102, 106, 110, 115, 127, 190

Kennebunk Plains Preserve, 144–145, *144, 145*, 148, *149*

Kerchner, Theresa, 132

Kimber, Bob, 152

Knox, ME, 60

land clearing for farms, 3, 67, 101–102, *102*, 104–117, *113, 114*, 132–134, 224–225, 233

land development for vacation homes, 205, *206*; and impacts on Maine forests, 210–211; projections for the future, 233

land ownership identity and forest management in Maine, 106–110, 115–116, 120–121, 124, 127, 135–138, 208–209, *209, 210, 211*; and impacts on Maine forests, 210–211; projections for the future, 233

land surveys in Maine: methods, 56–61, *59, 60*; presettlement forest patterns revealed, 61–66, *62, 63, 65*; tension zones revealed, 61–64, *63, 65*; frequency of natural disturbance revealed, 64–66

Lansky, Mitch, 127

larch. *See* tamarack

Likens, Gene, 233

Lindbladh, Matts, 37

linden. *See* basswood

Linnaeus, Carl, 143

Livermore Falls, ME, 25

Livingston, Bill, 221, 225, 232–233

logging in Maine, 131–132, 134; during Colonial times, 104, 106–107; harvest levels, 106–107, 109–110, 112–116, *116*, 123–124, 127–129, 135, 136–137; impacts on forests, 106–107, 109–110, 113, 115, 121–123, 124, 127–129, 132–139, 210–211; innovations, 112–116, 120–124, 127–129; for local uses, 109–110; lumber vs. paper, 120–124, *121*, 127–128; pine vs. spruce, 112–116, *116*; for pine, 108–110, 112–116, *116*; for shipbuilding, 111; for spruce, 113, 115, *116*, 120–121. *See also* clear-cutting; paper industry

Longfellow, Henry Wadsworth, 50

Lorimer, Craig, 65

lynx, Canada, 7, 118, 152, 163, 221, 228, 229

Machias, ME, 104

Maine Department of Inland Fisheries and Wildlife, 5, 123, 144, 146, 175

Maine Forest Service, 135, 136, 197, 199

Maine Geological Survey, 1, 24

Maine Indian Basketmakers Alliance, 199

Maine Land Use Regulatory Commission, 6

Maine Natural Areas Program, 80–81, 146, 165, 169, 177

mammoth, woolly, 41–42, 42

Manomet Center for Conservation Sciences, 207, 208, 209, 211

maple, mountain, 72

maple, Norway, 199–200, 202, 203

maple, red: in current forests, 72, 132–134, 135, 138, 151, 169, 174, 199; in past forests, 39, 61, 62

maple, striped, 62, 72

maple, sugar: in current forests, 48, 152, 158, 174; and dieback, 192, 196, 200; in future forests, 224, 233; harvest and use of, 111; in old-growth forests, 70, 72, 72, 78, 80, 82, 88; in presettlement forests, 56–57, 61, 62, 132, 135

maple: in current forests, 5, 48, 103, 132–134, 135, 138, 151–152, 158, 169–171, 174; and dieback, 192, 196, 199; in future forests, 199–200, 202, 203, 224, 233; harvest and use of, 111; in old-growth forests, 70, 72, 78, 80, 82, 88; in post-glacial forests, 34, 35, 39; in presettlement forests, 6, 54–59, 60, 61, 62, 66–68, 101, 132, 135

Marsh, George Perkins, 121

marten, American, 118, 163, 221, 226, 227

Martin, Paul, 43

Massachusetts Bay Colony, 57, 106

Matawaska, ME, 56

Mathews, Stephen, 224–225

McCullough, David, 44

McKibben, Bill, 4, 133

McMahon, Janet, 164–165, 173–174, 189–190, 216; and ecoregional system for Maine, 164–169

McNair, Wesley, vii

Medway, ME, 25

megafauna, extinction of, 42

Merriam, C. Hart, 161

Midwest, opening of, 4, 102–103, 117, 120

migration of tree species in post-glacial times. See forests of Maine: changes due to to climate change

Milankovitch Cycles, 22–24, 22. See also climate change

Milankovitch, Milutin, 22–23

Miller, Eric, 196

Millinocket, ME, 120–121, 128

Mitchener, Lori, 65

Monhegan Island, 104, 197, 204

moose, 43–45, 118, 122, 126, 133, 221

Moosehead Lake, 3, 5–6, 5, 18, 27, 205

Morrill, ME, 38, 56

Motzkin, Glenn, 189

Mount Abraham, 56, 101, 175

Mount Agamenticus, 4, 154–155, 155, 171

Mount Katahdin, 1, 6, 18–19, 26, 145, 147, 150, 153, 167, 169, 172

Mount Kineo, 26, 27

mountain ash, 56, 62, 72, 161

mountain laurel, 155

mountain lion, 118, 126

Muir, John, 121–122

Muscongus Bay, 189–190

Native Americans. See Indians

Natural Resources Council of Maine, 125, 127

Nature Conservancy, 125, 129, 148, 150, 160, 176–177, 204, 209. See also Basin

Preserve; Big Reed Forest Reserve;
Kennebunk Plains Preserve; St.
John River Preserve
new forest in Maine, 129, 138
New Vineyard, ME, 56
nitrogen saturation, 192–197
no-analog climates and communities: in the past, 46; in the future, 231–232
non-native invasive organisms, 115; diseases, 124–125, 197–198, *198*, 201, 204, 232–233; likely impacts, 204, 232–233; insects, 197–201, *198*, 204, 225–226, 232–233; plants, 197, 199–201, *202*, *203*, 204; the tens rule, 201, 204
Northeastern Mixed Forest Province, 160
northern blazing star, 145–147, *145*, 146
Northern Forest Lands Study, 128
northern hardwood forest, 7, 85, 103, 151–152, *153*, 165–166, 169, 199–200, 221
Nurse, Andrea, 15–16

oak, chestnut, 153, 155, 161, 171
oak, red, 55, 56, 61, 62, 66, 71, *80*, 111, 132, 166, 171, 189
oak, scrub, 88, 144, 167, 169
oak, white, 61, 62, *62*, 133, 161, *162*, 166, 171, 220, 222
oak: in current forests, 6, 15–16, 132–134, *135*, 144, 154–155, 161, *162*, 166–169, 171, 189; in future forests, 220–221, 222; harvest and use of, 106, 110–111; in old-growth forests, 71, *81*; in post-glacial forests, 15–16, 34–36, *34*, *35*, 45; in presettlement forests, 15–16, 50–51, 54–58, 61–64, *62*, *63*, *64*, 66–68, 82, 85, 88–89, 101, 132–134, *135*; and sudden death, 198, *198*

old-growth forests in Maine: Big Reed Forest Reserve old growth landscape, *49*, 68–81, *69*, *70*, *71*, *72*, *73*, *74*, *77*, *79*; disturbance and dynamics of, 73–81, 86–89; small old-growth areas, *49*, *80*, 81–84, *83*, *84*; species in, 70–73, *80*, 81–85, 88–89; structure of, 74, *80*, 81–85, 88–89
Orono Bog Boardwalk, 150–151, *151*
Orono, ME, 69, 110
Orwig, David, 199
otter, 118, 126
Oxford, ME, 63

paleoecology: evidence for presettlement forests, 66–68; and reconstruction of past environments, 15–16, 29–40, *32*, *33*, *34*, 46, 49
paper industry, 102, 104, 120–124, *121*, 127–131, 134, *135*, 136–138
passenger pigeon, 118
peatlands. *See* bogs and fens
Pemaquid, ME, 104
Penobscot Bay, 54, *54*, 64, 165
Penobscot River, 3, 5, 55, 106, 110, 115, 165
Perham: family and farm, 100–103; Hannah, 102; Lemuel, 56, 101; Mary Ann Hobbs, 102; Silas, 59, 100–101; Silas Decater, 102; Silas French, 103
pesticide spraying in Maine forests, 127, 129, 197
Phillips, ME, 18
Phippsburg, ME, *4*, 15–16, *30*, 50, *51*, 67, 88
Pinchot, Gifford, 122
pine, jack, 7, 29, 161, *163*
pine, pitch, *4*, 50, 54, 61–64, *62*, 67, 144, 166–169, 171
pine, red, 29, 62, 192
pine, white: in current forests, 36, 38–39, 72, *80*, 82, 132–135, 138, 166,

221, 225; and dieback, 221, 225; in future forests, 220, 221, 225; harvest and use of, 3, 53, 104, 106, 107–116, *107*; in old-growth forests, *80*, 82; in post-glacial forests, 16, 34–36, *35*, 38–39, 68; in presettlement forests, 50, 53–55, 61–64, *62*, 67, 85, 88, 93–94

pioneer species, 29

Piscataqua River, 3, 106

plate tectonics, 17, 18; and Maine, 18–20, *19*

Plum Creek Timber Company development proposal, 6, 205

plum, wild, *62*

pollen cores. *See* paleoecology

Popham Colony, 48–51, *49*, 88, 104

Popham, George, 48–50, 54–55

poplar, balsam, 2, 6, *66*, 161

popple. *See* aspen

population of Maine, 4, 44, 102–103, 104–106, 108, *108*, 112, 117–119, 204

Portland, ME, 1–2, *2*, 3, 4, *5*, 151, 211

Prasad, Anantha, 224–225

preservation of Maine forests, 4–6, 68–84, 121–122, 125, 127–131, *130*, 134, 175–179, 209

Presumpscot River, 3, 110

Pring, Martin, 54

Province of Maine, 57

public lands in Maine, 122, 175–179, *176*, *177*, 209

pulp industry. *See* paper industry

Quimby, Roxanne, 131

railroad, 4, 102, 110, 112–113, 117, 123

rare species and communities, 4, 69, 118, 144–145, 147–149, 153–155, 164, 166–167, 171, 205–206; Maine's rarity ranking system, 146–147

Real Estate Investment Trust (REIT), 128, 208

recreation in Maine's forests, 122, 127, 130, 211, 233

refugia, for tree species during glaciation, 37–39

Revolutionary War and settlement of Maine. *See* American Revolution, and settlement of Maine

Richards, Charles, 153

Richmond Island, ME, 104

Rockland, ME, 190

Roorbach, Bill, 152

Rosier, James, 53–55

Rowland, Erika, 79

Rumford, ME, 61, *120*

Russell, Emily, 46, 134

Saco River, 3, 41, 54, 110

Saddleback Mountain, 152

Salmon Falls River, 104

salmon, 118

Sandy River, 57, 100, 101, *101*, 103, 106, 113

Sanger, David, 44

sassafrass, 154

Schauffler, Molly, 37, 67, 79

science: challenges of forecasting, 191, 230; evidence and uncertainty, 231; strength of modern science, 230–231; theory, 17–18

sea level change, 2, 25, 41, 213, 220

sea mink, 43

seal: bearded, 41; harp, 42; ringed, 42

Sebois Plantation, ME, 58

Secord, Theresa, 199

Seymour, Bob, 68, 129

Sferra, Nancy, 144–145, 160, 204

shadbush, *62*, 72

ships masts, of white pine, 3, 38, 53–55, 106, 109, 111

skidder, 123, 127

Smith, Captain John, 142, 154

Smith, David C., 104, 123

Somes Sound (Mt. Desert Island), 27

South Berwick, ME, 104

South Branch Pond, 67

South Brewer, ME, 113

Spiess, Arthur, 44

spruce budworm, 66, 76–77, 77, 86, 104, 127–129, 137, 232

spruce-fir forest: dieback, 192–193, 232; future, 221, 226; harvest and use of, 127, 129, 137; modern, 5, 129, 152–154, 154, 158, 166–171, 170, 192–193, 216; old-growth, 80; post-glacial, 32–35, 33, 37, 44; presettlement, 51, 66

spruce, black, 15, 37, 66–67, 72, 152, 161, 174

spruce, red: in current forests, 37, 134, 152, 158, 160–161, 162, 170, 189, 192; in future forests, 220, 222; in old-growth forests, 72, 76–77, 80, 84; in past forests, 37, 48, 50–51, 59, 62, 66, 88

spruce, white, 2, 2, 37, 72, 161

spruce: and acid rain, 192–193, 194; in current forests, 5–6, 5, 15, 33, 38, 48, 88, 134–135, 138, 152–154, 160–161, 162, 166–171, 170, 174, 189, 192–193, 204, 216; in future forests, 216, 220–221, 222, 232; harvest and use of, 86, 103–104, 110–112, 115, 116, 120–121, 124, 127, 129, 134–138; in old-growth forests, 67, 72, 76–77, 80, 82, 84; in post-glacial forests, 2, 2, 16, 29, 32–37, 32–34, 44–46, 89, 132; in pre-settlement forests, 6, 16, 46, 50–56, 59, 60, 61–62, 62, 64–68, 65, 78–79, 80, 86, 135

St. John River Preserve (The Nature Conservancy), 66, 129, 152–154, 154, 209

St. John River, 3, 5, 56, 110, 112, 115, 120, 166, 171, 174, 177, 178. See also St. John River Preserve

Stolzenberg, William, 43

stone walls, 4, 83, 102–103, 102, 109, 109, 189

Sugarloaf Mountain, 101

Sumner, ME, 56

superorganism idea in ecology, 40–41

swamp, 7, 28, 56, 61, 72, 78, 85, 155, 169, 171

Talbot, George, 122

tamarack, 29, 61, 62, 71, 111, 171, 174

temperate deciduous forest biome, 37, 40, 154–156, 155, 160–163, 162, 163

Temple Stream, ME, 107

Thompson, Woody, 1, 24

Thoreau, Henry David, 1, 6, 110, 121

thrush, Bicknell, 206, 228

Timber Investment Management Organization (TIMO), 128, 208

total energy gradient and ecological diversity in Maine. See geography and ecological diversity in Maine

tourism in Maine forests, 3, 117, 122, 124, 127, 211

Tredici, Peter del, 204

Tumbledown Mountain, 26, 170, 171

tupelo, 62, 62, 161

turkey, wild, 118, 126, 190

University of Maine Climate Change Institute, 15, 37, 38, 67

Unorganized Territory, xi, 7, 38, 205

Verrazano, Giovanni da, 53

Vickery, Barbara, 148–149

Vickery, Peter, 148–149

Walker, John, 53

walnut, white. See butternut

warbler, prairie, *162*, 163

warbler, Wilson's, *162*, 163

Watershed High School, 190

Waymouth, George, 48, 53–55, *54*

Wayne, ME, 132

Wegener, Alfred, 17

Wells, Jeff, 216

Wells, ME, 106

wetlands, 28, 66–67, 71, 118–119, 150–151, *151*, 166, 171, *202*, 204. *See also* bogs and fens; swamp

White, Eric, 205

Whitman, Andrew, 208

Whittaker, Robert, 40

wildfire: in current forests, 122, 134, 144–145, *145*, 148–149, 156, 161, 167, 171, 174; in future forests, 220, 232–233; in old-growth forests, 76, 86–89; in post-glacial forests, 30, 35–36, 38, 43, 46; in presettlement forests, 58–59, 64–67, 79, 86–89

Wildlands and Woodlots project, 177

wildlife: extirpation, 104, 110, 117–119, 122–123, 134; return of, 126, 133–134; legislation affecting, 122–123, 125

Williams, Jack, 46, 232

willow, 29

wind energy in Maine, 205–207

witness trees. *See* land surveys in Maine

wolf, gray, 117–118, 126, 134

woodpecker, pileated, 118, 133

woodpecker, red-bellied, *221*

wool industry and sheep, 102, 108–109, 111–112, 117

Yamasaki, Mariko, 119

Young, Aaron, 64

Younger Dryas, 26, 34–35, 43, 45. *See also* Indians: and Younger Dryas climate change

Zachos, James, 20

Zielinski, Gregory, 173